I0063689

Smart machines, Remote Sensing, Precision Farming, Processes, Mechatronic, Materials and Policies for Safety and Health Aspects

Special Issue Editors

Andrea Colantoni
Danilo Monarca
Massimo Cecchini
Vincenzo Laurendi
Mauro Villarini
Filippo Gambella

MDPI • Basel • Beijing • Wuhan • Barcelona • Belgrade

MDPI

Special Issue Editors
Andrea Colantoni, Danilo Monarca, Massimo Cecchini, Mauro Villarini
University of Tuscia
Italy

Vincenzo Laurendi
National Institute for Insurance against Accidents at Work (INAIL)
Italy

Filippo Gambella
Department of Agricultural Science, University of Sassar
Italy

Editorial Office
MDPI
St. Alban-Anlage 66
Basel, Switzerland

This edition is a reprint of the Special Issue published online in the open access journal *Agriculture* (ISSN 2077-0472) from 2017–2018 (available at: http://www.mdpi.com/journal/agriculture/special_issues/safety_health).

For citation purposes, cite each article independently as indicated on the article page online and as indicated below:

Lastname, F.M.; Lastname, F.M. Article title. *Journal Name* **Year**, *Article number*, page range.

First Edition 2018

ISBN 978-3-03842-865-7 (Pbk)
ISBN 978-3-03842-866-4 (PDF)

Table of Contents

About the Special Issue Editors

Andrea Colantoni studied Agricultural Mechanization and wrote his thesis on the "Study and development of innovative technologies for small and medium sized companies for the use of renewable energy sources". He is a member of the Italian Association of Agricultural Engineers, the European Society of Agricultural Engineers, the International Commission of Agricultural Engineering (CIGR), and the Italian Association of Scientific Agricultural Societies (AISSA). He is involved in the biomasses analysis and characterization of the physical chemical properties of different biomaterials using both experimental and computational tools. He is involved in international projects concerning the gasification processes for electric energy production by biomass sources. He is also developing national projects on the small scale mechanization in agriculture for sloped fields, focusing on the safety improvement of workers.

Danilo Monarca is Full Professor of Agriculture Mechanics at the Tuscia University. Author of over 320 scientific papers, he has been President of the AIIA (Italian Association of Agricultural Engineering) in the quadriennium 2014–2017. Member of the EurAgEng Council, of the Club of Bologna, of the Georgofili Academy and of the National Academy of Agriculture, actually is coordinator of the PhD board in Engineering for Energy and Environment of the Tuscia University.

Massimo Cecchini is an Associate Professor in "Agricultural, Forestry and Biosystem Engineering". He is the Vice President of the 5th Section (Ergonomics and Work Organization) of the Italian Association of Agricultural Engineering (AIIA). He was Director of the online Master in "Quality and Safety in Food Production" and a Master in "Management of Agritourism Enterprises". From 2012 to 2016, he was the Coordinator of a Ph.D. course on "Engineering for Agricultural and Forestry Systems" at the University of Tuscia (Italy). From 2014 to 2016, he was a member of the "Working Group on Safety Issues Related to the Maintenance of Tall Trees" at INAIL. He was also a Scientific Coordinator of the agreement between the "Agency for the Development of Public Administration" of the Lazio Region and the GEMINI Department (known today as DAFNE). He is a member of the Technical Commission within the framework of INAIL and the University of Tuscia's collaboration. He has been a speaker at numerous conferences, and has authored more than 270 works in national and international journals and conference proceedings.

Vincenzo Laurendi is a researcher at the National Institute for Insurance against Accidents at Work. Inail is an Italian public non-profit entity safeguarding workers against physical injuries and occupational diseases. He is a member of ISO—International Organization for Standardization, CEN—European Committee for Standardization, MDWG—Machinery Directive Working Group of the European Commission, AdCo Machinery of the European Commission, the National Academy of Agriculture and of the Georgofili Academy (of which he is a member of the Consultative Commission for Occupational Safety). He is also the scientific coordinator of the main research activities of Inail, on the safety of agricultural and forestry machinery.

Mauro Villarini, was born in Rome, Italy, in 1979. He received his B.S. and M.S. degrees in mechanical engineering from the University of Sapienza, Rome, in 2003 and his Ph.D. degree in Energy from the same university, in 2007. From 2007 to 2012, he has been a Research Assistant at the Sapienza University. Since 2012, he has been an Assistant Professor at the DAFNE Department, Tuscia University of Viterbo. He is the author of three books and more than 20 articles. His research interests include thermal and electrical processes of the energy system (solar, thermal, biomass, gasification and photovoltaic) with particular focus on agricultural applications.

Filippo Gambella, PhD is senior researcher of Agriculture Mechanics, in the Department of Agricultural Science University of Sassari, Italy. He is member of the AIIA (Italian Association of Agricultural Engineering) and Director of Center for Precision Farming (CIRAP). He is Author or Co-author of over 50 scientific papers published in national and international journals and proceedings in national and international conferences on topics related to the scientific sector AGR09. He had scientific interest in different areas of agricultural engineering and in particular, adoption of precision farming in agriculture by drones, image analysis, RGB, NIR and fluorimetric sensors. In the mechanization and harvesting of saffron's flowers. In the development of new hand held equipment and interception system for the harvest of table olive.

Acknowledgement

Editors would like to express their special thanks to all the colleagues who gave the opportunity to realize this Special Issue. A special acknowledgement to the National Institute for Insurance against Accidents at Work (INAIL) for funding the project PROMOSIC.

agriculture

MDPI

Editorial

Smart Machines, Remote Sensing, Precision Farming, Processes, Mechatronic, Materials and Policies for Safety and Health Aspects

Andrea Colantoni [1],*, Danilo Monarca [1], Vincenzo Laurendi [2], Mauro Villarini [1], Filippo Gambella [3] and Massimo Cecchini [1]

[1] Department of Agricultural and Forestry Sciences (DAFNE), University of Tuscia, Via S. Camillo De Lellis, 1-01100 Viterbo, Italy; monarca@unitus.it (D.M.); mauro.villarini@unitus.it (M.V.); cecchini@unitus.it (M.C.)

[2] National Institute for Insurance against Accidents at Work (INAIL), Via Fontana Candida, 1-00078 Monte Porzio Catone (RM), Italy; v.laurendi@inail.it

[3] Department of Agriculture, University of Sassari, Viale Italia, 39-07100 Sassari, Italy; gambella@uniss.it

* Correspondence: colantoni@unitus.it; Tel.: +39-076-1357-356

Received: 12 March 2018; Accepted: 16 March 2018; Published: 23 March 2018

Abstract: The purpose of this Special Issue is to publish high-quality research papers, as well as review articles, addressing recent advances on systems, processes, and materials for work safety, health, and environment. Original, high-quality contributions that have not yet been published, or that are not currently under review by other journals or peer-reviewed conferences, have been sought. The main topics have been the protection system aimed to agricultural health and safety especially applied to mechanization sector (harvester, chippers), often involved in accidents at work, in the context of Directive 2006/42/EC, and to other families of risk as the chemical one and issues pertinent to safety. Methodologies for gradual and sustainable safety improvements on farms have been investigated in the vision of preliminary applications. Furthermore, the application of technologies aimed to the improvement and facilitation of operations in the agriculture sector as monitoring, precision farming, internet of things, application of evolved networks and machines of new conception.

Keywords: Agriculture Engineering; mechatronic; sensors; safety engineering; precision farming

1. Introduction

The introduction of "smart machines" for agricultural operations will allow several advantages, such as an increase in their efficiencies, a reduction in environmental impacts and a reduction of work injuries. There are partially- and fully-automatic devices for most aspects of agricultural functions, from seeding and planting to harvesting and post harvesting, from spraying to livestock management, and so on [1–5]. Moreover "precision farming", using sensors and robotic technologies are applied to existing systems. Work health and safety are also linked to the use of modern technologies, e.g., the protection of machinery operators from crush, entanglement, and shearing by means of mechatronic solutions [6–8]. Another aspect is the use of robots and smart automation, which can also benefit from the gathering of operational data, such as machine condition and fleet monitoring, allowing preventive maintenance and improved fleet management [9]. Considerable advances in sensing hardware, information technologies, smart systems, and software algorithms, have led to significant new developments in the areas of equipment health monitoring, fault diagnosis, and prognosis. These advances enable industries to undergo a fundamental shift towards condition- based maintenance to improve equipment availability and readiness at reduced operating costs throughout the system life-cycle [10–12]. The emergence of sensor networks is also bringing the possibility of collective learning algorithms and decision-theoretic approaches to facilitate effective and scalable diagnostic/prognostic technology for widespread deployment of condition-based maintenance [9]. The mentioned technological development is applicable to the relevant context of safety engineering [13–16]. Furthermore, energy, safety and agriculture have an important role in reducing environmental emissions [17–20]. All the systems aimed at the management of energy, safety and environment are performed and optimized by means of innovative technologies, materials, processes, and methods [21–32].

The purpose of this Special Issue is to publish high-quality research papers, as well as review articles, addressing recent advances on systems, processes, and materials for work safety, health, and environment.

The objectives of this special issue are:

- study of man–machine dialogue systems;
- analysis on towed or carried machines: forestry chippers, manure spreaders, round balers and others;
- safety and health management system design and engineering;
- safety and health monitoring sensors and sensing;
- data-driven methods for anomaly detection, diagnosis and prognosis;
- precision farming;
- mechatronic;
- automotive and agriculture machinery applications;
- engineering of hybrid and integrated systems and their efficiency maximization, especially for safety and health purposes, aimed to injuries and accidents reduction;
- use of remote sensor and mechatronic systems applied in several aspects.

2. Papers in this Special Issue

The special issue "Smart machines, Remote Sensing, Precision Farming, Processes, Mechatronic, Materials and Policies for Safety and Health Aspects" brings together some of the latest research results in the field of smart machines connected with the safety and health aspects. It presents eighteen papers, which deal with a wide range of research activities.

We can divide the special issue in three parts, as follow.

2.1. Research Articles

The first contribution in this section explores the "Agricultural Health and Safety Survey in Friuli Venezia Giulia" by Sirio Rossano Secondo Cividino, Gianfranco Pergher, Nicola Zucchiatti and Rino Gubiani [33]. The work in the agricultural sector has taken on a fundamental role in the last decades, due to the still too high rate of fatal injuries, workplace accidents, and dangerous occurrences reported each year [34]. The average old age of agricultural machinery is one of the main issues at stake in Italy. Numerous safety problems stem from that; therefore, two surveys were conducted in two different periods, on current levels of work safety in agriculture in relation to agricultural machinery's age and efficiency, and to show the levels of actual implementation of the Italian legislation on safety and health at work in the agricultural sector [34,35]. The surveys were carried out, considering a sample of 161 farms located in the region Friuli Venezia Giulia (North-East of Italy). The research highlights the most significant difficulties the sample of farms considered have in enforcing the law. One hand, sanitary surveillance and workers' information and training represent the main deficiencies and weakest points in family farms. Moreover, family farms do not generally provide the proper documentation concerning health and safety at workplaces, when they award the contract to other companies. On the other hand, lack of maintenance program for machinery and equipment, and of emergency plans and participation of workers' health and safety representative, are the most common issues in farms with employees. Several difficulties are also evident in planning workers' training programs. Furthermore, the company physician's task is often limited to medical controls, so that he is not involved in risk assessment and training. Interviews in heterogeneous samples of farms have shown meaningful outcomes, which have subsequently been used to implement new databases and guidelines for Health and Safety Experts and courses in the field of Work Safety in agriculture. In conclusion, although the legislation making training courses for tractor operators and tractor inspections compulsory dates back to the years 2012 and 2015, deadlines have been prorogued, and the law is not yet fully applied, so that non-upgraded unfit old agricultural machinery is still being used by many workers, putting their health and their own lives at risk.

The second paper concerns the "Definition of a Methodology for Gradual and Sustainable Safety Improvements on Farms and Its Preliminary Applications" by Sirio Rossano Secondo Cividino, Gianfranco Pergher, Rino Gubiani, Carlo Moreschi, Ugo Da Broi, Michela Vello and Fabiano Rinaldi [36]. In many productive sectors, ensuring a safe working environment is still an underestimated problem and, especially, so in farming. A lack of attention to safety and poor risk awareness by operators represents a crucial problem, which results in numerous serious injuries and fatal accidents. The Demetra project, involving the collaboration of the Regional Directorate of INAIL (National Institute for Insurance against Accidents at Work), aims to devise operational solutions to evaluate the risk of accidents in agricultural work and analyze the dynamics of occupational accidents by using an observational method to help farmers ensure optimal safety levels. The challenge of the project is to support farmers with tools designed to encourage good safety management in the agricultural workplaces.

The third contribution is "Analysis of the Almond Harvesting and Hulling Mechanization Process: A Case Study" by Simone Pascuzzi and Francesco Santoro. The aim of this paper is the analysis of the almond harvesting system with a very high level of mechanization frequently used in Apulia for the almond harvesting and hulling process. Several tests were carried out to assess the technical aspects related to the machinery and to the mechanized harvesting system used itself, highlighting their usefulness, limits, and compatibility within the almond cultivation sector. Almonds were very easily separated from the tree, and this circumstance considerably improved the mechanical harvesting operation efficiency even if the total time was mainly affected by the time required to manoeuvre the machine and by the following manual tree beating. The mechanical pick-up from the ground was not effective, with only 30% of the dropped almond collected, which mainly was caused by both the pick-up reel of the machine being unable to approach the almonds dropped near the base of the trunk and the surface condition of the soil being unsuitably arranged for a mechanized pick-up operation. The work times concerning the hulling and screening processes, carried out at the farm, were heavily affected by several manual operations before, during, and after the executed process; nevertheless, the plant work capability varied from 170 to 200 kg/h with two operators.

The fourth paper entitled "Safety Improvements on Wood Chippers Currently in Use: A Study on Feasibility in the Italian Context" by Giorgia Bagagiolo, Vincenzo Laurendi and Eugenio Cavallo, following formal opposition by France on the harmonized safety standards regarding manually-loaded wood chippers (EN 13525:2005+A2:2009) which presumed compliance with the Essential Health and Safety Requirements (EHSR) required by the Machine Directive (Directive 2006/42/EC), have recently been withdrawn, and a new draft of the standard is currently under revision. In order to assess the potential impact of the expected future harmonized standards within the Italian context, this study has examined the main issues in implementing EHSRs on wood chippers already being used. Safety issues regarding wood chippers already in use were identified in an analysis of the draft standard, through the observation of a number of case studies, and qualitative analysis of the essential technical interventions. A number of agricultural and forestry operators and companies participated in the study, pointing out the technical and economic obstacle facing the safety features requested by the pending new standard. It emerged that the main safety issues concerned the implementation of the reverse function, the stop bar, and the protective devices, the infeed chute dimension, the emergency stop function, and the designated feeding area. The possibility of adopting such solutions mainly depends on technical feasibility and costs, but an important role is also played by the attitude towards safety and a lack of adequate information regarding safety obligations and procedures among users.

The fifth paper concerns "Phytotoxicity and Chemical Characterization of Compost Derived from Pig Slurry Solid Fraction for Organic Pellet Production" by Niccolò Pampuro, Carlo Bisaglia, Elio Romano, Massimo Brambilla, Ester Foppa Pedretti and Eugenio Cavallo. The phytotoxicity of four different composts obtained from pig slurry solid fraction composted by itself (SSFC) and mixed with sawdust (SC), woodchips (WCC) and wheat straw (WSC) was tested with bioassay methods. For each compost type, the effect of water extracts of compost on seed germination and primary root growth of cress (*Lepidium Sativum* L.) was investigated. Composts were also chemically analysed for total nitrogen, ammonium, electrical conductivity and heavy metal (Cu and Zn). The chemicals were correlated to phytotoxicity indices. The mean values of the germination index (GI) obtained were 160.7, 187.9, 200.9 and 264.4 for WSC, WCC, SC and SSFC, respectively. Growth index (GrI) ranged from the 229.4%, the highest value, for SSFC, followed by 201.9% for SC, and 193.1% for WCC, to the lowest value, 121.4%, for WSC. Electrical conductivity showed a significant and negative correlation with relative seed germination at the 50% and 75% concentrations. A strong positive correlation was found for water-extractable Cu with relative root growth and germination index at the 10% concentration. Water-extractable Zn showed a significant positive correlation with relative root growth and GI at the 10% concentration. These results highlighted that the four composts could be used for organic pellet production and subsequently distributed as a soil amendment with positive effects on seed germination and plant growth (GI > 80%).

The sixth paper illustred "A Study of the Lateral Stability of Self-Propelled Fruit Harvesters" by Maurizio Cutini, Massimo Brambilla, Carlo Bisaglia, Stefano Melzi, Edoardo Sabbioni, Michele Vignati, Eugenio Cavallo and Vincenzo Laurendi. Self-propelled fruit harvesters (SPFHs) are agricultural machines designed to facilitate fruit picking and other tasks requiring operators to stay close to the foliage or to the upper part of the canopy. They generally consist of a chassis with a variable height working platform that can be equipped with lateral extending platforms. The positioning of additional masses (operators, fruit bins) and the maximum height of the platform (up to three meters above the ground) strongly affect machine stability. Since there are no specific studies on the lateral stability of SPFHs, this study aimed to develop a specific test procedure to fill this gap. A survey of the Italian market found 20 firms manufacturing 110 different models of vehicles. Observation and monitoring of SPFHs under real operational conditions revealed the variables mostly likely to affect lateral stability: the position and mass of the operators and the fruit bin on the platform. Two SPFHs were tested in the laboratory to determine their centre of gravity and lateral stability in four different settings reproducing operational conditions. The test setting was found to affect the stability angle. Lastly, the study identified two specific settings reproducing real operational conditions most likely to affect the lateral stability of SPFHs: these should be used as standard, reproducible settings to enable a comparison of results.

The seventh article entitled "Development of a Variable Rate Chemical Sprayer for Monitoring Diseases and Pests Infestation in Coconut Plantations" by Grianggai Samseemoung, Peeyush Soni and Pimsiri Suwan shows an image processing-based variable rate chemical sprayer for disease and pest-infested coconut plantations was designed and evaluated. The manual application of chemicals is considered risky and hazardous to workers, and provides low precision. The designed sprayer consisted of a sprayer frame, motors, a power system, a chemical tank and pump, a crane, a nozzle with a remote monitoring system, and motion and crane controlling systems. As the target was confirmed, the nozzle was moved towards the target area (tree canopy) using the remote monitoring system. The pump then sprayed chemicals to the target at a specified rate. The results suggested optimal design values for 5–9 m tall coconut trees, including the distance between nozzle and target (1 m), pressure (1.5 bar), spraying rate (2.712 L/min), the highest movement speed (1.5 km/h), fuel consumption (0.58 L/h), and working capacity (0.056 ha/h). The sprayer reduced labor requirements, prevented chemical hazards to workers, and increased coconut pest controlling efficiency.

The eighth article is: "Analysis of the almond harvesting and hulling mechanization process". A case study by Simone Pascuzzi e Francesco Santoro, the aim of this paper is the analysis of the almond harvesting system with a very high level of mechanization frequently used in Apulia (Southern Italy). It is the leading Italian region for the production of olive oil (115×10^6 kg of oil/year), and the olive oil chain is really important from a business point of view. Currently, the extraction of olive oil is essentially performed by using a mechanical pressing process (traditional olive oil mills), or by the centrifugation process (modern olive oil mills). The aim of this paper is to evaluate in detail the noise levels within a typical olive oil mill located in the northern part of the Apulia region during olive oil extraction. The feasibility of this study focusing on the assessment of workers' exposure to noise was tested in compliance with the Italian-European Regulations and US standards and criteria. Several measurements of the noise emission produced by each machine belonging to the productive cycle were carried out during olive oil production. The results obtained were then used to evaluate possible improvements to carry out in order to achieve better working conditions. An effective reduction in noise could probably be achieved through a combination of different solutions, which obviously have to be assessed not only from a technical point of view but also an economic one. A significant reduction in noise levels could be achieved by increasing the area of the room allotted to the olive oil extraction cycle by removing all the unnecessary partition walls that might be present.

The ninth paper regards the "Monitoring and Precision Spraying for Orchid Plantation with Wireless WebCAMs" by Grianggai Samseemoung, Peeyush Soni and Chaiyan Sirikul, face up the processing images taken from wireless WebCAMs on the low altitude remote sensing (LARS) platform, this research monitored crop growth, pest, and disease information in a dendrobium orchid's plantation. Vegetetative indices were derived for distinguishing different stages of crop growth, and the infestation density of pests and diseases. Image data was processed through an algorithm created in MATLAB® (The MathWorks, Inc., Natick, MA, USA). Corresponding to the orchid's growth stage and its infestation density, varying levels of fertilizer and chemical injections were administered. The acquired LARS images from wireless WebCAMs were positioned using geo-referencing, and eventually processed to estimate vegetative-indices (Red = 650 nm and NIR = 800 nm band center). Good correlations and a clear cluster range were obtained in characteristic plots of the normalized difference vegetation index (NDVI) and the green normalized difference vegetation index (GNDVI) against chlorophyll content. The coefficient of determination, the chlorophyll content values (μmol m^{-2}) showed significant differences among clusters for healthy orchids ($R^2 = 0.985$–0.992), and for infested orchids ($R^2 = 0.984$–0.998). The WebCAM application, while being inexpensive, provided acceptable inputs for image processing. The LARS platform gave its best performance at an altitude of 1.2 m above canopy. The image processing software based on LARS images provided satisfactory results as compared with manual measurements.

The tenth paper is "Energy and Carbon Impact of Precision Livestock Farming Technologies Implementation in the Milk Chain: From Dairy Farm to Cheese Factory" by Giuseppe Todde, Maria Caria, Filippo Gambella and Antonio Pazzona speak of Precision Livestock Farming (PLF) is being developed in livestock farms to relieve the human workload and to help farmers to optimize production and management procedure. The objectives of this study were to evaluate the consequences in energy intensity and the related carbon impact, from dairy farm to cheese factory, due to the implementation of a real-time milk analysis and separation (AfiMilk MCS) in milking parlors. The research carried out involved three conventional dairy farms, the collection and delivery of milk from dairy farms to cheese factory and the processing line of a traditional soft cheese into a dairy factory. The AfiMilk MCS system installed in the milking parlors allowed to obtain a large number of information related to the quantity and quality of milk from each individual cow and to separate milk with two different composition (one with high coagulation properties and the other one with low coagulation properties), with different percentage of separation. Due to the presence of an additional milkline and the AfiMilk MCS components, the energy requirements and the related environmental impact at farm level were slightly higher, among 1.1% and 4.4%. The logistic of milk collection was also significantly reorganized in view of the collection of two separate type of milk, hence, it leads an increment of 44% of the energy requirements. The logistic of milk collection and delivery represents the process which the highest incidence in energy consumption occurred after the installation of the PLF technology. Thanks to the availability of milk with high coagulation properties, the dairy plant, produced traditional soft cheese avoiding the standardization of the formula, as a result, the energy uses decreased about 44%, while considering the whole chain, the emissions of carbon dioxide was reduced by 69%. In this study, the application of advance technologies in milking parlors modified not only the on-farm management but mainly the procedure carried out in cheese making plant. This aspect makes precision livestock farming implementation unimportant technology that may provide important benefits throughout the overall milk chain, avoiding about 2.65 MJ of primary energy every 100 kg of processed milk.

The eleventh papers is "Adoption of Web-Based Spatial Tools by Agricultural Producers: Conversations with Seven Northeastern Ontario Farmers Using the GeoVisage Decision Support System" by Daniel H. Jarvis, Mark P. Wachowiak, Dan F. Walters and John M. Kovacs. The paper reports the findings of a multi-site qualitative case study research project designed to document the utility and perceived usefulness of weather station and imagery data associated with the online resource GeoVisage among northeastern Ontario farmers. Interviews were conducted onsite at five participating farms (three dairy, one cash crop, and one public access fruit/vegetable) in 2014–2016, and these conversations were transcribed and returned to participants for member checking. Interview data was then entered into Atlas.ti software for the purpose of qualitative thematic analysis. Fifteen codes emerged from the data and findings center around three overarching themes: common uses of weather station data (e.g., air/soil temperature, rainfall); the use of GeoVisage Imagery data/tools (e.g., acreage calculations, remotely sensed imagery); and future recommendations for the online resource (e.g., communication, secure crop imagery, mobile access). Overall, weather station data and tools freely accessible through the GeoVisage site were viewed as representing a timely, positive, and important addition to contemporary agricultural decision-making in northeastern Ontario farming.

The twelfth article is "Safety-Critical Manuals for Agricultural Tractor Drivers: A Method to Improve Their Usability" by Maurizio Cutini, Giada Forte, Marco Maietta, Maurizio Mazzenga, Simon Mastrangelo and Carlo Bisaglia. This work sets out the planning phases adopted for the first time to put together a manual on injury and accident prevention in the use of farm tractors. The goal is to convey information more effectively than at present, while taking the end users' opinions into consideration. The manual was devised, created, and tested based on a human-centred design (HCD) process, which identified the operators' requirements using a participatory ergonomics (PE) strategy. The main topics of the manual were outlined by engaging the users in a qualitative research activity (i.e., focus groups and workshops with final users), and the contents were prioritized and labelled by way of a noun prioritization activity. The users were involved right up to the choice of graphics and print layout in order to orient the publication to the farming context. The research activity highlighted a divergence between the operators' requirements and the topics currently dealt with in the sector publications. The project resulted in the publication of the "Safe Tractor" manual, which features some innovations. The experience highlighted the need to adopt HCD processes to create innovative editorial products, which can help speed up the dissemination of safety culture in the primary sector.

The thirteenth paper face up the "Precision Farming in Hilly Areas: The Use of Network RTK in GNSS Technology" by Alvaro Marucci, Andrea Colantoni, Ilaria Zambon and Gianluca Egidi [37]. The number of GNSS satellites has greatly increased over the last few decades, which has led to increased interest in developing self-propelled vehicles. Even agricultural vehicles have a great potential for use of these systems. In fact, it is possible to improve the efficiency of machines in terms of their working uniformity, reduction of fertilizers, pesticides, etc. with the aim of (i) reducing the timeframes of cultivation operations with significant economic benefits and, above all; (ii) decreasing environmental impact. These systems face some perplexity in hilly environments but, with specific devices, it is possible to overcome any signal deficiencies. In hilly areas then, the satellite-based system can also be used to safeguard operators' safety from the risk of rollover. This paper reports the results obtained from a rural development program (RDP) in the Lazio Region 2007/2013 (measure project 1.2.4) for the introduction and diffusion of GNSS satellites systems in hilly areas.

The fourteenth article is "Identification of Optimal Mechanization Processes for Harvesting Hazelnuts Based on Geospatial Technologies in Sicily (Southern Italy)" by Ilaria Zambon, Lavinia Delfanti, Alvaro Marucci, Roberto Bedini, Walter Bessone, Massimo Cecchini and Danilo Monarca [38]. Sicily is a region located in the southern Italy. The typical Mediterranean landscape can be appreciated due to its high biodiversity [39–49]. Specifically in Sicily, hazelnut plantations have adapted in a definite area in Sicily (the Nebrodi Park, Sant'Agata Militello, Messina, Italy) due to specific morphological and climatic characteristics. However, many of these plantations are not used today due to adverse conditions, both to collect hazelnuts and to reach hazel groves. Though a geospatial analysis, the paper aims to identify which hazelnut contexts can be actively used for agricultural, economic (e.g., introduction of a circular economy) and energetic purposes (to establish a potential agro-energetic district) [40,42]. The examination revealed the most suitable areas giving several criteria (e.g., slope, road system), ensuring an effective cultivation and consequent harvesting of hazelnuts and providing security for the operators since many of hazelnut plants are placed in very sloped contexts that are difficult to reach by traditional machines. In this sense, this paper also suggests optimal mechanization processes for harvesting hazelnuts in this part of Sicily.

2.2. Review Articles

The first review is "Analysis of the Cause-Effect Relation between Tractor Overturns and Traumatic Lesions Suffered by Drivers and Passengers: A Crucial Step in the Reconstruction of Accident Dynamics and the Improvement of Prevention" by Carlo Moreschi, Ugo Da Broi, Sirio Rossano Secondo Cividino, Rino Gubiani, Gianfranco Pergher, Michela Vello and Fabiano Rinaldi. The evaluation of the dynamics of accidents involving the overturning of farm tractors is difficult for both engineers and coroners. A clear reconstruction of the causes, vectorial forces, speed, acceleration, timing and direction of rear, front and side rollovers may be complicated by the complexity of the lesions, the absence of witnesses and the death of the operator, and sometimes also by multiple overturns. Careful analysis of the death scene, vehicle, traumatic lesions and their comparison with the mechanical structures of the vehicle and the morphology of the terrain, should help experts to reconstruct the dynamics of accidents and may help in the design of new preventive equipment and procedures.

The second review is "Whole-Body Vibration in Farming: Background Document for Creating a Simplified Procedure to Determine Agricultural Tractor Vibration Comfort" by Maurizio Cutini, Massimo Brambilla and Carlo Bisaglia. The operator exposure to high levels of whole-body vibration (WBV) presents risks to health and safety and it is reported to worsen or even cause back injuries. Work activities resulting in operator exposure to whole-body vibration have a common onset in off-road work such as farming. Despite the wide variability of agricultural surface profiles, studies have shown that with changing soil profile and tractor speed, the accelerations resulting from ground input present similar spectral trends. While on the one hand such studies confirmed that tractor WBV emission levels are very dependent upon the nature of the operation performed, on the other, irrespective of the wide range of conditions characterizing agricultural operations, they led researchers to set up a possible and realistic simplification and standardization of tractor driver comfort testing activities. The studies indicate the usefulness, and the possibility, of developing simplified procedures to determine agricultural tractor vibration comfort. The results obtained could be used effectively to compare tractors of the same category or a given tractor when equipped with different seats, suspension, tyres, etc.

2.3. Technical Note

The first technical note is "Mechatronic Solutions for the Safety of Workers Involved in the Use of Manure Spreader" by Massimo Cecchini, Danilo Monarca, Vincenzo Laurendi, Daniele Puri and Filippo Cossio [50]. An internationally acknowledged requirement is to analyze and provide technical solutions for prevention and safety during the use and maintenance of manure spreader wagons. Injuries statistics data and specific studies show that particular constructive criticalities have been identified on these machines, which are the cause of serious and often fatal accidents. These accidents particularly occur during the washing and maintenance phases, especially when such practices are carried out inside the hopper when the rotating parts of the machine are in action. The current technical standards and the various safety requirements under consideration have not always been effective for protecting workers. To this end, the use of SWOT analysis (Strengths, Weaknesses, Opportunities, and Threats) allowed authors to highlight critical and positive aspects of the different solutions studied for reducing the risk due to contact with the rotating parts. The selected and tested solution consists of a decoupling system automatically activated when the wheels of the wagon are not moving. Such a solution prevents the contact with the moving rotating parts of the machine when the worker is inside the hopper. This mechatronic solution allowed to obtain a prototype that has led to the resolution of the issues related to the use of the wagon itself: in fact, the system guarantees the stopping of manure spreading organs in about 12 s from the moment of the wheels stopping [50].

The second technical note is "Innovative Solution for Reducing the Run-Down Time of the Chipper Disc Using a Brake Clamp Device" by Andrea Colantoni, Francesco Mazzocchi, Vincenzo Laurendi, Stefano Grigolato, Francesca Monarca, Danilo Monarca and Massimo Cecchini [51]. Wood-chippers are widely used machines in the forestry, urban and agricultural sectors. The use of these machines implies various risks for workers, primarily the risk of contact with moving and cutting parts. These machine parts have a high moment of inertia that can lead to entrainment with the cutting components. This risk is particularly high in the case of manually fed chippers. Following cases of injury with wood-chippers and the improvement of the technical standard (Comité Européen de Normalisation-European Norm) EN 13525:2005+A2:2009, the technical note presents the prototype of an innovative system to reduce risks related to the involved moving parts, based on the "brake caliper" system and electromagnetic clutch for the declutching of the power take-off (PTO). The prototype has demonstrated its potential for reducing the run-down time of the chipper disc (95%) and for reducing the worker's risk of entanglement and entrainment in the machine's feed mouth.

3. Conclusions

In summary, the papers of the special issue represent some of the latest and most promising research results in this new and exciting field, which continues to make significant impact on real-world applications. We are confident that this special issue will stimulate further research in this area.

Acknowledgments: We thank all authors of the special issue.

Author Contributions: The contribution to the programming and executing of this special must be equally divided by the authors.

Conflicts of Interest: The authors declare no conflicts of interest.

References

1. Febbi, P.; Menesatti, P.; Costa, C.; Pari, L.; Cecchini, M. Automated determination of poplar chip size distribution based on combined image and multivariate analyses. *Biomass Bioenergy* **2014**, *73*, 1–10. [CrossRef]
2. Moscetti, R.; Haff, R.P.; Monarca, D.; Cecchini, M.; Massantini, R. Near-infrared spectroscopy for detection of hailstorm damage on olive fruit. *Postharvest Biol. Technol.* **2016**, *120*, 204–212. [CrossRef]
3. Moscetti, R.; Monarca, D.; Cecchini, M.; Haff, R.P.; Contini, M.; Massantini, R. Detection of mold-damaged chestnuts by near-infrared spectroscopy. *Postharvest Biol. Technol.* **2014**, *93*, 83–90. [CrossRef]
4. Moscetti, R.; Saeys, W.; Keresztes, J.C.; Goodarzi, M.; Cecchini, M.; Monarca, D.; Massantini, R. Hazelnut quality sorting using high dynamic range short-wave infrared hyperspectral imaging. *Food Bioprocess Technol.* **2015**, *8*, 1593–1604. [CrossRef]

5. Stella, E.; Moscetti, R.; Haff, R.P.; Monarca, D.; Cecchini, M.; Contini, M.; Massantini, R. Recent advances in the use of non-destructive near infrared spectroscopy for intact olive fruits. *J. Near Infrared Spectrosc.* **2015**, *23*, 197–208. [CrossRef]

6. Pascuzzi, S.; Santoro, F. Evaluation of farmers' OSH hazard in operation nearby mobile telephone radio base stations. In Proceedings of the 16th International Scientific Conference "Engineering for Rural Development", Jelgava, Latvia, 24–26 May 2017; Volume 16, pp. 748–755.

7. Pascuzzi, S.; Santoro, F. Exposure of farm workers to electromagnetic radiation from cellular network radio base stations situated on rural agricultural land. *Int. J. Occup. Saf. Ergon. (JOSE)* **2015**, *21*, 351–358. [CrossRef] [PubMed]

8. Pascuzzi, S. The effects of the forward speed and air volume of an air-assisted sprayer on spray deposition in "tendone" trained vineyards. *J. Agric. Eng.* **2013**, *44*, 125–132. [CrossRef]

9. Villarini, M.; Cesarotti, V.; Alfonsi, L.; Introna, V. Optimization of photovoltaic maintenance plan by means of a FMEA approach based on real data. *Energy Convers. Manag.* **2017**, *152*, 1–12. [CrossRef]

10. Manetto, G.; Cerruto, E.; Pascuzzi, S.; Santoro, F. Improvements in Citrus Packing Lines to Reduce the Mechanical Damage to Fruit. *Chem. Eng. Trans.* **2017**, *58*, 391–396.

11. Bianchi, B.; Tamborrino, A.; Santoro, F. Assessment of the energy and separation efficiency of the decanter centrifuge with regulation capability of oil water ring in the industrial process line using a continuous method. *J. Agric. Eng.* **2013**, *44*, 278–282. [CrossRef]

12. Russo, G.; Verdiani, G.; Anifantis, A.S. Re-use of agricultural biomass for nurseries using proximity composting. *Contemp. Eng. Sci.* **2016**, *9*, 1151–1182. [CrossRef]

13. Pascuzzi, S.; Santoro, F. Analysis of the almond harvesting and hulling mechanization process: A case study. *Agriculture* **2017**, *7*, 100. [CrossRef]

14. Pascuzzi, S.; Santoro, F. Analysis of possible noise reduction arrangements inside olive oil mills: A case study. *Agriculture* **2017**, *7*, 88. [CrossRef]

15. Pascuzzi, S. A multibody approach applied to the study of driver injures due to a narrow-track wheeled tractor rollover. *J. Agric. Eng.* **2015**, *46*, 105–114.

16. Pascuzzi, S.; Blanco, I.; Anifantis, A.S.; Scarascia Mugnozza, G. Hazards assessment and technical actions due to the production of pressured hydrogen within a pilot photovoltaic-electrolyser-fuel cell power system for agricultural equipment. *J. Agric. Eng.* **2016**, *47*, 88–93. [CrossRef]

17. Bocci, E.; Villarini, M.; Vecchione, L.; Sbordone, D.; Di Carlo, A.; Dell'Era, A. Energy and economic analysis of a residential Solar Organic Rankine plant. *Energy Procedia* **2015**, *81*, 558–568. [CrossRef]

18. Anifantis, A.S.; Pascuzzi, S.; Scarascia Mugnozza, G. Geothermal source heat pump performance for a greenhouse heating system: An experimental study. *J. Agric. Eng.* **2016**, *47*, 164–170. [CrossRef]

19. Marucci, A.; Zambon, I.; Colantoni, A.; Monarca, D. A combination of agricultural and energy purposes: Evaluation of a prototype of photovoltaic greenhouse tunnel. *Renew. Sustain. Energy Rev.* **2018**, *82*, 1178–1186. [CrossRef]

20. Zambon, I.; Monarca, D.; Cecchini, M.; Bedini, R.; Longo, L.; Romagnoli, M.; Marucci, A. Alternative energy and the development of local rural contexts: An approach to improve the degree of smart cities in the Central-Southern Italy. *Contemp. Eng. Sci.* **2016**, *9*, 1371–1386. [CrossRef]

21. Anifantis, A.S.; Colantoni, A.; Pascuzzi, S.; Santoro, F. Photovoltaic and hydrogen plant integrated with a gas heat pump for greenhouse heating: A mathematical study. *Sustainability* **2018**, *10*, 378. [CrossRef]

22. Anifantis, A.S. Performance assessment of photovoltaic, ground source heat pump and hydrogen heat generator in a stand-alone systems for greenhouse heating. *Chem. Eng. Trans.* **2017**, *58*, 511–516.

23. Pascuzzi, S.; Anifantis, A.S.; Blanco, I.; Scarascia Mugnozza, G. Electrolyzer performance analysis of an integrated hydrogen power system for greenhouse heating a case study. *Sustainability* **2016**, *8*, 629. [CrossRef]

24. Carlini, M.; Mosconi, E.M.; Castellucci, S.; Villarini, M.; Colantoni, A. An economical evaluation of anaerobic digestion plants fed with organic agro-industrial waste. *Energies* **2017**, *10*, 1165. [CrossRef]

25. Zambon, I.; Colantoni, A.; Carlucci, M.; Morrow, N.; Sateriano, A.; Salvati, L. Land quality, sustainable development and environmental degradation in agricultural districts: A computational approach based on entropy indexes. *Environ. Impact Assess. Rev.* **2017**, *64*, 37–46. [CrossRef]

26. Anifantis, A.S.; Colantoni, A.; Pascuzzi, S. Thermal energy assessment of a small scale photovoltaic, hydrogen and geothermal stand-alone system for greenhouse heating. *Renew. Energy* **2017**, *103*, 115–127. [CrossRef]

27. Zambon, I.; Colantoni, A.; Cecchini, M.; Mosconi, E.M. Rethinking sustainability within the viticulture realities integrating economy, landscape and energy. *Sustainability* **2018**, *10*, 320. [CrossRef]

28. Salerno, M.; Gallucci, F.; Pari, L.; Zambon, I.; Sarri, D.; Colantoni, A. Costs-benefits analysis of a small-scale biogas plant and electric energy production. *Bulg. J. Agric. Sci.* **2017**, *23*, 357–362.

29. Colantoni, A.; Zambon, I.; Cecchini, M.; Marucci, A.; Piacentini, L.; Feltrin, S.; Monarca, D. Greenhouses plants as a landmark for research and innovation: The combination of agricultural and energy purposes for a more sustainable future in Italy. *Chem. Eng. Trans.* **2017**, *58*, 469–474.

30. Pascuzzi, S.; Cerruto, E. An innovative pneumatic electrostatic sprayer useful for tendone vineyards. *J. Agric. Eng.* **2015**, *3*, 123–127. [CrossRef]

31. Pascuzzi, S.; Cerruto, E.; Manetto, G. Foliar spray deposition in a "tendone" vineyard as affected by airflow rate, volume rate and vegetative development. *Crop Prot.* **2016**, *91*, 34–48. [CrossRef]

32. Baldoin, C.; Balsari, P.; Cerruto, E.; Pascuzzi, S.; Raffaelli, M. Improvement in pesticide application on greenhouse crops: Results of a survey about greenhouse structures in Italy. *Acta Hortic.* **2008**, *801*, 609–614. [CrossRef]

33. Cividino, S.R.S.; Pergher, G., Zucchiatti N.; Gubiani, R. Agricultural health and safety survey in Friuli Venezia Giulia. *Agriculture* **2018**, *8*, 9. [CrossRef]

34. Cecchini, M.; Colantoni, A.; Monarca, D.; Cossio, F.; Riccioni, S. Survey on the status of enforcement of European directives on health and safety at work in some farms of central Italy. *Chem. Eng. Trans.* **2017**, *58*, 103–108.

35. Cecchini, M.; Cossio, F.; Marucci, A.; Monarca, D.; Colantoni, A.; Petrelli, M.; Allegrini, E. Survey on the status of enforcement of European directives on health and safety at work in some Italian farms. *J. Food Agric. Environ.* **2013**, *11*, 595–600.

36. Cividino, S.R.S.; Pergher, G.; Gubiani, R.; Moreschi, C.; Da Broi, U.; Vello, M.; Rinaldi, F. Definition of a methodology for gradual and sustainable safety improvements on farms and its preliminary applications. *Agriculture* **2018**, *8*, 7. [CrossRef]

37. Marucci, A.; Colantoni, A.; Zambon, I.; Egidi, G. Precision farming in hilly areas: The use of network RTK in GNSS technology. *Agriculture* **2017**, *7*, 60. [CrossRef]

38. Zambon, I.; Delfanti, L.; Marucci, A.; Bedini, R.; Bessone, W.; Cecchini, M.; Monarca, D. Identification of optimal mechanization processes for harvesting Hazelnuts based on geospatial technologies in Sicily (Southern Italy). *Agriculture* **2017**, *7*, 56. [CrossRef]

39. Cecchini, M.; Zambon, I.; Pontrandolfi, A.; Turco, R.; Colantoni, A.; Mavrakis, A.; Salvati, L. Urban sprawl and the 'olive' landscape: Sustainable land management for 'crisis' cities. *GeoJournal* **2018**, 1–19. [CrossRef]

40. Colantoni, A.; Delfanti, L.; Recanatesi, F.; Tolli, M.; Lord, R. Land use planning for utilizing biomass residues in Tuscia Romana (central Italy): Preliminary results of a multi criteria analysis to create an agro-energy district. *Land Use Policy* **2016**, *50*, 125–133. [CrossRef]

41. Colantoni, A.; Ferrara, C.; Perini, L.; Salvati, L. Assessing trends in climate aridity and vulnerability to soil degradation in Italy. *Ecol. Indic.* **2015**, *48*, 599–604. [CrossRef]

42. Colantoni, A.; Mavrakis, A.; Sorgi, T.; Salvati, L. Towards a 'polycentric' landscape? Reconnecting fragments into an integrated network of coastal forests in Rome. *Rend. Lincei* **2015**, *26*, 615–624. [CrossRef]

43. Duvernoy, I.; Zambon, I.; Sateriano, A.; Salvati, L. Pictures from the other side of the fringe: Urban growth and peri-urban agriculture in a post-industrial city (Toulouse, France). *J. Rural Stud.* **2018**, *57*, 25–35. [CrossRef]

44. Monarca, D.; Cecchini, M.; Colantoni, A. Plant for the production of chips and pellet: Technical and economic aspects of an case study in the central Italy. In *International Conference on Computational Science and Its Applications*; Springer: Berlin/Heidelberg, Germany, 2011; Volume 6785, pp. 296–306.

45. Mosconi, E.M.; Carlini, M.; Castellucci, S.; Allegrini, E.; Mizzelli, L.; di Trifiletti, M.A. Economical assessment of large-scale photovoltaic plants: An Italian case study. In *International Conference on Computational Science and Its Applications*; Springer: Berlin/Heidelberg, Germany, 2013; pp. 160–175.

46. Mosconi, E.M. Opportunity and function of energy wholesale market in Italy. *Riv. Giurdica Dell'Ambient.* **2003**, *18*, 1101–1110.

47. Ruggieri, A.; Braccini, A.M.; Poponi, S.; Mosconi, E.M. A meta-model of inter-organisational cooperation for the transition to a circular economy. *Sustainability* **2016**, *8*, 1153. [CrossRef]

48. Ruggieri, A.; Mosconi, E.M.; Poponi, S.; Silvestri, C. Digital innovation in the job market: An explorative study on cloud working platforms. *Lect. Notes Inf. Syst. Organ.* **2016**, *11*, 273–283.

49. Zambon, I.; Benedetti, A.; Ferrara, C.; Salvati, L. Soil Matters? A multivariate analysis of socioeconomic constraints to urban expansion in Mediterranean Europe. *Ecol. Econ.* **2018**, *146*, 173–183. [CrossRef]

50. Cecchini, M.; Monarca, D.; Laurendi, V.; Puri, D.; Cossio, F. Mechatronic Solutions for the Safety of Workers Involved in the Use of Manure Spreader. *Agriculture* **2017**, *7*, 95. [CrossRef]

51. Colantoni, A.; Mazzocchi, F.; Laurendi, V.; Grigolato, S.; Monarca, F.; Monarca, D.; Cecchini, M. Innovative solution for reducing the run-down time of the chipper disc using a brake clamp device. *Agriculture* **2017**, *7*, 71. [CrossRef]

agriculture

Article

Identification of Optimal Mechanization Processes for Harvesting Hazelnuts Based on Geospatial Technologies in Sicily (Southern Italy)

Ilaria Zambon * , Lavinia Delfanti, Alvaro Marucci, Roberto Bedini, Walter Bessone , Massimo Cecchini and Danilo Monarca

Department of Agricultural and Forestry Sciences, DAFNE Tuscia University, Via San Camillo de Lellis snc, 01100 Viterbo, Italy; laviniadelfanti@unitus.it (L.D.); marucci@unitus.it (A.M.); r.bedini@unitus.it (R.B.); walter.bessone@regione.piemonte.it (W.B.); cecchini@unitus.it (M.C.); monarca@unitus.it (D.M.)
* Correspondence: ilaria.zambon@unitus.it; Tel.: +39-076-135-7356

Academic Editor: Ole Wendroth
Received: 19 June 2017; Accepted: 6 July 2017; Published: 9 July 2017

Abstract: Sicily is a region located in the southern Italy. Its typical Mediterranean landscape is appreciated due to its high biodiversity. Specifically, hazelnut plantations have adapted in a definite area in Sicily (the Nebroidi park) due to specific morphological and climatic characteristics. However, many of these plantations are not used today due to adverse conditions, both to collect hazelnuts and to reach hazel groves. Though a geospatial analysis, the present paper aims to identify which hazelnut contexts can be actively used for agricultural, economic (e.g., introduction of a circular economy) and energetic purposes (to establish a potential agro-energetic district). The examination revealed the most suitable areas giving several criteria (e.g., slope, road system), ensuring an effective cultivation and consequent harvesting of hazelnuts and (ii) providing security for the operators since many of hazelnut plants are placed in very sloped contexts that are difficult to reach by traditional machines. In this sense, this paper also suggests optimal mechanization processes for harvesting hazelnuts in this part of Sicily.

Keywords: hazelnuts; spatial analysis; mechanization processes; precision farming; rural landscape; Sicily

1. Introduction

The rural landscapes of Mediterranean Europe are characterized by their peculiar crops, whose agricultural practices have led to different land use changes [1]. In recent years, there has been a strong abandonment of agricultural areas [2,3], supporting a consequent reforestation development [1,4].

Hazelnuts represent ones of most produced nut crops in the Mediterranean contexts, as in Italy [5], since as agricultural products have relevant nutritional and economic value [6]. Given their profitability, they are also grown on unsuitable ground, due to the absence of land use policies (as in Langhe region in Italy) [2,7]. For example, Turkey imposed specific regulations for cultivating hazelnuts in given areas, where the maximum elevation is 750 m, the slope is more than 6% and IV or upper class of LCC [8]. According to such government regulations, potential hazelnut areas can be mapped with specific criteria (e.g., slope, elevation, and land use–land cover) using GIS technology [9]. Consequently, their detection may be useful to observe landscape changes, providing greater support to national and international institutions in the assessment of rural agriculture policies [10] and their latent consequences on local society, landscape, and production [11–13].

Defining hazelnut areas is possible through maps and satellite images by advanced computer programs such as Geographical Information Systems (GIS) and Remote Sensing (RS) technologies, which offer benefits in data management and acquisition [6,14]. In recent decades, GIS and RS have been appreciated within rural applications linked to resources at several spatial scales [9,15]. GIS presents a suitable tool for processing, analyzing, and collecting spatial information [7,16,17]. Spatial analysis reveals elevation, aspect, slope, and soil data using GIS methods and even investigates environmental situations, soil attributes, and topographic changes [6,9,18]. From RS technologies, land cover classification is regularly achieved by a multi-class scenery and supervised arrangement of textural or spectral characteristics at pixel level [19,20]. Remote sensing imagery permits to provide data about hazelnuts from satellite images [21], which can be then integrated to other database in GIS with the aim of securing sustainable development of rural areas [6,22–24]. Therefore, through Geographical Information Systems (GIS) and remote sensing with multi-temporal high-resolution satellite data, land use changes, vegetation cover, soil degradation, and further issues can be monitored integrally [25,26].

Remote identification of hazelnuts is not reasonably straightforward [2,7,27,28]. However, it is necessary (i) to optimize harvesting methods and (ii) to distinguish rural landscape dynamics and socio-economic and land use changes to achieve sustainable development [29,30]. Their detection usually takes place through a visual interpretation of very high resolution remote sensing imagery to exploit spectral and textural features, due to the absence of an automated method [20]. However, few studies have focused on mapping hazel groves with high resolution imagery [7,20,31–33]. Vegetation variables appear continuous and difficult to distinguish, e.g., biomass, fraction of vegetation cover, or leaf area index [28,34]. For instance, NDVI values appear very close for hazel groves and further woody vegetation [20]. In fact, it is difficult distinguishing hazelnuts from forest areas and other similar crops (such as olives) that are also typical of the Mediterranean landscape [35]. Their identification from other areas can decrease the inventory expenses by saving money and time [35]. The existence of vegetation maps, performed through Geographic Information Systems (GIS), can be useful for both qualitative and quantitative assessments of natural resources in a definite context [36–40].

The importance of having analytical parameters is essential to find hazelnut plants. The latter are usually located at an altitude of 500 and 1000 m [41]. Their typical altimetry is motivated by the degree of humidity and climate, with a slope between 6% and 30% [6]. Furthermore, the cultivation of hazelnuts is not recommended on steep slopes, since they are not able to prevent and hinder potential soil erosion processes [42,43].

Hazelnut production is frequently characterized by irregular plantations and inconstant density, from steep slopes and rough terrain environments [44]. There are several mechanization methods for collecting hazelnuts, aiming to rationalize costs and harvest production using appropriate existing technologies [45,46]. Several research activities have been launched to assess the collection of hazelnuts, minimizing the risks for the operators in the field (e.g., risk of overturning) [44]. Hazelnuts are usually planted in rows along which herbicides are distributed during the year on the herbaceous vegetation for improving mechanical operation during the harvest [47]. The major problem during the hazelnut collection concerns the situations of high slopes and terraces in addition to the risk of roll-over problems [44]. Furthermore, the intense hazelnut harvesting can lead to negative consequences (e.g., soil erosion) [47] and it is therefore necessary to evaluate how to optimize the collection depending on the soil characteristics.

The purpose of this paper is to identify hazelnuts with the aim of proposing strategies and optimizing mechanization systems through geo-spatial technologies. The case study focuses on 10 municipalities in the Sicily region, which are part of the National association of hazelnuts. In these contexts, many hazelnut plantations appear to be woods. Hazelnuts have well-adapted in the Nebrodi mountains [38], but very often are in problematic areas to reach and work in safety. The present paper aims to recognize the areas that really can contribute to the primary sector in economic terms, estimating the potential hazelnut cultivation, ensuring opportunities for cultivation and the security for operators during the harvesting according to the intrinsic characteristics of such context. In this framework, an optimization of collection and mechanization processes, depending on geo-morphological and territorial characteristics and avoiding possible pollution, was reached. This first examination estimates the biomass obtained from suitable hazelnut plants pruning, as a real solution to produce energy through thermo-chemical processes, i.e., combustion, gasification, and pyrolysis [48]. Finally, the work aims to suggest the consolidation of an agro-energetic district in this context. The latter provides several benefits, as it strengthens the local economy linked to the cultivation of hazelnuts and can start a reality based on the circular economy with the purpose of re-using agricultural residues for energy purposes.

2. Materials and Methods

2.1. Context of Study

Sicily is a Southern Italian region with many forests and fields designed for agricultural activities. Among these, hazelnuts have settled as one of the most visible crops in the north-eastern part of Sicily (along the Nebrodi mountains), given the confident morphological and climatic features [38]. The cultivation of hazelnuts in Sicily covers a surface area of 16,482 hectares, producing each year around 204,306 quintals. The diffusion of hazelnut trees in this context took place in 1890, after the crisis of gelsiculture. Today, thanks to their ease of adaptation, dense root system and profitable productivity, hazelnuts are the predominant yield of the Nebrodi agrarian landscape [38]. In this regard, the municipalities of the province of Messina of Castell'Umberto, Montalbano Elicona, Sant'Angelo di Brolo, Raccuja, Santa Domenica Vittoria, San Piero Patti, San Salvatore Fitalia, Sinagra, Tortorici and Ucria are part of the National Association of Hazel Towns ('Associazione Nazionale Città della Nocciola'), representing the region of Sicily.

2.2. Data Analysis and Materials

ESRI ArcGIS software was used to integrate data and accomplish spatial analysis [6]. GIS technology is decisive to spatial surveys for examining the context of the study. As computer-based system, it allows to capture, storage, recovery, analyze and display geographic data [17]. In this study, GIS techniques were used to overlay maps (vegetation map of Sicily, Corine Land Cover (CLC), and other geospatial data, as well as road system), to make elaborations examining where the hazelnuts are located and to hypothesize mechanization processes focusing on some of their morphological characteristics: DTM, slope, aspect and curvature. The National Terrain Model (DTM) map is the representation of the interpolation of orographic data from the map of the Military Geographic Institute. The resulting product is a 20 m regular step matrix, whose elements (pixels) show the values of the quotas. The Slope identifies the maximum rate of change in value from that cell to its neighbors. Principally, the maximum change in elevation over the distance (among the cell and its eight neighbors) finds the steepest downhill descent from the cell. The Curvature displays the shape or curvature of the slope and is calculated by computing the second derivative of the surface. The curvature, parallel to the slope, indicates the direction of maximum slope. A part of a surface can be concave or convex, by looking at the curvature value. It affects the acceleration and deceleration of flow across the surface: (i) a negative value indicates that the surface is upwardly convex at that cell, and flow will be decelerated, (ii) a positive profile indicates that the surface is upwardly concave at that cell, and the flow will be accelerated, and (iii) a value of zero indicates that the surface is linear. As the slope direction, aspect displays the downslope direction of the maximum rate of change in value from each cell to its neighbors. The values of each cell in the output raster designate the compass direction that the surface faces at such location, measured in degrees from 0 (north) to 360 (again north). Having no downslope direction, flat areas assume a value of -1.

The vegetation map was used as the base for the land use. It represents a convenient combination of the vegetal landscape, whose complex diversity reproduces the greatest physiographic, geomorphological, lithological, and bioclimatic variability of this region. In fact, the vegetation map is characterized by 36 phytocoenotic categories. As a result of years of research, it gives a summary of the widespread phytosociological and cartographic literature in Sicily. It was performed through Geographic Information Systems (GIS) at several scales (1:50,000, 1:25,000, and 1:10,000) and provides both qualitative and quantitative assessments of natural resources [36–40].

The vegetation map of Sicily was prepared at a 1:250,000 scale according several stages: (i) preparation of a GIS project (1:10,000 scale) with an inclusive database and thematic layers with georeferenced materials, (ii) photo-interpretation of the vegetation with satellite images (e.g., Landsat TM), orthophotos and digital data on the Technical Map of Sicily; (iii) validations with other maps, such as land use, vegetation or geology, (iv) validation of the photo-interpretation through field survey and verification, (v) digitization of the outcomes and further data, and (vi) phytosociological classification of the mapped types, categorized by 36 phytocoenotic classes [38]. Therefore, the vegetation map identifies all the existing crops in Sicily in a precise and detailed way. For instance, hazelnuts (identified with the code 202) occupy a surface area of about 9500 hectares.

2.3. Mechanization Framework

The cultivation of hazelnuts is characterized by several factors that make it difficult and dangerous to use mechanization systems for operators, at all stages of cultivation, especially in the harvesting phase. Some of these factors are predominantly irregular plantations, a high degree of acclimatization of the slopes (which also reach 35 degrees), uneven ground conditions, a lack or absence of business and interpersonal viability, presence of obstacles to the passage of machines, and unusual soil management with the abandonment of pruning residues on the ground.

The north-eastern part of Sicily along the Nebrodi mountains has seen the spontaneous diffusion of hazelnuts, which have easily adapted [38]. Despite their potential productivity, hazelnut plants are placed in very problematic environments, especially for the harvest phase, and therefore most of them are abandoned. Traditional vehicles have difficulty reaching these contexts (e.g., steep slopes that make it unsafe for operator intervention). Therefore, the currently-mechanization methodologies are equal to zero. In fact, harvesting is still by hand-picking in the few cultivated areas.

Focusing the prototype tested by [44], the present work suggests using a similar device that is self-propelled and easily transportable for harvesting in areas with poor or absent roads between farms (Figure 1). The device can move even under critical slope circumstances (even up to 30–35%) and overcoming substantial difficulties (e.g., terraces, where can be assemble the harvester to a mini crawler with hydraulic or hydrostatic transmission). In this manner, mechanization can be introduced in principally disadvantaged areas, with consequences in terms of safety for operators and a cost-benefit decrease. In operational stages, however, their prototype collects in a stationary position with the assistance of a suction line, permitting operation on highly sloped surfaces (more than 20%). The prototype tested by [44] is ideal for this Sicilian context, avoiding problems linked to steep slopes and movement among hazelnut groves.

Figure 1. Photo of the small-scale machine for nuts harvesting proposed in the study of [44].

3. Results

By means of a first GIS processing, the municipalities chosen are counted on a surface area of almost 4970 hectares (representing 52% of the hazels in the Sicily region) (Table 1). Comparing the vegetation map with the CLC, the accuracy of the first map was confirmed (Figure 2). While the CLC considered hazelnuts as forests, the vegetation map of the Sicily region highlighted their presence as hazelnuts (code 202). By comparing the two maps, 63% of the hazelnuts identified as "orchards" in CLC, while 22% are categorized as "deciduous forests" in CLC. As a first clarification, the CLC tends to aggregate hazelnuts in the category "orchards". Nonetheless, many fields of hazelnuts (22%) visually appeared as forests.

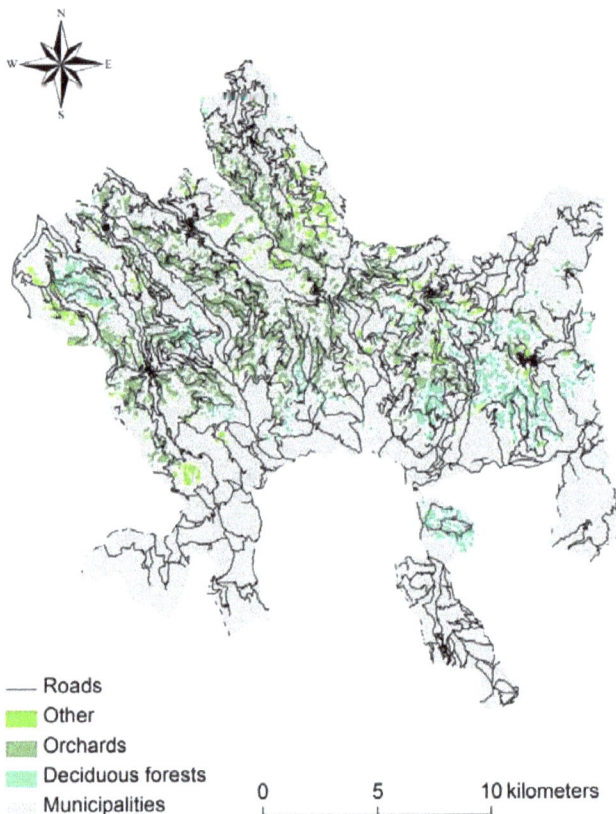

Figure 2. Hazelnut areas identified by the vegetation map of the region of Sicily. Each plot corresponds to the land use observed in CLC. Source: own elaboration.

Table 1. Hectares of hazelnuts belonging to vegetation map overlapped to CLC classes for each municipality.

Hectares of Hazelnuts Belonging to CLC Classes	Continuous Urban Fabric	Discontinuous Urban Fabric	Arable Crops in Non-Irrigated Areas	Orchards and Minor Fruits	Olive Grove	Annual Crops Associated with Permanent Crops	Complex Crop and Systems	Mainly Occupied Areas	Deciduous Forests	Areas with Natural Pasture at High Altitude	Areas Affected by Fires	Total of Hazelnuts
Code of CLC	111	112	211	222	223	241	242	243	311	321	334	
Castell'Umberto		20.88		85.16	17.95				0.24	0.00		124.23
Montalbano Elicona		5.16	4.07	91.99	1.47			35.49	457.65	16.82	0.20	612.85
Raccuja		8.46		315.92	15.40			21.75	25.37	4.64		391.53
San Piero Patti	3.32			423.81	66.77			91.16	205.15	17.86		815.28
San Salvatore di Fitalia			7.21		5.07		19.72		138.41	43.77		289.47
Santa Domenica Vittoria		14.49		65.52				2.50	118.95	0.00		118.95
Sant'Angelo di Brolo	0.22		0.71	471.31	26.53	23.51	2.18	38.21	19.68	82.73		659.21
Sinagra		1.87		455.59	44.39			3.12	1.26	9.75		515.98
Tortorici		11.97	0.47	669.69				0.56	92.44	63.45		838.59
Ucria	3.54	7.17		535.19	0.19			1.63	54.85	4.32		603.35
Total	3.54	70.00	12.45	3114.18	177.78	23.51	21.90	194.42	1114.00	243.33	0.20	4969.43
	0%	1%	0%	63%	4%	0%	0%	4%	22%	5%	0%	100%

Source: own elaboration.

Representing 43% of the hazelnuts considered, San Piero Patti and Tortorici are the municipalities that recorded the highest presence of hazel trees within their administrative boundaries. Hazelnuts plants prefer high altitudes (Table 2). In fact, 84% of them can be observed between 500 and 1000 m above sea level (with an average of 755 m by examining municipalities in analysis). Castell'Umberto, San Salvatore di Fitalia, Sant'Angelo di Brolo, and Sinagra are the municipalities that identified a larger percentage of hazelnuts at a moderate altimeter than the others, i.e., between 500 and 750 m. The municipality of Santa Domenica Vittoria demonstrated that 86% of its hazelnuts are above 1000 m, although it recorded, in quantitative terms, a reduced surface area for hazelnuts compared to other study contexts.

Table 2. Surface area (hectares) (top) and percentage (bottom) of hazelnuts for each municipality depending on DTM classes (meters).

	DTM classes (meters)					
	<250	250–500	500–750	750–1000	>1000	% Area Compared to Total
Castell'Umberto	0.0	10.7	85.5	12.9	0.0	2%
Montalbano Elicona	0.0	4.8	158.8	353.1	42.5	12%
Raccuja	0.0	0.5	143.1	195.0	18.3	8%
San Piero Patti	0.0	56.9	309.3	340.5	42.6	16%
San Salvatore di Fitalia	3.1	71.1	149.1	45.3	0.0	6%
Santa Domenica Vittoria	0.0	0.0	0.0	15.5	93.4	2%
Sant'Angelo di Brolo	0.8	70.5	346.9	191.8	0.0	13%
Sinagra	5.8	133.9	214.0	117.9	6.6	10%
Tortorici	0.0	31.9	313.1	368.4	58.2	17%
Ucria	0.0	24.9	173.0	317.6	43.3	12%
	<250 (%)	250–500 (%)	500–750 (%)	750–1000 (%)	>1000 (%)	Average DTM
Castell'Umberto	0%	10%	78%	12%	0%	627
Montalbano Elicona	0%	1%	28%	63%	8%	825
Raccuja	0%	0%	40%	55%	5%	802
San Piero Patti	0%	8%	41%	45%	6%	754
San Salvatore di Fitalia	1%	26%	56%	17%	0%	595
Santa Domenica Vittoria	0%	0%	0%	14%	86%	1088
Sant'Angelo di Brolo	0%	12%	57%	31%	0%	671
Sinagra	1%	28%	45%	25%	1%	614
Tortorici	0%	4%	41%	48%	8%	779
Ucria	0%	4%	31%	57%	8%	797
						755

Through the GIS program, maps concerning DTM, slope (classified in percentage terms), aspect, and curvature were produced (Figure 3). Starting from a DTM map the slope of the roads, which must be driven by the vehicles, and the specific slope of each hazelnut areas were calculated. Ambiguous contexts (such as a high degree of slope or altitude) have been assessed in a parallel analysis through available orthophoto investigation, even if the vegetation map detected hazelnut plants. Within this operation, the elevate degree of correctness of the vegetation map of Sicily can be confirmed.

Aspect   Slope   Curvature

DTM (in meters)

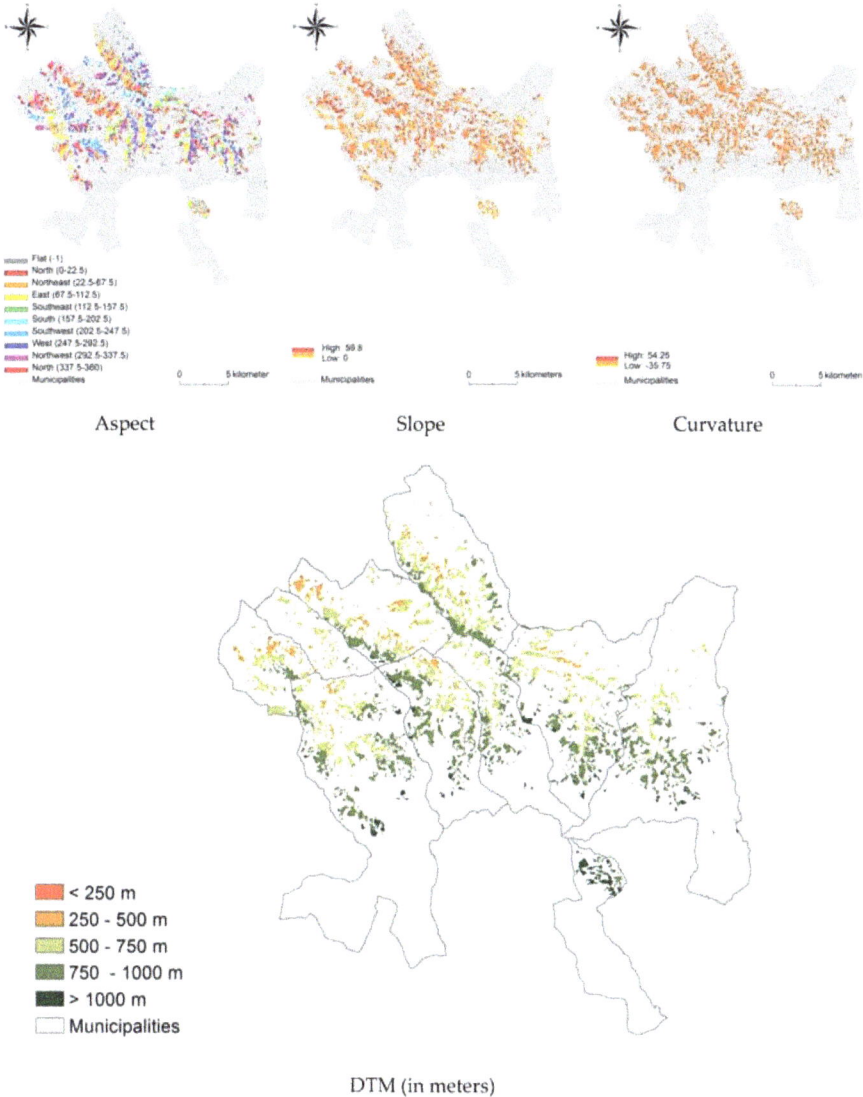

Figure 3. Morphological structures of the territory using GIS program. Source: own elaboration.

The slope of hazelnut areas was classified into seven classes: '1': 0%; '2': 1–10%; '3': 11–20%; '4': 21–30%; '5': 31–40%; '6': 41–50%; '7': >50%. The slope of the streets in the ten municipalities was classified into 17 classes: '1': 0%; '2': 1–2%; '3': 3–4%; '4': 5–6%; '5': 7–8%; '6': 9–10%; '7': 11–12%; '8': 13–14%; '9': 15–16%; '10': 17–18%; '11': 19–20%; '12': 21–22%; '13': 23–24%; '14': 25–26%; '15': 27–28%; '16': 29–30%; '17': >30%. Zones with a steep slope (>30%) and high altitude (>1000 m) are the ones to avoid for mechanized harvesting as it results in increased risk for operators when they should collect hazelnuts.

Through the raster calculator tool using GIS program, the territory was analyzed observing the most suitable places to introduce mechanization processes. Figure 4 displays the optimal contexts for hazelnuts (in legend with the label "0"), with minimal risk for operators, where the slope is minimal, with optimum altitudes to hazelnuts and ease in terms of mobility for the machines that need to reach such areas. There are also further favorable contexts for hazelnuts with good altitude and slopes. Finally, areas that should be avoided for greater risk for operators, due to their high altitude and slopes, discontinuous road system with strong slopes.

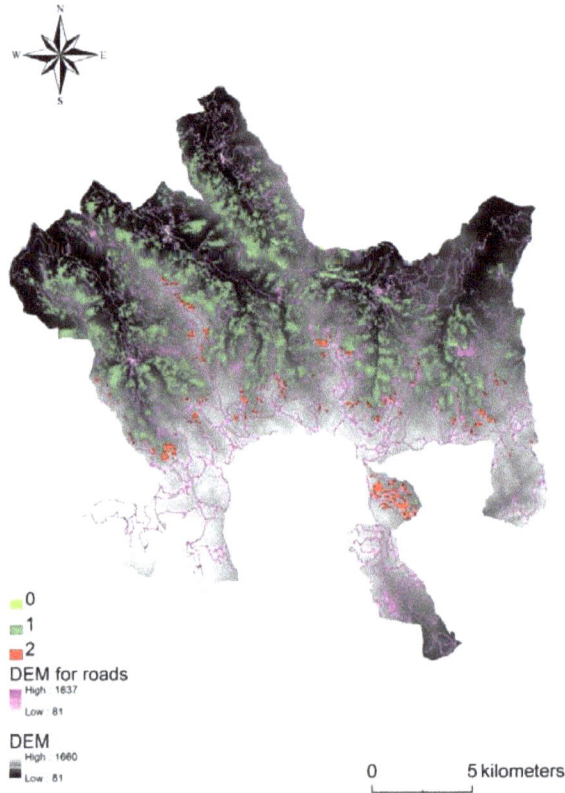

Figure 4. Possibility of mechanization. Legend 0: optimal areas for hazelnut, with minimal risk for operators; 1: favorable areas for hazelnut with good altitude and slopes. 2: areas to be avoided for greater risk for operators, including high altitude and slopes and road systems with strong slopes. Source: own elaboration.

Checking the results obtained, a region group elaboration was run using the GIS program. It identifies the degree of feasibility of cultivation and collection of hazelnuts depending on the morphological characteristics (Figure 5). Four groups of hazelnut areas can be observed. In this elaboration, the most optimal contexts emerge both to grow and manage the cultivation of hazelnuts and to provide the right security measures for the operators who must collect the hazelnuts (class "1"). In fact, in Figure 5, it is possible to clearly distinguish the southern zones, which are the ones that are higher in altitude (>1000 m), sloping (>30%) and mostly affect the safety of workers (class "4"). However, the best areas ("1") occupy only 430 hectares (about 9% compared to the total surface area of hazelnuts in the ten municipalities). Unsuitable contexts have a surface of 370 hectares (about 7% of the total surface area in analysis). The intermediate areas (classes "2" and "3" for the region group elaboration) are those that occupy the largest surface areas (almost 4200 hectares). Finally, 4600 hectares can be used as agro-energetic districts.

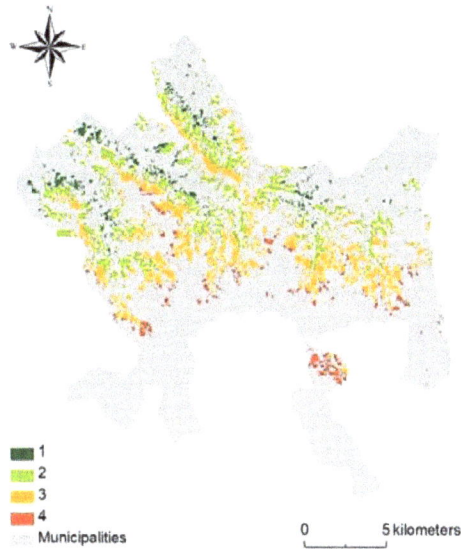

Figure 5. Region group elaboration using GIS program to identify the degree of feasibility of cultivation and collection of hazelnuts depending on the morphological and spatial characteristics. Source: own elaboration.

4. Discussion

Hazelnuts represents one of the major economic realities that constitute the primary sector of the Sicily region [49]. The latter is a unique Mediterranean context, given its climate, landscape, and peculiar characteristics [50,51]. Particularly, the Nebroidi park allows for the easy adaptability of hazelnuts [38]. However, high altitudes and the acclivity of slopes make it difficult to cultivate and harvest hazelnuts [42,43]. This study aims to identify the most favorable contexts to increase the growing and harvesting of hazelnuts using appropriate vehicles. First, the territorial characteristics should be considered, such as slope or the road system necessary to reach these contexts. Using and processing data through GIS technologies and databases obtained by remote sensing processes at local level was decisive.

Spatial data collection permitted the comparison of different databases. The vegetation map of Sicily has highlighted how a deep knowledge of the local contexts and the use of remote sensing and GIS technologies, in addition to a large bibliographic collection, allows for a detailed analysis, identifying several kinds of crops. In fact, limiting to a CLC map could causes an actual error in calculating surface areas destined for hazelnuts: only 63% of hazelnuts fall into the category of "orchards" in the CLC map. Data processing has confirmed the adequacy of the vegetation map of Sicily: most of the hazelnuts are found at slopes that are not too high (between 6% and 30%) [6] and at altitudes between 500 and 1000 m [41]. GIS processing has thus let to recognize the most appropriate areas for the hazelnuts, since their cultivation is not recommended on the steep slopes, since they cannot prevent and hinder environmental matters, as soil erosion processes [42,43]. Furthermore, when some contexts appeared uncertain (e.g., when the Vegetation map of Sicily detected hazelnut plantations along high degree of slope or altitude), a parallel analysis (orthophoto investigation of specific areas) assessed such outcomes, confirming the high correctness of the Vegetation map of Sicily.

Another issue that must be addressed in this paper concerns the collection of hazelnuts. In these contexts, traditional methods are still used, such as hand-picking. This makes the collection of hazelnuts expensive, wasteful, with high labor costs and long working hours. As a possible solution to the mechanization of harvesting the nuts in soils with planting distances and irregular with steep slopes (from 24 to 35%), the prototype, proposed by [44] can adapt to the most difficult conditions. It is not necessary to use other harvester machines. Very often, hazelnuts are located along inclined slopes or unconnected areas, where traditional means of mechanization fail to work optimally. The prototype (i) is smaller than existing machines, (ii) ensures agility in maneuvering and high stability in steep slopes, (iii) is easy to use and versatile, (iv) reduces capital amortization times, (v) is easy to be transported by simple means such as small trolleys or pickups, (vi) increases the capacity to collect hazelnuts, and (vii) improves working conditions (e.g., substantial reduction in the risk of biomechanical overload compared to manual harvesting). The prototype of [44] allows for simple collection, safeguarding the health and safety of workers, and reduces the time necessary for the hazelnut harvest (e.g., it separates hazelnuts from other elements such as weeds or leaves). Furthermore, as a work accessory, the prototype proposed by [44] fits to other machines depending on the working context.

Besides identifying the most suitable areas, the present paper also aims to offer a chance of sustainable development, such as increasing cultivation of hazelnuts, protecting the workers' safety and optimizing work times concerning picking hazelnuts given the intrinsic territorial adversity. The concepts of circular economy and agro-energy districts could be effectively applied in these territories [52,53]. From the point of view of agro-energetic districts, it is assumed that the former depends on several parameters, i.e., the cultivation type and site and the planting distance, defining the most appropriate use of residual biomass [48]. For intensive farming of hazelnut, the pruned biomass can reach about 1848 kg/ha [54]. Obtainable residual biomass from hazelnut trees pruning can be positively considered as an actual economic chance for this area. From our study, it is possible to estimate to get a biomass of 8500 kg (4600 hectares).

In conclusion, from the economic point of view, a greater cultivation of hazelnuts would also give more employment alternatives, increasing the employment status and leading to a valorization of local agriculture [55]. As in other region (e.g., Latium and Piedmont in Italy or in Turkey) where hazelnuts are important for the primary sector [7,8,56–58], they can be defined as an economic resource in Sicily since they could provide income opportunities in hilly and mountainous areas where other agricultural activities are limited by the hostile environment [55]. Potential revenue deriving from this kind of cultivation can be estimated depending on how many hectares are put back into culture [55]. Finally, hazelnuts are defined as one of the most profitable fruit, demonstrating a high degree of sustainability, mostly owing to the low input necessities for orchard management and the opportunity of using agricultural waste as potential biomass [48,52,54,55,58,59].

5. Conclusions

The present paper started from the collection and comparison of available materials. The GIS elaboration is decisive for analyzing the Sicilian context and discriminate the spatial database by choosing the most appropriate one. Using this method, the most suitable area for cultivation hazelnuts can be detected. Also, innovative mechanization processes should be employed since they are still undeveloped and can mitigate the physical obstacles to hazelnut production (e.g., discontinuous road system, high slope). Finally, a sustainable vision is offered with the aim to promote a circular economy and agro-energetic district in this Sicilian context based on hazelnut cultivation.

Acknowledgments: This study was supported by the SICILNUT Project founded by MIPAAF.

Author Contributions: Ilaria Zambon analyzed the data and wrote the paper; Lavinia Delfanti collected the materials (e.g., shapefile data) concerning Sicily; Roberto Bedini, Massimo Cecchini, and Danilo Monarca collected the materials concerning mechanization processes; Alvaro Marucci was involved in the critical review of the results obtained; and Walter Bessone revised the manuscript.

Conflicts of Interest: The authors declare no conflict of interest.

References

1. Bonet, A. Secondary succession of semi-arid Mediterranean old-fields in south-eastern Spain: Insights for conservation and restoration of degraded lands. *J. Arid Environ.* **2004**, *56*, 213–233. [CrossRef]
2. Godone, D.; Garbarino, M.; Sibona, E.; Garnero, G.; Godone, F. Progressive fragmentation of a traditional Mediterranean landscape by hazelnut plantations: The impact of CAP over time in the Langhe region (NW Italy). *Land Use Policy* **2014**, *36*, 259–266. [CrossRef]
3. Sitzia, T.; Semenzato, P.; Trentanovi, G. Natural reforestation is changing spatial patterns of rural mountain and hill landscapes: A global overview. *For. Ecol. Manag.* **2010**, *259*, 1354–1362. [CrossRef]
4. Sluiter, R.; de Jong, S.M. Spatial patterns of Mediterranean land abandonment and related land cover transitions. *Landsc. Ecol.* **2007**, *22*, 559–576. [CrossRef]
5. FAO. Food and Agricultural Commodities Production. Available online: http://faostat.fao.org/site/339/default.aspx (accessed on 1 July 2010).
6. Aydinoglu, A.C. Examining environmental condition on the growth areas of Turkish hazelnut (*Corylus colurna* L.). *Afr. J. Biotechnol.* **2010**, *9*, 6492–6502.
7. Reis, S.; Yomralioglu, T. Detection of current and potential hazelnut (*Corylus*) plantation areas in trabzon, north east Turkey using GIS and RS. *J. Environ. Biol.* **2006**, *27*, 653–659. [PubMed]
8. TURKSTAT. Turkish Statistical Institute. Available online: www.turkstat.gov.tr (accessed on 12 May 2001).
9. Sarıoğlu, F.E.; Saygın, F.; Balcı, G.; Dengiz, O.; Demirsoy, H. Determination of potential hazelnut plantation areas based GIS model case study: Samsun city of central Black Sea region. *Eurasian J. Soil Sci.* **2013**, *2*, 12–18.
10. London Economics. *Evaluation of the CAP Policy on Protected Designations of Origin (PDO) and Protected Geographical Indications (PGI)*; European Commission—Agriculture and Rural Development: Bruxelles, Belgium, 2008; p. 275.
11. Martinez-Casasnovas, J.A.; Ramos, M.C.; Cots-Folch, R. Influence of the EU CAP on terrain morphology and vineyard cultivation in the Priorat region of NE Spain. *Land Use Policy* **2010**, *27*, 11–21. [CrossRef]
12. Van Berkel, D.B.; Verburg, P.H. Sensitising rural policy: Assessing spatial variation in rural development options for Europe. *Land Use Policy* **2011**, *28*, 447–459. [CrossRef]
13. Westhoek, H.J.; van den Berg, M.; Bakkes, J.A. Scenario development to explore the future of Europe's rural areas. *Agric. Ecosyst. Environ.* **2006**, *114*, 7–20. [CrossRef]
14. Official Gazette. *The Regulation of the Law Planning Hazelnut Production and Determining Hazelnut Plantation Areas*; Official Gazette: Ankara, Turkey, 2009; pp. 27289.14.
15. Dengiz, O.; Ozcan, H.; Köksal, E.S.; Kosker, Y. Sustainable Natural Resource Management and Environmental Assessment in The Salt Lake (Tuz Golu) Specially Protected Area. *J. Environ. Monit. Assess.* **2010**, *161*, 327–342. [CrossRef] [PubMed]
16. Lioubimtseva, E.; Defourny, P. GIS based landscape classification and mapping of European Russia. *Landsc. Urban Plan.* **1999**, *44*, 63–75. [CrossRef]
17. Longley, P.A.; Goodchild, M.F.; Maguire, D.J.; Rhind, D.W. *Geographic Information Systems and Science*; Bath Press: London, UK, 2001.
18. Bolca, M.; Kurucu, Y.; Dengiz, O.; Nahry, A.D.H. Terrain characterization for soils survey of Kucuk Menderes plain, South of Izmir, Turkey, using remote sensing and GIS techniques. *Zemdirb. Agric.* **2011**, *98*, 93–104.
19. Wilkinson, G. Results and implications of a study of fifteen years of satellite image classification experiments. *IEEE Trans. Geosci. Remote Sens.* **2005**, *43*, 433–440. [CrossRef]
20. Reis, S.; Taşdemir, K. Identification of hazelnut fields using spectral and Gabor textural features. *ISPRS J. Photogramm. Remote Sens.* **2011**, *66*, 652–661. [CrossRef]
21. Lillesand, T.M.; Kiefer, R.W. *Remote Sensing and Image Interpratation*; The Lehigh Press: New York, NY, USA, 2000.
22. Cohen, Y.; Shoshany, M. A national knowledge-based crop recognition in Mediterranean environment. *Int. J. Appl. Earth Observ. Geoinf.* **2002**, *4*, 75–87. [CrossRef]
23. Grauke, L.J.; Thompson, T.E. Rootstock development in temperate nut crops. Genetics and breeding of tree fruits and nuts. *Acta Horticult.* **2003**, *622*, 553–566. [CrossRef]
24. Yomralioglu, T.; Inan, H.I.; Aydinoglu, A.C.; Uzun, B. Evaluation of initiatives for spatial information system to support Turkish agriculture policy. *Sci. Res. Essay* **2009**, *4*, 1523–1530.

25. Mundia, C.N.; Aniya, M. Analysis of land use/cover changes and urban expansion of Nairobi city using remote sensing and GIS. *Int. J. Remote Sens.* **2005**, *26*, 2831–2849. [CrossRef]

26. Yuan, F.; Sawaya, K.E.; Loeffelholz, B.; Bauer, M.E. Land cover classification and change analysis of the Twin Cities (Minnesota) metropolitan area by multi temporal Landsat remote sensing. *Remote Sens. Environ.* **2005**, *98*, 317–328. [CrossRef]

27. Franco, S. Use of remote sensing to evaluate the spatial distribution of hazelnut cultivation: Results of a study performed in an Italian production area. *Acta Horticult.* **1997**, *445*, 381–388. [CrossRef]

28. Kavzoglu, T. Increasing the accuracy of neural network classification using refined training data. *Environ. Model. Softw.* **2009**, *24*, 850–858. [CrossRef]

29. De Aranzabal, I.; Schmitz, M.F.; Aguilera, P.; Pineda, F.D. Modelling of landscape changes derived from the dynamics of socio-ecological systems: A case of study in a semiarid Mediterranean landscape. *Ecol. Indic.* **2008**, *8*, 672–685. [CrossRef]

30. Tzanopoulos, J.; Jones, P.J.; Mortimer, S.R. The implications of the 2003 Common Agricultural Policy reforms for land-use and landscape quality in England. *Landsc. Urban Plan.* **2012**, *108*, 39–48. [CrossRef]

31. Fabi, A.; Varvaro, L. Remote sensing in monitoring the dieback of hazelnut on the 'Monti Cimini' district (Central Italy). *Acta Horticult.* **2009**, *845*, 521–526. [CrossRef]

32. Taşdemir, K. Exploiting spectral and spatial information for the identification of hazelnut fields using self-organizing maps. *Int. J. Remote Sens.* **2012**, *33*, 6239–6253. [CrossRef]

33. Yalniz, I.; Aksoy, S. Detecting regular plantation areas in satellite images. In Proceedings of the IEEE 17th Signal Processing and Communications Applications Conference, Antalya, Turkey, 9–11 April 2009.

34. Kimes, D.S.; Nelson, R.F.; Manry, M.T.; Fung, A.K. Attributes of neural networks for extracting continuous vegetation variables from optical and radar measurements. *Int. J. Remote Sens.* **1998**, *19*, 2639–2663. [CrossRef]

35. Aslan, Ü.; Özdemir, İ. Separation of Agricultural Aimed Plantations from the Forest Cover by Using the LANDSAT-5TM and SPOT-4 HRVIR Data in Turkey. International Archives of Photogrammetry. *Remote Sens. Spat. Inf. Sci.* **2004**, *36*, 324–327.

36. Biondi, E.; Calandra, R. La cartographie phytoécologique du paysage. *Écologie* **1998**, *29*, 145–148.

37. Biondi, E.; Catorci, A.; Pandolfi, M.; Casavecchia, S.; Pesaresi, S.; Galassi, S.; Pinzi, M.; Vitanzi, A.; Angelini, E.; Bianchelli, M.; et al. Il Progetto di "Rete Ecologica della Regione Marche" (REM), per il monitoraggio e la gestione dei siti Natura 2000 e l'organizzazione in rete delle aree di maggiore naturalità. *Fitosociologia* **2007**, *44*, 89–93.

38. Gianguzzi, L.; Papini, F.; Cusimano, D. Phytosociological survey vegetation map of Sicily (Mediterranean region). *J. Maps* **2016**, *12*, 845–851. [CrossRef]

39. Pedrotti, F. *Cartografia Geobotanica*; Pitagora Editrice: Bologna, Italy, 2004; p. 248.

40. Rivas-Martínez, S. Notions on dynamic-catenal phytosociology as a basis of landscape science. *Plant Biosyst.* **2005**, *139*, 135–144. [CrossRef]

41. Duran, C. Drought and vegetation analysis in Tarsus River Basin (Southern Turkey) using GIS and Remote Sensing data. *J. Hum. Sci.* **2015**, *12*, 1853–1866. [CrossRef]

42. Ozturk, I.; Tanik, A.; Seker, D.Z.; Levent, T.B.; Ovez, S.; Tavsan, C.; Ozabali, A.; Sezgin, E.; Ozdilek, O. *Technical Report on the Land-Use Methodology Being Tested and Draft Land-Use Plans, Testing of Methodology on Spatial Planning for ICZM; Akçakoca District Pilot Project*; ITU: Istanbul, Turkey, 2007.

43. Tanik, A.; Seker, D.Z.; Ozturk, I.; Tavsan, C. GIS based sectoral conflict analysis in a coastal district of Turkey. *Int. Arch. Photogramm. Remote Sens. Spat. Inf. Sci.* **2008**, *37*, 665–668.

44. Monarca, D.; Cecchini, M.; Colantoni, A.; Bedini, R.; Longo, L.; Bessone, W.; Caruso, L.; Schillaci, G. Evaluation of safety aspects for a small-scale machine for nuts harvesting. In Proceedings of the MECHTECH 2016 Conference—Mechanization and New Technologies for the Control and Sustainability of Agricultural and Forestry Systems, Alghero, Italy, 29 May–1 June 2016; pp. 32–35.

45. Monarca, D.; Cecchini, M.; Massantini, R.; Antonelli, D.; Salcini, M.C.; Mordacchini, M.L. Mechanical harvesting and quality of "marroni" chestnut. *Acta Horticulturae* **2005**, *682*, 1193–1198. [CrossRef]

46. Formato, A.; Scaglione, G.; Ianniello, D. Application of software for the optimization of the surface shape of nets for chestnut harvesting. *J. Agric. Eng.* **2013**, *44*. [CrossRef]

47. Recanatesi, F.; Ripa, M.N.; Leone, A.; Luigi, P.; Luca, S. Land use, climate and transport of nutrients: Evidence emerging from the Lake Vicocase study. *Environ. Manag.* **2013**, *52*, 503–513. [CrossRef] [PubMed]

48. Bilandzija, N.; Voca, N.; Kricka, T.; Matin, A.; Jurisic, V. Energy potential of fruit tree pruned biomass in Croatia. *Span. J. Agric. Res.* **2012**, *10*, 292–298.

49. Cotugno, L. Territory and Population—Demographic Dynamics in Sicily. *Rev. Hist. Geogr. Toponomast.* **2011**, *6*, 81–91.

50. Barbera, G.; Cullotta, S. An inventory approach to the assessment of main traditional landscapes in Sicily (Central Mediterranean Basin). *Landsc. Res.* **2012**, *37*, 539–569. [CrossRef]

51. Colantoni, A.; Ferrara, C.; Perini, L.; Salvati, L. Assessing trends in climate aridity and vulnerability to soil degradation in Italy. *Ecol. Indic.* **2015**, *48*, 599–604. [CrossRef]

52. Colantoni, A.; Delfanti, L.M.P.; Recanatesi, F.; Tolli, M.; Lord, R. Land use planning for utilizing biomass residues in Tuscia Romana (central Italy): Preliminary results of a multi criteria analysis to create an agro-energy district. *Land Use Policy* **2016**, *50*, 125–133. [CrossRef]

53. Colantoni, A.; Longo, L.; Gallucci, F.; Monarca, D. Pyro-Gasification of Hazelnut Pruning Using a Downdraft Gasifier for Concurrent Production of Syngas and Biochar. *Contemp. Eng. Sci.* **2016**, *9*, 1339–1348. [CrossRef]

54. Cecchini, M.; Monarca, D.; Colantoni, A.; Di Giacinto, S.; Longo, L.; Allegrini, E. Evaluation of biomass residuals by hazelnut and olive's pruning in Viterbo area. In Proceedings of the International Commission of Agricultural and Biological Engineers, Section V. CIOSTA XXXV Conference "From Effective to Intelligent Agriculture and Forestry", Billund, Denmark, 3–5 July 2013.

55. Cerutti, A.K.; Beccaro, G.L.; Bagliani, M.; Donno, D.; Bounous, G. Multifunctional ecological footprint analysis for assessing eco-efficiency: A case study of fruit production systems in Northern Italy. *J. Clean. Prod.* **2013**, *40*, 108–117. [CrossRef]

56. Gönenc, S.; Tanrıvermis, H.; Bülbül, M. Economic assessment of hazelnut production and the importance of supply management approaches in Turkey. *J. Agric. Rural Dev. Trop. Subtrop.* **2006**, *107*, 19–32.

57. Petriccione, M.; Ciarmiello, L.F.; Boccacci, P.; De Luca, A.; Piccirillo, P. Evaluation of 'Tonda di Giffoni' hazelnut (*Corylus avellana* L.) clones. *Sci. Horticult.* **2010**, *124*, 153–158. [CrossRef]

58. Di Giacinto, S.; Longo, L.; Menghini, G.; Delfanti, L.M.P.; Egidi, G.; De Benedictis, L.; Salvati, L. A model for estimating pruned biomass obtained from *Corylus avellana* L. *Appl. Math. Sci.* **2014**, *8*, 6555–6564. [CrossRef]

59. Zambon, I.; Colosimo, F.; Monarca, D.; Cecchini, M.; Gallucci, F.; Proto, A.R.; Colantoni, A. An innovative agro-forestry supply chain for residual biomass: Physicochemical characterisation of biochar from olive and hazelnut pellets. *Energies* **2016**, *9*, 526. [CrossRef]

agriculture

MDPI

Article

Precision Farming in Hilly Areas: The Use of Network RTK in GNSS Technology

Alvaro Marucci, Andrea Colantoni, Ilaria Zambon * and Gianluca Egidi

Department of Agricultural and Forestry Sciences (DAFNE), University of Tuscia, Via S. Camillo de Lellis, 01100 Viterbo, Italy; marucci@unitus.it (A.M.); colantoni@unitus.it (A.C.); egidi.gianluca@yahoo.it (G.E.)
* Correspondence: ilaria.zambon@unitus.it; Tel.: +39-0761357356

Academic Editor: Les Copeland
Received: 21 June 2017; Accepted: 19 July 2017; Published: 20 July 2017

Abstract: The number of GNSS satellites has greatly increased over the last few decades, which has led to increased interest in developing self-propelled vehicles. Even agricultural vehicles have a great potential for use of these systems. In fact, it is possible to improve the efficiency of machining in terms of their uniformity, reduction of fertilizers, pesticides, etc. with the aim of (i) reducing the timeframes of cultivation operations with significant economic benefits and, above all, (ii) decreasing environmental impact. These systems face some perplexity in hilly environments but, with specific devices, it is possible to overcome any signal deficiencies. In hilly areas then, the satellite-based system can also be used to safeguard operators' safety from the risk of rollover. This paper reports the results obtained from a rural development program (RDP) in the Lazio Region 2007/2013 (measure project 1.2.4) for the introduction and diffusion of GNSS satellites systems in hilly areas.

Keywords: precision farming; hilly areas; Network RTK; GNSS technology

1. Introduction

Precision agriculture has provided a remarkable positive contribution to the primary sector globally at various levels [1–4]. Unlike conventional agricultural methods, it can adapt crop yields by considering the local variability of the physical, chemical, and biological characteristics of soils, as well as the application time through the development of technological and computer support [5–7]. Precision agriculture employs machines equipped with "intelligent systems", which can measure production factors in relation to the real needs of a plot and of different homogeneous areas within it [8]. According to [9], several goals can be achieved, such as: (i) enhancing yields with the same total inputs; (ii) reducing inputs with equal yield; (iii) increasing yields by reducing inputs at the same time. Furthermore, when applied to conservative soil methods, the principles of precision agriculture can (i) maintain environmental benefits, (ii) improve corporate income, and (iii) rationalize the use of machines [10,11].

Analyzing precision farming from environmental, economic, and management points of view, several positive impacts emerge, which should not be underestimated [10,12–14]. Focusing on economic benefits, an overall optimization of crop interventions can be detected, together with a quantitative reduction of distributed chemicals [13] and an improved operational and safety capacity for operators in their workplace. It should also be stressed that the above-mentioned economic benefits can also be derived from two reasons: firstly, for the probable growth in costs of the production factors; secondly, on the other hand, for the likely reduction in the cost of purchasing technical equipment, which occurs as demand and technological evolution rise.

The modern approach to precision agriculture is based on flexible customized and equipped technology solutions with extensive interoperability. This method is essential as it is able to manage the wide variability of usage conditions in typical Italian farms, which are characterized by their (often highly fragmented) land capital [15–19]. The diffusion of highly-innovative techniques, favored by their adaptability to each agricultural reality, would substantially contribute to the modern transformation of productive processes in the Italian primary sector [20]. Nowadays, the use of a service platform for product data capture related to production is widely available in several areas through local devices. The latter include sensors, automatic guidance systems with Global Navigation Satellite Systems (GNSS), and central processing systems. Specifically, GNSS technology, as a European global navigation satellite system, ensures greater positioning accuracy and reliability compared to GPS, as it is designed to provide real-time positioning services [21–23]. The use of the signals of GNSS was initially anticipated in the late 1980s [24,25] and experimentally confirmed in the early 1990s [26]. Regardless of this initial approach to GNSS techniques, the initial efforts to evaluate soil moisture from reflectivity dimensions arise in 2003 [27]. GNSS, such as the Global Positioning System (GPS) technology, has grown and developed as a controlling atmospheric remote sensing tool able to provide precise observations of atmospheric parameters. The success of GPS has stimulated additional progress of GNSS technology. Nevertheless, these two systems can be defined as complementary [28]. An additional value of these kind of technologies is that they can perform in all weather conditions. This feature of GNSS offers useful information also during cloudy and rainy days, which are still uncertain blocks to radar systems and low Earth orbiting satellites [29].

Potential technical enhancements must be set in the background of processing power and sophisticated technology previously incorporated into GPS receivers [13]. These devices (as GNSS technology) permit land mapping. Through the latter, a geo-referenced data survey, allows observation of the characteristics of the field to be cultivated (e.g., size and perimeter) and defines subsequent planning phases. An information flux between rural machines and the farm's management can be performed. In this way, through these devices, it is possible to optimize automation with vehicles, which can repeat (or exclude) a path already done. Lastly, an appropriate technical analysis points out (i) the enhancement points and agronomic practices to be implemented or, simply, (ii) a unique business database reporting the surfaces of the cultivated areas and the activities carried out on them.

GNSS technologies can rapidly get involved when critical events emerge due to a specific incident or agent. This smart advice could a save of products and reduce environmental impact, without (i) having to maximize the use of technical vehicles (e.g., seeds and fertilizers) and (ii) taking account of the tangible need for crops or the (qualitative and quantitative) presence of weeds. Even if these techniques are mainly spreading in flat areas, increasing attention to site-specific management in precision farming is emerging in hilly contexts [30]. In the latter areas, precision agriculture continues to be very limited due to the effort required during the assessment. Hills can be a relevant cause for missing observations [31]. One of their main limits is the highest error probability, even where hills are small, compared to the flatter regions [23,32].

The positioning system (as GPS) can continuously record both the in-field and the correct position of a vehicle in use [33]. All the collected data can then be processed through a Geographic Information System program (GIS). The latter can produce valid agro-technological assessments and multifaceted examinations of rural fields [34].

Measuring the effective distribution of such systems in less-favored areas, a project was carried out under the rural development program (RDP) in the Lazio Region 2007/2013 measure 1.2.4. The latter aims to assess financial farming conditions. European support might be granting merely the supplementary financing needed to implement the project in more attractive farms (in terms of local sources or banking system) [35].

The experimental activities envisaged in the project concerned:

- the employment of a permanent station network (Network RTK), exportable as a connection model to other users, (i) improving the efficiency of applications in precision agriculture and (ii) exporting processing data in the areas (to be tested) in a corporate GIS program;
- the execution of experimental assessments on defined areas, evaluating how to reduce the use of technical vehicles in three different ways of driving (manual, assisted, automatic);

- the identification of areas with special orographic constraints, using the GIS program, sending warnings to operators both at entry and exit from the area, and even in cases of excessive inclination of the vehicle during its use (avoiding potential risk of overturning).

2. Materials and Methods

For the execution of experimental tests (at the stage of weeding), three farms have been identified. Achieving the main purpose of this work, a homogeneous environment was chosen for the selected farms, where only one seasonal crop was present (wheat). The three farms are representative of the agricultural reality in Tuscia. The territory of Tuscia is an area that often coincides with the province of Viterbo, located in the central part of Italy. The satellite-navigation systems have been installed on all the vehicles of the selected farms, including an electric steering wheel (Figure 1), control monitor, and GNSS antenna (Figure 2).

Figure 1. Steering wheel with automatic guidance device.

Figure 2. Monitor and antenna installed on the tractor.

In these instruments, calibrations were performed for automatic guidance and configuration for the network connection of permanent RTK stations.

The network of permanent RTK stationary stations was developed for the distribution of differential corrections for geodesy and topography, enable the performance topographic surveys with only one receiver connected to one of the available networks.

This technology and methodology, if used even in precision agriculture applications, can: (i) avoid installing and using the "base", where it is possible to operate with only one receiver on the vehicle; (ii) operate across the area covered by the network and border areas, without having the problem of distance from the reference station, always with the same position precision; (iii) gain greater system reliability and integrity, due to the systematic error management resulting from the overall network calculation; (iv) dispel all doubt and ambiguity compared to the traditional RTK technique; (v) know the spatial position of the vehicle in a single and controlled reference system; (vi) have a high degree of repeatability, over time, both of the positioning and operation of the vehicle; (vii) receive distribution across the internet (NTRIP protocol) of differential corrections; and (viii) operate in areas with low GPRS/UMTS coverage through local solutions (RTK Bridge).

Differential correction occurs via the Virtual Reference Station (VRS), whose operating principles are:

- the receiver on the agricultural vehicle estimates its approximate location (error of a few meters) through the GPS and GLONASS satellites only and sends it to the control center thanks to the active connection of its modem;
- the control center generates a dedicated (virtual) base station for the potential user and sends differential corrections in real time;
- the receiver on the agricultural vehicle corrects its estimated position with the data sent by the network and achieves a centimeter accuracy.

The subsequent experimental activity concerned the evaluation of reducing the number of technical vehicles employed. Such activity, envisaged for an entire crop cycle, was addressed to every single agronomic practice from soil preparation up to harvesting.

The test area extends over as surface of approximately three hectares, divided into three equal portions, following their physical and geo-morphological characteristics. By way of example, the results in the next paragraph are related to the weeding operation. This choice is surely one of the most beneficial activities from satellite navigation systems.

The transaction data was recorded and downloaded from the system for consequent analysis. The execution times for each individual processing have been scheduled, by tracking the consumption of the product used and verifying the correct functioning of the satellite system in the intended modes of use. During the tests, three driving systems were verified: manual, assistive, and automatic.

In the first mode, cultivation operations were carried out by an operator without any technological aid. In this case, the system recorded only the path of the machine and other technical data (e.g., position, time, feed rate, path).

In the second mode, the same operations were carried out using the assisted guidance GNSS system. In this situation, the operator was facilitated more by the onboard computer system. The latter reported the path and the deviation from the trajectory to be followed in a monitor/video.

The third mode involved the implementation of a total automation processes. The tractor followed specific paths, defined on a map, through the automatic pilot. The operator was limited to monitoring the correct operation of devices and operating machines. This compares the traditional operating modes with precision agriculture, in terms of products used, execution times, and job safety. The reduction of environmental emissions into the atmosphere and the inadequacies in the implementation of agronomic practices were also assessed.

Regarding the latest research goal, a fleet management device was not immediately visible from the cabin (as it is an alarm device in case of theft). The system has many configuration options and functions, as well as serving as an aid to the management and maintenance of the vehicle itself. It is also possible to define a "limit zone" in the system: if the vehicle exits or enters this area, at uninterrupted time intervals, even when the engine is off, the device sends SMS or e-mail alerts.

For areas with poor Internet coverage, a device called an RTK-Bridge has been tested (Figure 3).

The device was placed in an area where there was a low Internet coverage. It has been connected to the permanent station network, sending its approximate position. It downloaded the differential corrections and then redistributed all the media operating through a low power radio. The width of the radio coverage depends on the territorial profile and the presence of obstacles. Like all radio devices, it is better to place the device at the highest point of the working area. In addition, for increasing operator safety, a device has been tested to send alarms via SMS and to switch on an led in the cabin, when one or more previously set inclination values of the vehicle have been reached. Two limit corners were set (a first angle of 15° and a second one of 45°). This latter angle greatly exceeded the safety limits for crawler tractors, and thus represented an alarm for probable overturning of the vehicle.

Figure 3. RTK-Bridge device.

3. Results

The experimentation for the use of permanent stations in precision agriculture has achieved several significant results. It showed that (i) the correction sent by permanent stations via the Network RTK to the GNSS system, concerning the vehicle in motion, was uniform and then (ii) the degree of precision varied with respect to the satellite constellation, visible during use and according to GPRS coverage, with which the system connected firstly to the network and then to the server. The system has changed, passing from "FIXED" (the system is connected to the network and receives a centimeter-accurate correction) to "FLOAT" (connected to the network but receiving a sub-metric precision correction) or to "STD" (the system is not connected to the network and collects only satellite correction with metric accuracy). However, the system can maintain the predetermined trajectory of the agricultural vehicle for a few seconds and then it emits a warning alert due to the loss of centimeter accuracy. This allowed us to assess whether the loss of precision was owed to a reduced GPRS signal or to an insufficient satellite constellation. Therefore, it is possible to act accordingly. It is necessary to wait (i) for the satellite signal, which allows a calculation and correction with the "FIXED" solution, or (ii) for the automatic system reconnection to the server. In addition, all data concerning the work carried out from the selected representative farms has been performed in the GIS program. A map about the treated terrains for each individual processing can be achieved. The resulting map quantifies the size of the operational surface areas for subsequent treatment scheduling. This management method allowed a seasonal check of the tasks that need to be performed and a planning of crop operations based on business choices achieved with the help of the GIS system.

Considering the three driving methods (manual, assisted, and automatic), experimental examinations on test areas checked the incidence of the human factor in the management of normal agronomic practices. Several benefits derived from new technologies, optimizing production and ensuring decreases in production times, environmental impacts, and fuel consumption.

The comparison between traditional operating methods and those tested with precision agriculture revealed significant progress, comparing the three driving methods.

In Figure 4, the steps of the agricultural vehicle during the action of weeding are shown (bar 12 m). The GNSS system graphically represented the positioning of satellite acquisitions along the path followed by the tractor. The assisted and automatic guidance systems allowed a more homogeneous soil treatment compared to the manual one. The covered surface was much more homogeneous, with the almost total disappearance of untreated areas and overlapping areas. All this is accompanied by a high driving comfort and better ease of execution during the operational activities.

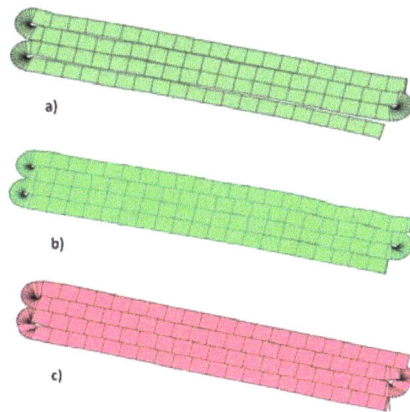

Figure 4. Operation crop weed control: GIS display in manual (**a**), assisted (**b**), and automatic guide (**c**).

The data for the activities carried out during the experimental field tests are shown in the Table 1. The cultivation operations concerned: (i) the soil preparation by 30–35 cm deep plowing carried out at the end of August without the GNSS system; (ii) fertilization, assuming GNSS aid; (iii) harrowing for the preparation of the sowing bed carried out in mid-November, without the GNSS system; (iv) sowing, following harrowing, with a seed density of about 300 kg/ha, without the GNSS system; (v) phytosanitary treatment, with chemical and fungicide weeding in mid-January; (vi) chemical weed control (ARIANE + AXIAL + activator) in February–March; (vii) chemical and fungicidal fertilization in April; collection of the product, with a grain yield of about 6 t/ha and 3.6 t/ha of straw, at the end of June–July, without the GNSS system.

Table 1. Primary data collected during experimental field trials.

	Type of Crop: Wheat				
				Modality	
Primary data			Manual	Assisted	Automatic
	Seeding density	kg/ha	300	257	254
	Grain yields	t/ha	6	5.1	5.1
	Quantity yield straw	t/ha	3.6	3.1	3.1
Cultivation operations	Plowing (30–35 cm)	August/September			
	Fertilization before sowing	Panfertil (Phosphate Biammonics 18–46) Dose: kg/ha	250	214	212
	Harrowing	November			
	Sowing	November			
	Chemical fertilization in January	Panfertil (Phosphate Biammonics 18–46) Dose: kg/ha	200	171	170
	Chemical disinfection + fungicide in January	GLEEN (kg/ha)	0.030	0.026	0.025
		Fungicida Zantara (Bayer) (l/ha)	1.5	1.28	1.27
	Chemical disinfection in February/March	ARIANE for wide leaf (l/ha)	3	2.6	2.5
		AXIAL for little leaf (l/ha)	1	0.86	0.85
		Activator (l/ha)	1	0.86	0.85
	Fungicidal chemical fertilization in April	AMIDAS urea (kg/ha)	200	171	170
		Fungicida Zantara (Bayer) (l/ha)	1.5	1.28	1.27
	Collection period	June/July			

In addition to a much better uniformity of treatment, significant reductions of products used should also be noted over time. Table 2 reports the different parameters chosen: time spent during the operations (expressed in minutes); product used, i.e., products which are used during the operations of weeding (expressed in liters); rural surface area (expressed in hectares); and distance travelled, otherwise the length of the path carried out by the vehicle (expressed in kilometers). The parameters were then detected according to the three-driving controls observed in the present work. With the same coverage area and distance travelled, the time employed and the amount of product used were considerably reduced in assisted and automatic guidance compared to manual driving. More specifically, time duration was reduced by 3.30% and the product used by 14.37% in assisted driving (when respectively they are 6.13% and 15% in automatic guidance). This result revealed that the transition from the assisted driving to the automatic mode did not affect the product used. For example, the overlapping or non-treated areas are almost non-existent in both modes. Instead, concerning the time spent, the automatic driving mode can be substantially preferred. The field tests carried out immediately provided divergent results, based on the parameters and the three driving types considered. For this reason, the data collected has not been combined even with statistical techniques.

Table 2. Comparison of the parameters found in the three types of soil processing.

Parameters Signed	Manual Guide	Assisted Guide		Automatic Guide	
Time spent (min)	7:04	6:50	−3.30%	6:38	−6.13%
Product used (l)	355	304	−14.37%	301	−15.21%
Surface area (ha)	1.25	1.25	-	1.25	-
Distance travelled (km)	1.10	1.10	-	1.10	-

Lastly, the introduction and installation of a telematic control system in common farms enabled the monitoring of their work and the performance of their vehicles. This can prevent risks in terms of unauthorized use of material goods, thefts, unplanned work, and unexpected movements. Above all, telematic control systems can constantly monitor the job conditions of workers, avoiding hazardous situations inside the passenger compartment on the agricultural vehicle.

The test performed confirmed the correct operation of the device installed in all its operating modes. Utilizing this method of monitoring and controlling agricultural operations, any type of operational risk can be avoided, from theft and damage of agricultural resources, to greater dangers—without excluding the overturning of agricultural vehicles—with the possibility of saving human lives.

4. Discussion

Despite an increasing awareness in hilly situations [30,36], techniques for precision farming are essentially used in flat areas. One of the major problems of precision agriculture concerns the difficulty of properly monitoring and managing hilly areas due to the high probability of error and loss of important information during detection operations [23,31,32]. This trouble can become critical if precision agriculture is applied to the Italian context. The latter presents a very varied territory, in which there is a strong presence of hillsides. Moreover, Italy has a fragmented rural landscape [20], which may complicate territorial analysis operations. Few studies have focused and compared GPS and GNSS technologies, demonstrating their validity in hilly areas [37].

In this paper, attempts have been made to minimize possible errors derived from GNSS technology. Devices as GNSS technology permit land mapping, optimizing machining operations in which rural vehicles can repeat (or exclude) a path already done. The innovative element of GNSS technology is the opportunity of putting rural machines and the farm management in communication. This interaction allows a timely technical analysis that points out the degree of improvement, further agronomic practices that need to be implemented, or a unique business database containing all the surface areas cultivated and the activities carried out on them. All the collected data can be processed through a Geographic Information System program (GIS), allowing rural processes and suitable agro-technological assessments [34]. GIS and remote sensing are layer-based systems, giving their users the flexibility to superimpose different levels of reality and find the best model for more accurate agricultural practice. An interpolated map describing the soil type, slope, and aspect in hilly areas can simulate the yield of crops with various variety groups and other agricultural inputs. With specific devices (RTK bridge), it is possible to overcome any signal deficiencies. Finally, in rugged hilly terrain areas, using the satellite navigation system can safeguard operators' safety from potential risk concerning the overturning of agricultural vehicles.

The present work gives new insights for research activities in hilly environments. In fact, with simple field trials at the stage of weeding, significant results have been achieved, without requiring statistical comparison. Achieving the main purpose, a homogeneous environment was chosen for the selected farms, where only one seasonal crop was present (wheat). Future research development may deal with heterogeneous environments having different crop areas and more than one crop season cycle.

5. Conclusions

Based on the results obtained with a RDP in the Lazio Region 2007/2013 measure 1.2.4, this paper reveals the actual possibilities of utilizing satellite guidance systems for agricultural vehicles in hilly areas, which are certainly less optimal than areas with more regular orography. The results obtained showed suitable possibilities of using these systems even in hilly environments. Therefore, the research topic proposed in this paper gives innovative insights, especially for researchers and even for farm producers. Considerable benefits have been achieved in terms of uniformity of machining, potential reduction of (chemical) products used, and operator safety during the working time.

Acknowledgments: Research carried out under the RDP Project Measure 124 entitled "Introduction of Sustainable Processes and Methods using RTK Satellite Technologies and Corporate GIS for the Development of Precision Agriculture in Tuscia" financed by the Lazio Region with a measure of concession of aids n. 31/124/10 of December 12th, 2014; Public Notice DGR n.76/2014 and s.m.i.

Author Contributions: Alvaro Marucci performed the experiments, analyzed the data, and contributed materials and tools; Andrea Colantoni designed the experiments; Ilaria Zambon and Gianluca Egidi wrote the paper; Ilaria Zambon was involved in the critical review of the results obtained.

Conflicts of Interest: The authors declare no conflict of interest.

References

1. Zhang, N.; Wang, M.; Wang, N. Precision agriculture—A worldwide review. *Comput. Electron. Agric.* **2002**, *36*, 113–132. [CrossRef]
2. Florax, R.J.G.M.; Voortman, R.L.; Brouwer, J. Spatial dimensions of precision agriculture: A spatial econometric analysis of millet yield on Sahelian coversands. *Agric. Econ.* **2002**, *27*, 425–443. [CrossRef]
3. McBratney, A.; Whelan, B.; Ancev, T.; Bouma, J. Future directions of precision agriculture. *Precis. Agric.* **2005**, *6*, 7–23. [CrossRef]
4. Pretty, J.; Sutherland, W.J.; Ashby, J.; Auburn, J.; Baulcombe, D.; Bell, M.; Campbell, H. The top 100 questions of importance to the future of global agriculture. *Int. J. Agric. Sustain.* **2010**, *8*, 219–236. [CrossRef]
5. Berti, A.; Zanin, G. Density Equivalent: A method for forecasting yield losses caused by mixed weed populations. *Weed Res.* **1994**, *34*, 326–332. [CrossRef]
6. Berti, A.; Borin, M.; Giupponi, C.; Morari, F.; Zanin, G.; Duso, C.; Furlan, L.; Rizzo, S.; Sartori, L.; Nardi, S.; et al. *Potenzialità Applicative dell'agricoltura di Precisione Nell'ambiente Veneto*; Veneto Agricultura: Milano, Italy, 2000.
7. Pierce, F.J.; Sadler, E.J. *The State of Site Specific Management for Agriculture*; ASA Publ.: Madison, WI, USA, 1997.
8. Verhagen, J.; Bouma, J. Modeling soil variability. In *The State of Site Specific Management for Agriculture*; Pierce, F.J., Sadler, E.J., Eds.; ASA Publ.: Madison, WI, USA, 1997.
9. Robert, P.C. Characterisation of soil conditions at the field level for soil specific management. *Geoderma* **1993**, *60*, 57–72. [CrossRef]
10. Bongiovanni, R.; Lowenberg-DeBoer, J. Precision agriculture and sustainability. *Precis. Agric.* **2004**, *5*, 359–387. [CrossRef]
11. Adamchuk, V.I.; Hummel, J.W.; Morgan, M.T.; Upadhyaya, S.K. On-the-go soil sensors for precision agriculture. *Comput. Electron. Agric.* **2004**, *44*, 71–91. [CrossRef]
12. Auernhammer, H. Precision farming—The environmental challenge. *Comput. Electron. Agric.* **2001**, *30*, 31–43. [CrossRef]
13. Stafford, J.V. Implementing precision agriculture in the 21st century. *J. Agric. Eng. Res.* **2000**, *76*, 267–275. [CrossRef]
14. Oliver, M.; Bishop, T.; Marchant, B. *Precision Agriculture for Sustainability and Environmental Protection*; Routledge: London, UK, 2013.
15. Lund, P.J.; Hill, P.G. Farm size, efficiency and economies of size. *J. Agric. Econ.* **1979**, *30*, 145–158. [CrossRef]
16. Van Dijk, T. Scenarios of Central European land fragmentation. *Land Use Policy* **2003**, *20*, 149–158. [CrossRef]
17. Alvarez, A.; Arias, C. Technical efficiency and farm size: A conditional analysis. *Agric. Econ.* **2004**, *30*, 241–250. [CrossRef]

18. Gorton, M.; Davidova, S. Farm productivity and efficiency in the CEE applicant countries: A synthesis of results. *Agric. Econ.* **2004**, *30*, 1–16. [CrossRef]
19. Galluzzo, N. Technical and economic efficiency analysis on Italian smallholder family farms using Farm Accountancy Data Network dataset. *Stud. Agric. Econ.* **2015**, *117*, 35–42. [CrossRef]
20. Pisante, M.; Stagnari, F.; Grant, C.A. Agricultural innovations for sustainable crop production intensification. *Ital. J. Agron.* **2012**, *7*, 40. [CrossRef]
21. Spiller, J.; Tapsell, A.; Peckham, R. Planning of future satellite navigation systems. *J. Navig.* **1999**, *52*, 47–59. [CrossRef]
22. Divis, D.A. Galileo enthusiasm and money propel Europe's GNSS. *GPS World* **1999**, *10*, 12–16.
23. Crespi, M.; Mazzoni, A.; Brunini, C. Assisted Code Point Positioning at Sub-meter Accuracy Level with Ionospheric Corrections Estimated in a Local GNSS Permanent Network. In *Geodesy for Planet*; Springer: Berlin, Germany, 2012.
24. Sánchez, N.; Alonso-Arroyo, A.; González-Zamora, A.; Martínez-Fernández, J.; Camps, A.; Vall-llosera, M.; Pablos, M.; Herrero-Jiménez, C.M. Airborne GNSS-R, thermal and optical data relationships for soil moisture retrievals. In Proceedings of the Geoscience and Remote Sensing Symposium (IGARSS), Milan, Italy, 26–31 July 2015; pp. 4785–4788.
25. Hall, C.D.; Cordey, R.A. Multistatic Scatterometry. In Proceedings of the IEEE International Geoscience and Remote Sensing Symposium, IGARSS, Edinburgh, UK, 12–16 September 1988; Volume 1, pp. 561–562.
26. Auber, J.-C.; Bibaut, A.; Rigal, J.-M. Characterization of Multipath on Land and Sea at GPS Frequencies. In Proceedings of the 7th International Technical Meeting of the Satellite Division of The Institute of Navigation (ION GPS 1994), Salt Lake City, UT, USA, 20–23 September 1994; pp. 1155–1171.
27. Zavorotny, V.U.; Masters, D.; Gasiewski, A.; Bartram, B.; Katzberg, S.; Axelrad, P.; Zamora, R. Seasonal polarimetric measurements of soil moisture using tower-based GPS bistatic radar. In Proceedings of the IEEE International Geoscience and Remote Sensing Symposium, IGARSS, Toulouse, France, 21–25 July 2003; pp. 781–783.
28. Choy, S.; Fu, F.; Dawson, J.; Jia, M.; Kuleshov, Y.; Chane-Ming, F.; Chuan-Sheng, K.; Yeh, T.K. Application of GNSS Atmospheric Sounding for Climate Studies in the Australian Region. In Proceedings of the FIG Working Week, Sofia, Bulgaria, 17–21 May 2015; pp. 1–13.
29. Awange, J.L. *Environmental Monitoring Using GNSS: Global Navigation Satellite Systems*; Springer Science & Business Media: Berlin, Germany, 2012.
30. Wu, W.; Liu, H.B.; Dai, H.L.; Li, W.; Sun, P.S. The management and planning of citrus orchards at a regional scale with GIS. *Precis. Agric.* **2011**, *12*, 44–54. [CrossRef]
31. Brown, N.; Kaloustian, S.; Roeckle, M. Monitoring of open pit mines using combined GNSS satellite receivers and robotic total stations. In Proceedings of the 2007 International Symposium on Rock Slope Stability in Open Pit Mining and Civil Engineering; ACG: Perth, Australia, 2007.
32. Ressl, C.; Pfeifer, N.; Mandlburger, G. Applying 3D affine transformation and least squares matching for airborne laser scanning strips adjustment without GNSS/IMU trajectory data. In Proceedings of the ISPRS workshop laser scanning, Calgary, Canada, 29–31 August 2011; pp. 1682–1777.
33. Neményi, M.; Mesterházi, P.Á.; Pecze, Z.; Stépán, Z. The role of GIS and GPS in precision farming. *Comput. Electron. Agric.* **2003**, *40*, 45–55. [CrossRef]
34. Pecze, Z.; Neményi, M.; Mesterházi, P.Á.; Stépán, Z. The function of the geographic information system (GIS) in precision farming. *IFAC Proc. Volumes* **2001**, *34*, 15–18. [CrossRef]
35. Dono, G.; Ceccarelli, L. Assessing the financial viability of agricultural investment: Indicators for projects submitted to the PSR (2007–2013) of the Lazio Region. *Riv. Econ. Agraria* **2010**, *65*, 465–485.
36. Fu, M.; Zhang, J. Construction Standard of Farmland Landscape Pattern in China Based on Precision Agriculture. In *Computer and Computing Technologies in Agriculture II, Volume 1: The Second IFIP International Conference on Computer and Computing Technologies in Agriculture (CCTA2008)*; Springer: Beijing, China, 2009.
37. Filip, A.; Bazant, L.; Mocek, H.; Cach, J. GPS/GNSS based train position locator for railway signalling. *WIT Trans. Built Environ.* **2000**, *50*, 16.

agriculture

MDPI

Article

Safety-Critical Manuals for Agricultural Tractor Drivers: A Method to Improve Their Usability

Maurizio Cutini [1,*], **Giada Forte** [2], **Marco Maietta** [2], **Maurizio Mazzenga** [2], **Simon Mastrangelo** [2,3] **and Carlo Bisaglia** [1]

1   Consiglio per la ricerca in agricoltura e l'analisi dell'economia agraria, Research Centre for Engineering and Agro-Food Processing (CREA-IT), via Milano 43, Treviglio 24047, Italy; carlo.bisaglia@crea.gov.it
2   Ergoproject Srl, Via Antonio Pacinotti 73/B, Roma 00146, Italy; g.forte@ergoproject.it (G.F.); m.maietta@ergoproject.it (M.M.); m.mazzenga@ergoproject.it (M.M.); s.mastrangelo@ergoproject.it (S.M.)
3   PAN-PAN Edizioni Srl, in Via Luciano 15, Milano 20156, Italy
*   Correspondence: maurizio.cutini@crea.gov.it; Tel.: +39-0363-49603

Received: 4 July 2017; Accepted: 31 July 2017; Published: 4 August 2017

Abstract: This work sets out the planning phases adopted for the first time to put together a manual on injury and accident prevention in the use of farm tractors. The goal is to convey information more effectively than at present, while taking the end users' opinions into consideration. The manual was devised, created, and tested based on a human-centred design (HCD) process, which identified the operators' requirements using a participatory ergonomics (PE) strategy. The main topics of the manual were outlined by engaging the users in a qualitative research activity (i.e., focus groups and workshops with final users), and the contents were prioritized and labelled by way of a noun prioritization activity. The users were involved right up to the choice of graphics and print layout in order to orient the publication to the farming context. The research activity highlighted a divergence between the operators' requirements and the topics currently dealt with in the sector publications. The project resulted in the publication of the "Safe Tractor" manual, which features some innovations. The experience highlighted the need to adopt HCD processes to create innovative editorial products, which can help speed up the dissemination of safety culture in the primary sector.

Keywords: focus group; editorial design; health and safety; usability

1. Introduction

Agricultural work is one of the most hazardous occupations as it ranks among the top jobs in work injury statistics. The fatality rate of such injuries is six times higher than the rate of all industries combined. In addition, concern about the growing number of leisure-related farm injuries is arising as well.

Figures concerning the burden of these injuries in the EU countries (EU15) show an average mortality rate of 13 deaths per 100,000 farm workers; this is confirmed also in the United States, where an average rate of 22 deaths per 100,000 workers was recorded. In both regions, peaks of more than 30.0 deaths per 100,000 workers were recorded as well [1,2]. Statistics referring to Italy reported tractors as the main cause both of injuries and deadly accidents; as a matter of fact, 56.5% of the total number of accidents in agriculture and forestry has been related to operating tractor tasks. Within this framework, and considering only the deadly accidents in agriculture and forestry operations, concern arises as 51% of these happened while workers were operating tractors (75% located on field and 25% while driving on roads). As far as accident dynamics are concerned, machine rollover represents the 77% of the accidents, while those involving the cardan shaft account for 0.7%, but 66% of cases result in the death of the operator [3].

Children often live, play, or even help on farms, and they are exposed to the dangers of tractors, machinery, and livestock [4]. Among the actions underway to lower the number and seriousness of accidents, effective training methods are increasingly being sought for agricultural machinery operators. Indeed, the "experience" factor alone does not seem to be significant, seeing as, in Italian statistics, fatalities mainly concern "senior" users; 40% of accidents are with operators aged over 50 [5]. As a result, institutions are intervening with information and worker training initiatives, both owing to new national regulatory obligations [6] and as an awareness-raising and divulgation activity. Furthermore, the design activities should also be more end user-oriented through what is now called a human-centred approach. This means that all designable components of a system have to be fitted to the characteristics of the intended users rather than selecting or adapting humans to fit the system [7]. This work sets out the planning phases adopted for the first time to put together a manual on injury and accident prevention in the use of farm tractors. The Human-Centred Design (HCD) process has also been formalized in the ISO-standard 13407-1999 human-centred design processes for Interactive Systems [8], currently revised by ISO 9241-210, 2010 [9]. These standards state the following key principles:

1. the active involvement of users and clear understanding of user and task requirements;
2. an appropriate allocation of function between user and system;
3. iteration of design solutions;
4. multi-disciplinary design teams.

HCD is a broad term to describe design processes in which end users influence how a design takes shape. It is both a broad philosophy and variety of methods [10]. It came to the fore through Norman and Draper [11], who focused on users' needs, carrying out an activity/task analysis, performing early testing and evaluation, and designing iteratively. According to Norman [12], the role of the designer is to facilitate the task for the user and to make sure that they are able to make use of the product as intended, with a minimum effort to learn how to use it. Norman notes that often the manuals that accompany products are not user-centred; consequently some design principles are needed to guide the design.

HCD requires the full exploration of the user's needs and the intended uses of the product. The need to involve actual users, often in the environment in which they would use the product being designed, is a natural evolution in the field of user-centred design [13]. Their involvement leads to more effective, efficient, and safer products and contributes to the acceptance and success of products [14]. The main methods used in user-centred design are as follows [15]:

- field studies (including contextual inquiry);
- user requirements analysis;
- iterative design;
- usability evaluation;
- task analysis;
- focus groups;
- formal heuristic evaluation;
- user interviews;
- prototype without user testing;
- surveys;
- informal expert review;
- card sorting;
- participatory design.

Before any usability design can begin, it is necessary to understand the context of use for the product, i.e., the goals of the user community, the main user, the task, and the environmental characteristics of the situation in which it will be operated [16]. The work set out here focuses on designing a manual on the health, safety, use and maintenance of farm tractors. When drawing up these publications, generally the contents they need to include are taken into account but not how they are read and used by the end reader. This research used a method that is "user-oriented" in all its phases of realization [17]. Every phase of the manual's definition envisaged the farm workers' active and direct participation, in particular concerning the choice of contents and their depiction in graphic form.

2. Materials and Methods

The first phase of the project set out to analyse the context in question. As a result, existing manuals were acquired to identify which topics needed to be dealt with and to assess their current presentation, organization, and reception. What emerged from the current publications was used as the basis for the workshops with the farm workers (WS1). They were all Italian and from the Bergamo area of the Lombardy region. During the workshops, the farm workers discussed and outlined their knowledge requirements, which were then put forward and tested in subsequent focus groups [18,19] involving a larger number of users. Once the context as well as the workers' real knowledge requirements had been defined, the next step was to devise the layout of the new publication. The iterative design phase began with a noun prioritization approach. Noun prioritization is the process of assigning priorities to things or tasks [20]. Prototypes were tested through expert observation and user workshops (WS2), until the final publication design was reached.

2.1. Issued Manual Evaluation: Definition of the Context in Question

The activity began by analysing a sample of equivalent publications to the planned one [21]. An analysis was made of nineteen publications by public institutions, the main focus of which was the safety of farm equipment (see list in Appendix A).

The aims were to catalogue the editorial features and to single out the best/worst practices in terms of clarity, immediacy in getting the message across, impossibility of misinterpretation, and completeness. Possible interesting topics and gaps in the communication process were identified in order to reach a first hypothesis on how to develop the new publication and to test it by involving final users, who were able to recognize the most and least mentioned topics and therefore to identify the ones not mentioned.

2.2. Preliminary Screening: Individual Interviews with Experienced Users

The first activities were individual interviews with three expert workers (41 to 52 years old, 20 to 40 years of experience). They were asked questions on their habitual use of safety manuals and user and maintenance manuals, their knowledge requirements, the information that they think a sector worker should know, and what they expect from a new publication on health and safety in the sector.

By comparing the results of this activity with what emerged from the previous benchmarking on other publications, it was possible to define a first structure (i.e., selection of topics, style of the illustrations) of the new publication.

2.3. Workshop 1: Further Investigations to Outline which Topics to Include in the New Publication and on Health and Safety Manual User Experiences

A workshop was organized with a larger group of operators (five experienced and three less experienced users, average age of 41, standard deviation of 6.37). The goal of the workshop was to obtain confirmation of what had emerged in the interviews. The experienced operators had 28.5 years of experience, while the less experienced operators counted 14.6 years. As a result, a further investigation could be made of their health and safety manual user experiences in order to more precisely define the topics to include in the new publication.

The results were compared with the initial hypothesis made after analysing the manuals, leading to the outline of a first layout (i.e., division into chapters, selection of topics) for the new publication.

2.4. Focus Group: Definition of the New Publication

Contrary to group interviews, which collect data from several people at the same time, focus groups explicitly use group interaction as the core of the method. This means that instead of the researcher asking to each person a question in turn, participants are encouraged to discuss and exchange comments and questions on experiences and points of view [19]. Focus groups are helpful "when insights, perceptions and explanations are more important than actual numbers" [22].

The aims of the focus group were:

- to further investigate how the users interact with the manuals;
- to devise the layout of the new publication.

It was possible to recruit twenty-two experienced farmers:

- landowners, agricultural contractors, farm labourers, and seasonal workers;
- aged between 20 and 60 (average age 36, SD = 12.46);
- all men.

The users were divided into three focus groups, each lasting one and a half hours (one with eight participants, two with two). A set of over 15 users is usually considered sufficient to perform a categorization activity such as noun prioritization [23].

They were attended by a moderator, who conducted the activities while promoting the participation of all and asking some open questions, and an observer who took notes and made additional questions in the final phase of the discussion. As agreed with the users, all of the sessions were video recorded.

Once the context as well as the workers' real knowledge requirements had been defined, the layout of the new publication was devised and then checked using noun prioritization. This last is an effective method for representing users' implicit mental models, which reveals their expectations on how the contents should be categorized. Once the mental models and implicit categorization are known, the information can be organized so that it is easier to find and use. This technique consists of showing the users cards with the name or description of a particular content. The users are then asked to divide the cards into groups and to put the topics into set categories. The participants were invited to categorize the topics and put them in order from the most to the least important. The categories of topics presented in the "Safe Tractor" publication were selected by researchers following the results from issued manual evaluation and input collected from users involved in the pre-focus group participatory activities (i.e., workshops and interviews). There were three set categories: (i) information, (ii) training, and (iii) action.

Sixteen topics (Table 1) had to be put into categories (six in each of the information and training categories and four in the action category), which had been selected from the results of the analysis of the topics dealt with in the existing publications and in WS1 (topics considered more/less interesting by the participants).

Table 1. The categories and the relevant topics proposed to the participants of the focus group.

Category	Topic	Description
Information	driving in the field	risks of injuries and accidents that can take place while driving in the field
	on-road driving	risks of injuries and accidents that can take place while driving on roads
	regulations	presentation of the health and safety regulations concerning agricultural sector workers
	environmental risks	the impact of work operations on the biological and/or atmospheric conditions
	risks linked to use of the farm vehicle	dangers connected to use during farming and maintenance operations
	statistics	e.g., number of workers in the sector, number of accidents/injuries
Training	work environment	basic notions on environmental and organizational aspects affecting work activities
	checklist	a 'paper and pencil' list that workers can use to check that the work is done in safety
	maintenance	information on the ordinary maintenance activities to perform on the machine every day
	work postures	indications on the postures to assume to prevent muscle/bone problems linked to the farm work
	first approach to the vehicle	indications on how to work safely designed for users new to the sector
Action	safety signs driving in the field on-road driving	
	injuries from use of the vehicle	information on the most likely injuries linked to the use of farm vehicles
	first aid	basic first aid procedures

2.5. Workshop 2: Discussion of the Draft with Experienced Users

The iterative design phase began after the focus group and noun prioritization activities. The authors of the manual were involved in drawing up the contents of the new publication, with the request to follow the indications below in order to create a text built around real users' requirements: "Always remember who your readers are: not just your boss or whoever has to review your texts, but the end users" [24]. The prototype was tested with the users in workshops (WS2) using the "walk-through/talk-through" method [25,26]. The users had the tasks of reading the contents, marking any passages that were difficult to understand and/or not very clear, and deciding the best solution for the graphical layout and for the order in which the topics were presented. During WS2, the publication was read out loud, followed by a free discussion. During this second session, the discussion was oriented in order to verify and further investigate subjective dimensions [27], such as:

- attractiveness: assessing whether the handbook is interesting enough to attract and keep the users' interest, with particular attention to the aspects/elements that they prefer and/or have most caught their interest;
- comprehension: assessing whether the transmitted message is clear and how this is understood by users;
- acceptance: ensuring that the handbook does not contain offensive or unpleasant elements and that it actually reflects users' convictions and beliefs;
- personal involvement: checking if users perceive that the instrument was designed precisely for their needs;
- persuasion: assessing whether the handbook convinces users to implement the proposed behaviour.

3. Results

3.1. Issued Manual Evaluation

A vast theoretical overview of the topic of "farm vehicle and equipment safety" but little information/practical training (e.g., how to get into and out of a tractor safely even though this is one of the most widespread causes of injuries) occured in 57.9% of the analysed publications. Table 2 shows the frequency with which the topics came up.

Table 2. Topics seen in the publications and their frequency.

Topic	Frequency (%)
Work environments (theory)	89.5
Safety signs	78.9
Farm vehicle safety (theory)	73.7
Statistics on the frequency of accidents	57.9
Checklist	31.6
Driving the farm vehicle on the road	26.3
Maintenance operations	15.8
First approach to the farm vehicle	15.8
First aid	10.5

Almost 85% of the analysed publications were divided into a range of six to eight chapters, further divided into paragraphs and subparagraphs. They were set out in the style of a compilation, with quotations from the regulations in force and technical descriptions of the risks to an operator when using a tractor. The images accompanying the texts were technical drawings taken from the regulations or drawn following the same type of model. 10% of the publications examined give a practical and concise presentation of the main dangers connected to the use of tractors as well as the solutions to be adopted in an emergency; the texts are accompanied by examples in the form of non-technical images. Lastly, the remaining 5% contain notes on the regulations and give information on day-to-day work activities; each topic is presented according to information/training sheets with example illustrations accompanying the text, subsections, and summary tables. Through this analysis, it was possible to identify the institutions' training/information priorities, which were then compared with the user requirements previously collected.

3.2. Preliminary Screening

The users interviewed individually (*n* = 3) underlined the need for publications on safety that give practical and concrete indications on "what to do", "what not to do", and "what to know" to avoid or prevent accidents. They highlighted that farm sector workers have a great need for training since they consider the existing publications unsuitable and incomplete.

The operators interviewed expressed information/training needs in contrast with what was found in the analysis of the existent publications. From the interviews, it also emerged that user and maintenance manuals are given little use both because there is not always a space designed for them and because the texts are particularly long and not very clear when those consulting them would instead like to find quick solutions to problems on the spot. Even when a tractor is used for the first time, the tendency is to ask more experienced colleagues.

3.3. Workshop 1

The workshop with a larger group of operators (five experienced and three less experienced, average age of 41, SD of 6.37) resulted in the classification of the topics that the users thought most useful (obtained from the average of the topics considered most useful by the five experienced users and the average of those considered most useful by the three less experienced users). These are presented separately depending on the amount of experience of the operators (Table 3).

Table 3. Topics in order of usefulness for the interviewed users, separated according to experience.

Classification	Experienced Users	Less Experienced Users
1	Checklist	First approach to the farm vehicle
2	On-road driving	Work environments (practical information)
3	Maintenance operations	Checklist
4	First aid	On-road driving
5	Farm vehicle safety (practical information)	Maintenance operations
6	-	First aid

The comparison (Figure 1) between the topics quoted by the operators as useful and those treated as such in the publications highlights the divergence ($r = -0.69$) between what the users would like to read and what they find in the manuals.

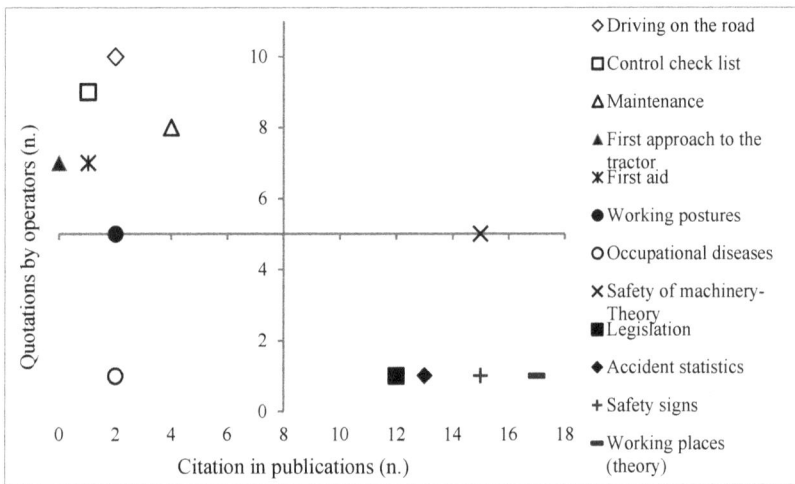

Figure 1. The *x*-axis shows the topics dealt with in the publications, the *y*-axis shows the topics quoted by the operators.

3.4. Focus Groups

3.4.1. User and Maintenance Manuals

The users only questioned user manuals to find out what error codes mean or the maximum admissible load weight. The vehicle owners only consult the manual when they are dealing with a new model, while they state that during the tractor's life span they ask either the dealer where they bought it or colleagues with the same model. The users complained about the fact that the information in the publications is not very comprehensible because they are written in a complex or superficial manner. Another problem is that these texts are not very accurate translations into Italian of instructions written in other languages. Examples of users' answers were: It only gets opened when it's absolutely necessary; It's written for clever clogs; The information's there, but we can't put it into practice; At times the only solution they give is to go and consult an authorized dealer. The participants pinpointed the length of the texts as the main factor discouraging them from reading and claimed that, in previous years, the dealers had suggested to the manufacturers that they divide the manuals into several leaflets precisely to make them easier to read.

3.4.2. Safety Manuals

All the operators admitted that they only read safety publications when they had to; for example, during a course held on the farm. They said they had leafed through them while waiting in the offices of some institution or other or because they came with the sector journals. They deemed them all to be very similar to one another, outdated and not really responding to their actual work conditions. Examples of operators' comments resulted: I'm not going to read them until I get sent on a course; When you go to the Italian farmers' union to sit in a queue, you leaf through them because that's all there is in the waiting room; It didn't say anything new. It was a waste of time. They did not express the need for clearer publications because they appeared skeptical and almost annoyed by the aims of these publications; they talked about them as containing lists of "obligations", which involved costs of varying proportions in order to meet them. "They're a cost, not a gain." The data categorized and summed up shows that the participants are sensitive to health/safety problems, but in practice they have difficulty in applying suitable measures in their everyday working conditions.

Nevertheless, it emerged that they wanted to be informed on these topics so long as those doing the information and training activities (public institutions, manufacturers, lawmakers) accounted for their real working and personal needs (e.g., clearer instructions, basic and useful information).

3.4.3. Noun Prioritization

The same people took part in the noun prioritization as took part in the focus groups. Of the 22 people present in the focus group, 19 took part in this second trial (average age 38, SD was 12.39). The preferences marked by the users were shown as the total of the degree of preference attributed to the topic (from 1, agree with most, to 5, agree with least) divided by the number of participants. For every category, the average of the user preferences given to each topic was compared with the average presence of the same topics in the 19 texts analysed during the issued manual evaluation activity. The results (Figures 2–4) show an inverse polarization on almost all the topics proposed; the topics deemed most interesting by the participants are those dealt with least in the publications analysed.

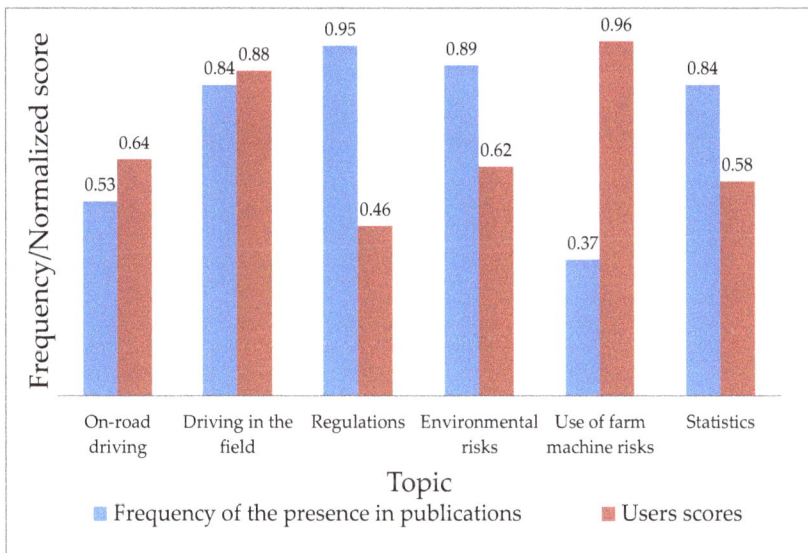

Figure 2. *Cont.*

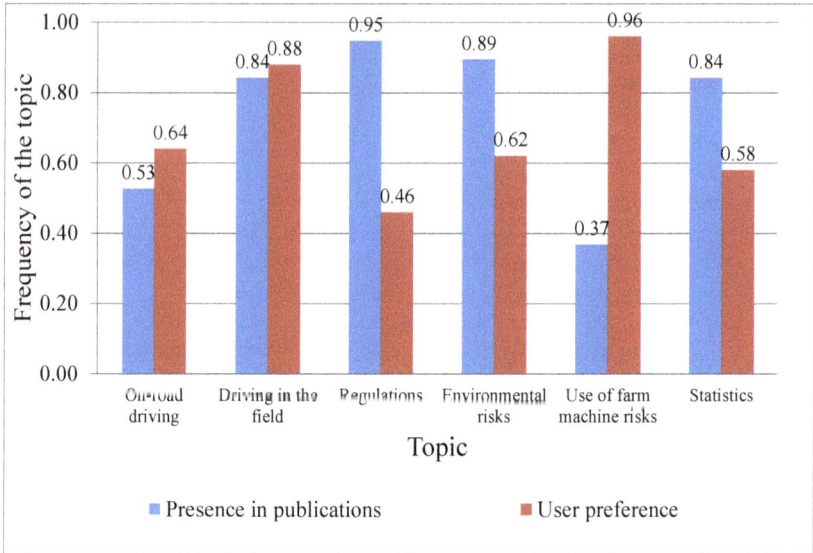

Figure 2. Frequencies of the topics dealt with in the sample of the 19 analysed publications (blue) and the normalized average scores expressed by the 19 operators (red) in the Information category.

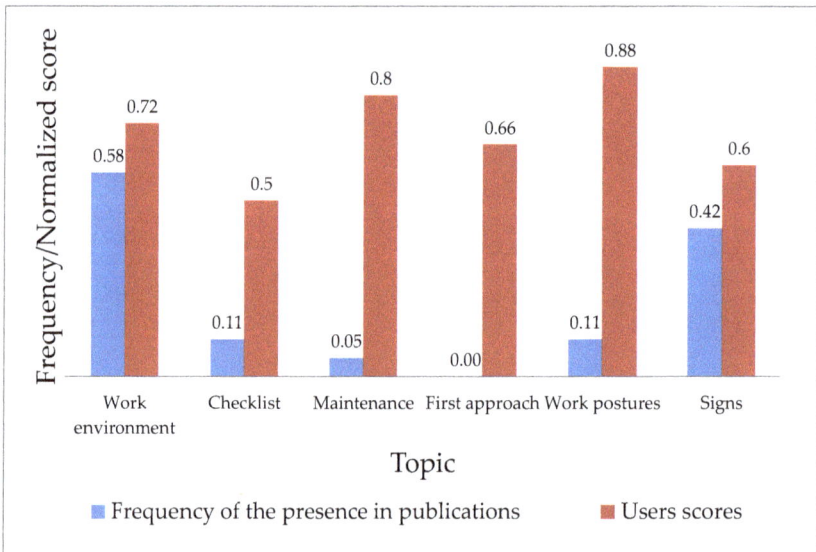

Figure 3. *Cont.*

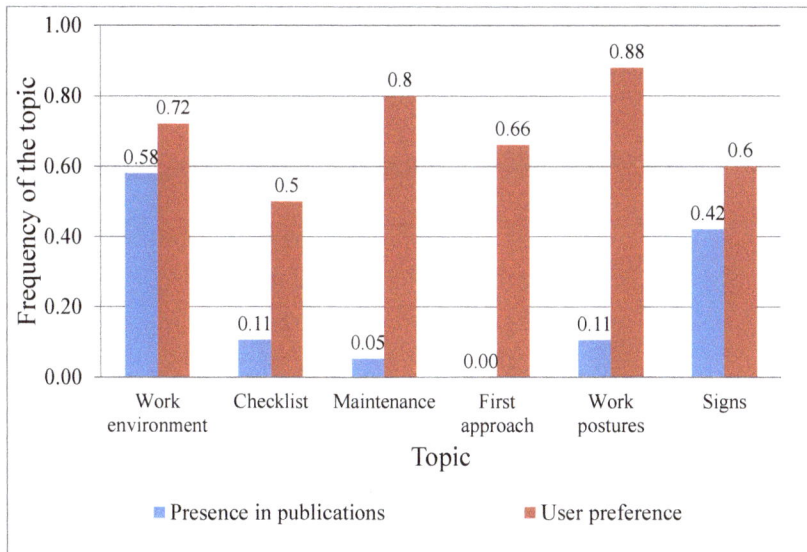

Figure 3. Frequencies of the topics dealt with in the sample of the 19 analysed publications (blue) and the normalized average scores expressed by the 19 operators (red) in the Training category.

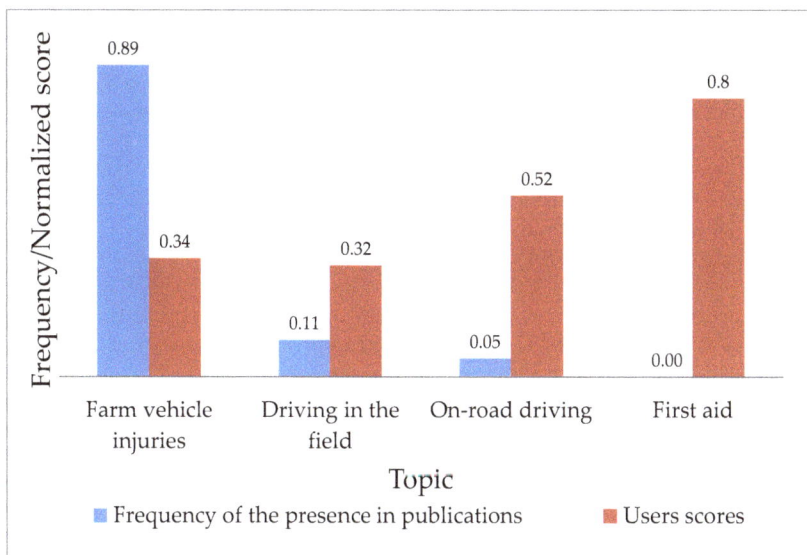

Figure 4. *Cont.*

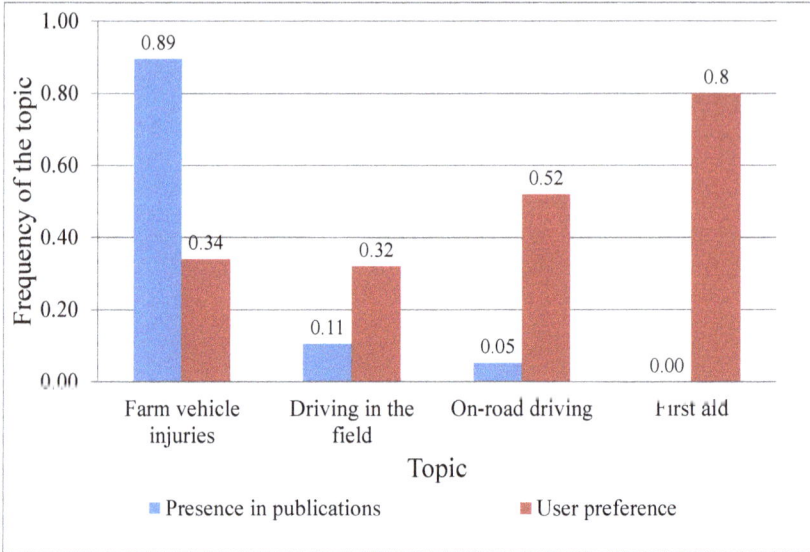

Figure 4. Frequencies of the topics dealt with in the sample of the 19 analysed publications (blue) and the normalized average scores expressed by the 19 operators (red) in the Action category.

At the end of the noun prioritization activity, the participants were asked to comment in groups on the topics they considered most significant or they found more difficult to place.

Regulations, belonging to the Information category were ranked among the bottom positions by the absolute majority of the participants. The users maintain that, in the publications, there is usually just a list of regulations; therefore they are considered of little interest. A topic that caused perplexity was the Checklist in the Training category; the users attributed little importance to the topic, not out of a lack of interest but because they thought that a tool of this kind (a list on paper to tick off with a pen) could be considered not very usable for the workers. Statistics, in the Information category, was the main topic of final discussion since half of the participants ranked it among the top positions, while the other half put it among the bottom. The users expressed their opinions by basing themselves on their previous experiences with sector manuals. The workers with most experience justified their choice by maintaining that using statistics on injuries and accidents that really happened could be a good way of raising awareness and convincing people of the need for safety when reading a publication; the users with less experience instead claimed that statistics are just seen as lists of numbers and ignored. The conclusion drawn was that users would declare greater interest in some topics if, in their experience, they had found them dealt with in a more explicit, concise, and, above all, less abstract manner (e.g., as far as statistics are concerned, they would prefer to find a smaller number of figures in the publications but ones linked to the most frequent accidents and injuries, rather than a mass of generic data). Another interesting piece of data that emerged from the discussion is that, when prompted to express personal expectations about the publications, the users admitted that they did not feel the need for a new publication, but that they would feel motivated to read it if the message was mainly put across through images and a few texts. In addition, they agreed in considering the division into areas (Information, Training, Action) as complete. They also asked for the topics to be dealt with in the new publication while taking their real working conditions into consideration. Furthermore, they explicitly requested that the excessive technicalities found in the previous publications be avoided.

3.5. Workshop 2: Iterative Design

After gathering the information, work began to make the manual prototype defining some elements such as the position of the illustrations on the page, the layout of the text in boxes, the thumb index at the side, and the print medium.

The first prototype of the manual (Figure 5) was presented to a group of experienced users, who were asked to assess the clarity of the contents and the coherence of the chosen graphics. From the preliminary investigation, it emerged that the areas the publication had been divided into (i.e., Information, Training, and Action) reflected the classification and institutional lexicon that had been identified as a worst practice during the manual evaluation.

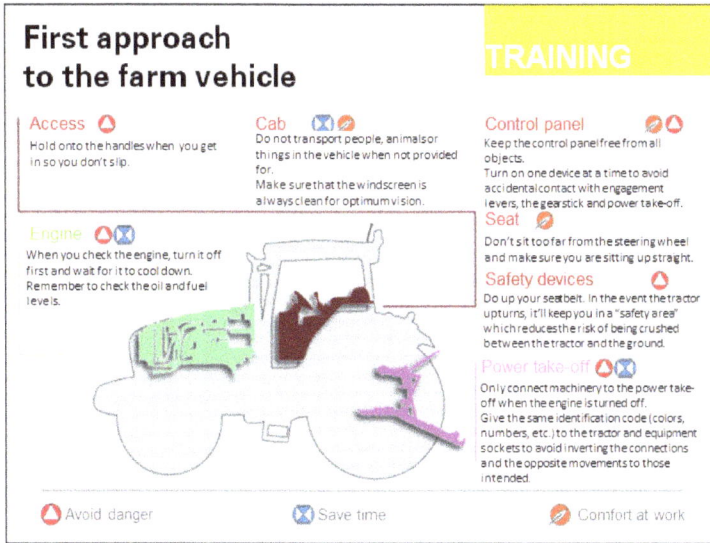

Figure 5. Prototype of manual page setting: example on the driver's position.

Hence, in order to ensure that the contents are conveyed effectively, the general organization of the text was reviewed to make it task-oriented. As a consequence, it was decided to separate the formal "regulations" from the "real" ways of using a vehicle according to the workflow defined with the users.

Therefore, the publication was divided into the chapters:

1. Vehicle at rest (what you need to know before you start to drive):

    *   Before you start work;
    *   Safety devices;
    *   Attaching equipment;
    *   Maintenance activities;
    *   Work shifts and stress.

2. Work in the field:

    *   Danger of overturning;
    *   Use of equipment;
    *   Presence of people around the machine.

3. On-road transport:

- Conduct on the road;
- How to behave with car drivers;
- Load quantities;
- Safety signs.

4. Annexes:

- What to do in the event of accidents;
- Maintenance checklist.

The group of experienced users then reviewed the texts according to the new layout for completeness, deeming it a good idea to also include a chapter on vocational illnesses, which had not been included in previous versions.

This additional chapter contains indications on:

- danger from chemical agents;
- danger from physical agents;
- biological risk.

Following the indications, the authors decided on further graphical elements, in particular the use of boxes and a review of the style of the illustrations (Figure 6).

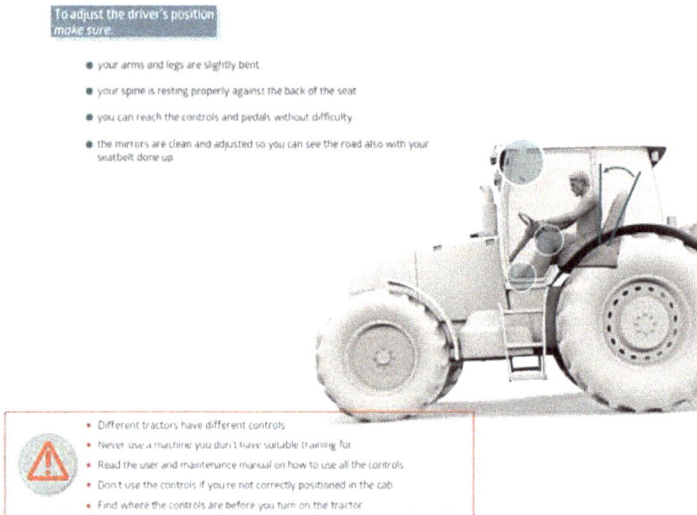

Figure 6. Intermediate draft of the section on the driving position.

To arrive at the final version of the publication, another four prototypes were made, which were assessed iteratively with direct user involvement. Participants were asked to read the prototype of the publication aloud. This allowed researchers to understand how workers wanted to use the manual for making meanings that are relevant to them, encouraging discussion, making connections with their lives, informing their view of themselves and others, asking themselves questions, and using the text to create personal storytelling [28].

Moreover, when read out loud, the participants found further issues, which they had not identified previously or pointed out in the feedback provided after reading the manual to themselves such as the presence of decorative elements and illustrations in partial contrast to the procedures described in the texts (e.g., the illustration of some personal protective equipment not set out in the text), complicated sentences (e.g., the presence of two negatives or long sentences with lots of clauses), and not very clear descriptions of activities (e.g., for attaching a winch and/or the procedure for making a U-turn on a slope). After this workshop, the authors came up with the final draft (Figure 7).

Figure 7. Final draft on the driving position.

3.6. The Final Manual

The final manual was drawn up by CREA-IT (Italian Agricultural Research Council) and ENAMA (Italian Institute for Agricultural Mechanization) experts, assessed and reviewed by experienced users, and, where necessary, simplified and updated to fit the needs expressed by the users. As far as the chosen graphics are concerned, the four colours of the different sections are included on the cover (Figure 8) to make the four areas that the manual is divided into easy to identify. Furthermore, the areas can be recognized inside the publication thanks to a vertical thumb index in the same colour as the area and page number.

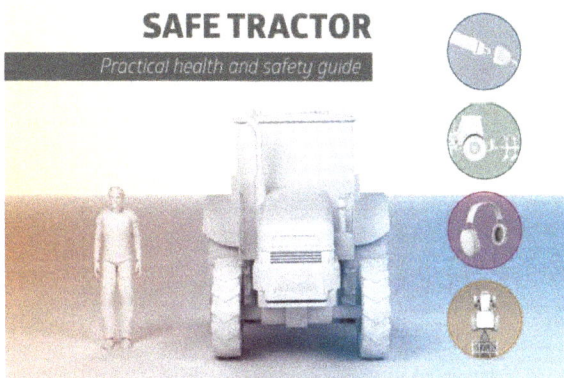

Figure 8. The cover with the four colours of the different sections included.

At the very beginning of the manual, some information on the latest data on accidents is visualized in graphic form (Figure 9). This helps to convey the information, make it easy to memorize and read, and to share the dimension of the problem.

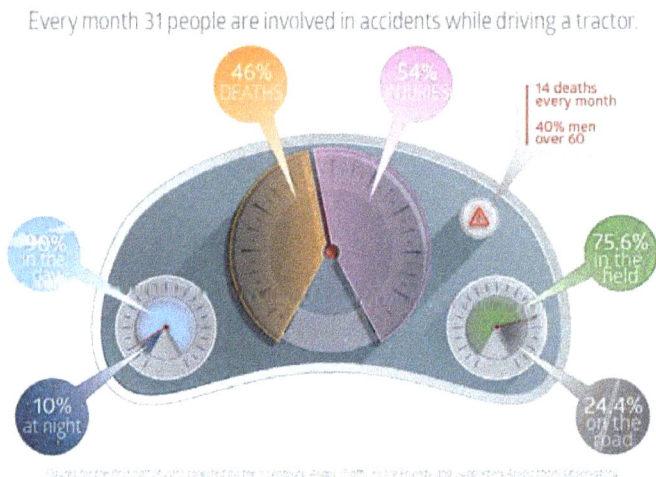

Figure 9. Example of graphic information on the latest data on accidents.

Every topic in the manual includes a short sentence (Figure 10) or figure (Figures 11 and 12) that summarizes real cases on fatalities to attract attention and get users to actively read it.

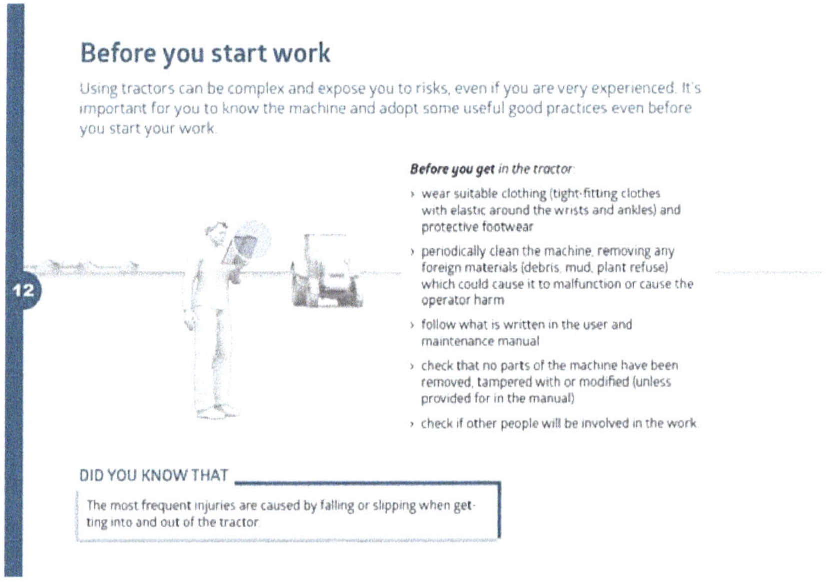

Figure 10. Example of a short sentence introducing every topic in the manual reports.

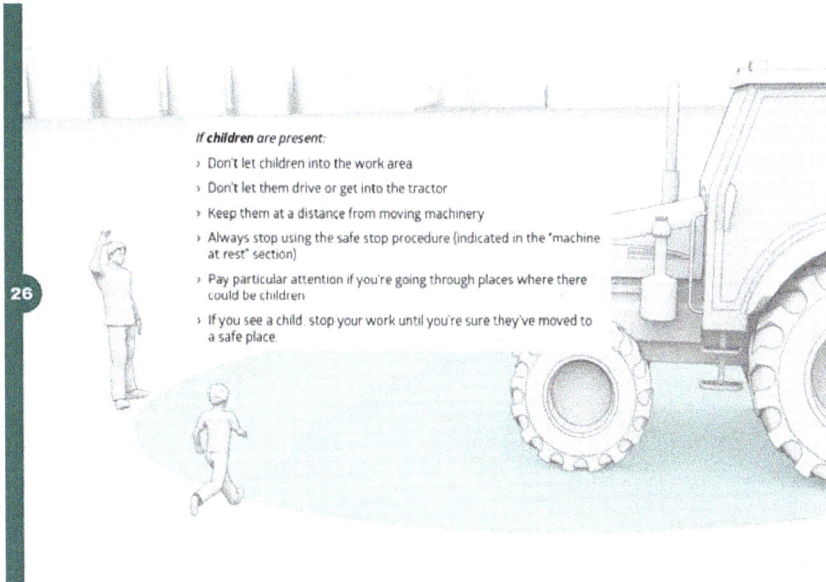

Figure 11. Example of a figure adopted to attract attention.

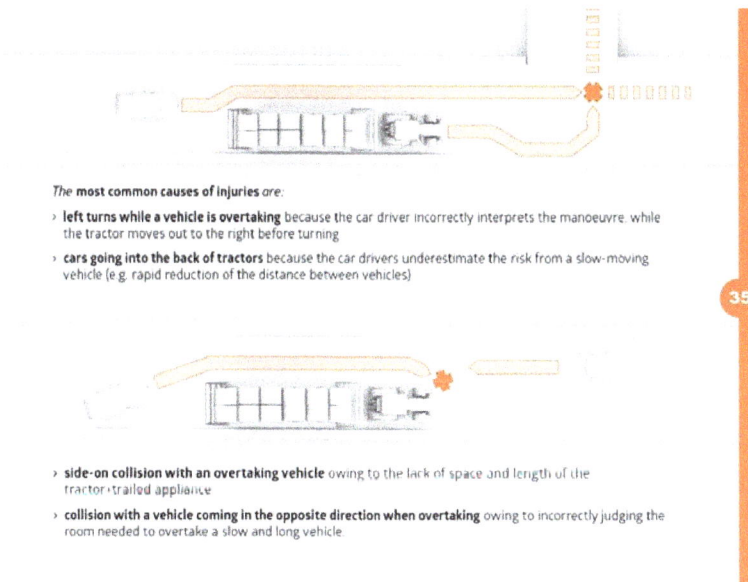

Figure 12. Example of a figure summarizing real cases of fatalities.

In order to only convey the information that the users consider useful for their particular work, every topic is set out over a maximum of two pages, which include short introductory texts and bullet points containing the main information to know or use to work in safety.

Some topics have boxes with further information. These have a dual function:

- the boxes labelled "Did you know that" contain additional information on frequent procedures and/or situations (Figure 10);
- the boxes labelled "Be careful" (in red, the colour used to underline situations of possible danger) contain further bullet points with actions to do and precautions to bear in mind before performing procedures (Figure 7).

The texts alternate with explanatory and practical illustrations (Figure 13) to use as concrete visual examples of what is relayed in the manual. Some pages contain colour spots to underline the behaviour to apply in the described operations.

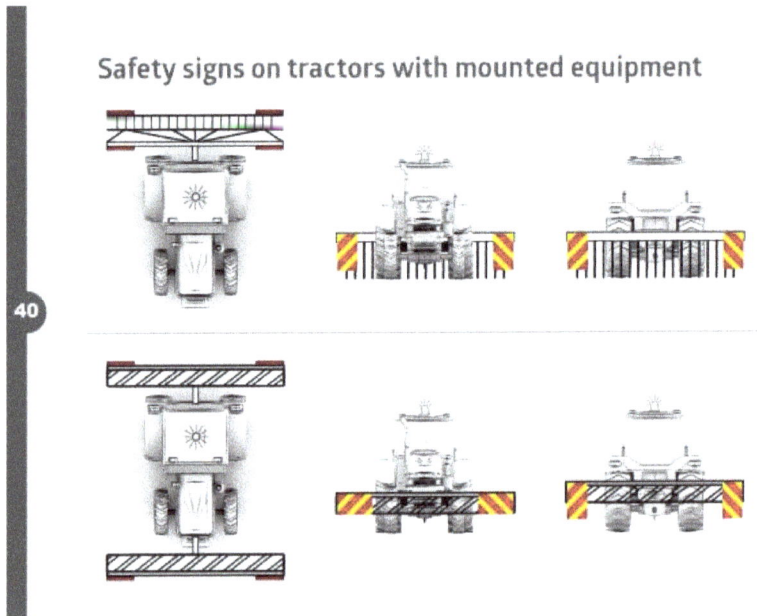

Figure 13. Example of practical illustrations.

4. Discussion

The "Safe Tractor" manual can be considered the first example of a publication in the farming sector created from its users' requirements. Through a participatory HCD approach, the users were involved in all the phases to create the manual.

A generational hierarchy was displayed in the interaction among the participants; the older users or those with more years of work experience were the ones who expressed their opinions with more conviction and more often, while the younger users intervened by confirming the opinions of their older colleagues or rejecting them, but only in part. The group leaders nevertheless played a positive role as they encouraged the others to participate actively in the discussion by responding promptly to the stimuli given by the moderator and proposing possibilities for change and improvement.

The final manual prototype has a new feature, that is, the print medium. Indeed, bearing in mind the conditions in which a tractor is normally used (high levels of damp, dust and at times dirt), which can damage the paper of a normal manual, and considering how the user consults it (in uncomfortable positions, in a hurry, sometimes wearing gloves), which can lead to whole or parts of pages being torn, it was decided to print the whole manual on a highly resistant artificial medium. The chosen material, which can be handled like ordinary paper, is biaxially oriented polypropylene film, or BOPP (Kartaplastic®, Tecnofoil srl, Azzano Decimo, Pordenone, Italy). This material is more resistant to ripping and more waterproof than paper but has a similar appearance and likewise can be printed on and recycled.

Right from the start of the research project, the users' support was deemed fundamental in selecting the topics to include in the publication and, more in general, in assessing their efficacy.

The result is a manual that includes information on what to know before getting into the driver's seat in "Vehicle at Rest". "Work in the Field" explains how to safely perform day-to-day activities; "On-road Transport" suggests how to avoid accidents with other vehicles' and "Occupational Illnesses" gives advice on how to prevent health problems.

The validation and assessment process envisaged by the HCD approach has not ended and may continue after the publication of the manual to iteratively improve it with every new edition. In this view, it is hoped that it will be possible to further the research by working on quantitative metrics as well by expanding the topics dealt with (e.g., first aid) and by involving other types of users (e.g., women, foreign workers). The authors should discuss the results and how they can be interpreted in light of previous studies and of the working hypotheses.

Acknowledgments: This study was funded by the INTRAC project, "Integration of ergonomic and safety aspects in agricultural tractors" funded by the Italian Ministry of Agricultural, Alimentary, and Forestry Policies, (MiPAAF; D.M. n. 12488/7303/11 of 09/06/2011).

Author Contributions: Carlo Bisaglia, Maurizio Cutini, and Simon Mastrangelo conceived and designed the experiments; Maurizio Cutini, Simon Mastrangelo and Giada Forte performed the experiments; Simon Mastrangelo, Giada Forte, Marco Maietta, and Maurizio Mazzenga analyzed the data; Marco Maietta and Maurizio Mazzenga followed the graphic design and Maurizio Cutini and Giada Forte wrote the paper.

Appendix

A list of the institutional publications analyzed by the authors to identify the editorial choices made to convey information on farm vehicle safety and maintenance to users.

- Adeguamento dei trattori agricoli o forestali (Technical adjustment of agricultural and forestry tractors)—INAIL, 2011.
- Agricoltura sicura (Safety agriculture), Azienda ULSS 20, Verona, 2010.
- Circolazione e sicurezza delle macchine agricole (traffic and safety of the agricultural machines), Egaf, 2003.
- Coltiviamo la cultura della sicurezza—La trattrice, (Let's cultivate the safety culture—The tractor) Veneto Agricoltura e INAIL, 2011.
- Come adeguare i trattori e le motoagricole usate (How to adjust technically used tractors and general purpose tractor), Regione Lombardia e INAIL, 2011.
- Compendio delle principali misure di sicurezza da applicare alle trattrici agricole (Summary of the main safety measures to apply to agricultural tractors), INAIL e Regione Lombardia, 2004.
- Ergonomics Checkpoints in Agriculture, International Labour Office in collaborazione con la International Ergonomics Association, 2012.
- Flyer trattrice, INAIL E ULSS Verona, 2009.
- Guida per l'adeguamento dei trattori agricoli e forestali (Guide for the technical adjustment of agricultural and forestry tractors)—Camera di Commercio Viterbo, 2008.
- I requisiti di sicurezza delle macchine irroratrici (Safety requirement of the sprayer machine)—ENAMA, 2002.
- L'installazione dei sistemi di ritenzione del conducente nei trattori agricoli o forestali (Installation of the driver's retention devices in agricultural or forestry tractors), Istituto Superiore per la Prevenzione e la Sicurezza del lavoro, 2009.

- L'installazione dei dispositivi di protezione in caso di ribaltamento nei trattori agricoli o forestali (Installation of the rollover protective system in agricultural or forestry tractor, ISPESL, 2008.
- La sicurezza delle macchine agricole (Agricultural machines safety), ENAMA, 2002.
- La sicurezza delle macchine agricole e degli impianti agro-industriali (Safety in agricultural machine and agro-industrial facilities), Associazione italiana di Ingegneria Agraria e ISPESL, 2002.
- Manuale della circolazione delle macchine agricole (Agricultural machines traffic manual), Regione Veneto e Polizia locale, 2009.
- Macchine semoventi 'Trattrici agricole a ruota' Sicurezza, (Self-propelled machines 'Agricultural wheeled tractor', Safety) CNR Torino, 2011.
- Manuale delle procedure di sicurezza (Safety procedure manual), Servizio sanitario regionale Emilia Romagna, 2011.
- Manuale per un lavoro sicuro in agricoltura (Manual for a safe work in agriculture), Regione Veneto, 2006.
- Sintesi delle norme di circolazione stradale riguardanti le macchine agricole (Summary of the rules of the road for agricultural machines), ENAMA, 2011.

References

1. European Commission (EC). *The Magnitude and Spectrum of Farm Injuries in the European Union Countries*; EC: Athens, Greece, 2004.
2. Health and Safety Authority (HSA). *Farm Safety Action Plan 2013–2015*; Health and Safety Authority (HSA): Dublin, Ireland, 2013.
3. Osservatorio INAIL Sugli Infortuni Nel Settore Agricolo e Forestale. In *Report Annuale Sugli Infortuni Mortali e Con Feriti Gravi Verificatisi Nel 2014 Nel Settore Agricolo e Forestale*; Dipartimento Innovazioni Tecnologiche e sicurezza degli Impianti, Prodotti e Insediamenti Antropici: Rome, Italy, 2015.
4. Pyykkonen, M.; Aherin, B. Occupational Health and Safety in Agriculture. In *Sustainable Agriculture*; Jakobsson, C., Ed.; Baltic University Press: Uppsala, Sweden, 2012; pp. 391–401.
5. ASAPS. *Asaps Associazione Sostenitori Amici Della Polizia Stradale, Report Primo Semestre 2013 su Incidenti ai Trattori Agricoli, Osservatorio Morti Verdi il Centauro*; ASAPS: Forlì, Italy, 2014.
6. EC. *Directive 2003/59/EC of the European Parliament and of the Council of 15 July 2003 on the Initial Qualification and Periodic Training of Drivers of Certain Road Vehicles for the Carriage of Goods or Passengers*; EUR-Lex: European Parliament, 2003; Available online: http://eur-lex.europa.eu/legal-content/EN/ALL/?uri=CELEX%3A32003L0059 (accessed on 3 August 2017).
7. ISO. *ISO 26800:2011: Ergonomics—General Approach, Principles and Concepts*; International Organization for Standardization: Geneva, Switzerland, 2011.
8. ISO. *ISO 13407: 1999 Human-Centred Design Processes for Interactive Systems*; International Organization for Standardization: Geneva, Switzerland, 1999.
9. ISO. *ISO 9241-210: 2010 Ergonomics of Human-System Interaction—Part 210: Human-Centred Design for Interactive Systems*; International Organization for Standardization: Geneva, Switzerland, 2010.
10. Abras, C.; Maloney-Krichmar, D.; Preece, J. User-Centred Design. In *Bainbridge 2004, W. Encyclopedia of Human-Computer Interaction*; Sage Publications: Thousand Oaks, CA, USA, 2004.
11. Norman, D.A.; Draper, S.W. *User-Centred System Design: New Perspectives on Human-Computer Interaction*; L. Erlbaum Associates Inc.: Hillsdale, NJ, USA, 1986.
12. Norman, D. *The Design of Everyday Things*; Doubleday: New York, NY, USA, 1988.
13. Sanders, E.B.N. From User-Centred to Participatory Design Approaches. In *Design and the Social Sciences 2002*; Frascara, J., Ed.; Taylor & Francis: Oxford, UK, 2002.
14. Preece, J.; Rogers, Y.; Sharp, H. *Interaction Design: Beyond Human-Computer Interaction*; John Wiley & Sons, Inc.: New York, NY, USA, 2002.
15. Vredenburg, K.; Mao, J.; Smith, P.; Carey, T. A Survey of User-Centred Design Practice. In Proceedings of the SIGCHI Conference on Human Factors in Computing Systems, Minneapolis, MN, USA, 20–25 April 2002.
16. Maguire, M. Context of use within usability activities. *Int. J. Hum. Comput. Stud.* **2001**, *55*, 453–483. [CrossRef]

17. UNI. *Technical Documentation of Product—Instructions for Use—Articulation and Exposition of The Content*; Ente Italiano di Normazione: Milan, Italy, 2000.
18. Langford, J.; McDonagh, D. *Focus Groups Supporting Effective Product Development*; Taylor & Francis: Oxford, UK, 2005.
19. Caplan, S. Using focus group methodology for ergonomic design. *Ergonomics* **1990**, *33*, 527–533. [CrossRef]
20. Fincher, S.; Tenenberg, J. Making sense of card sorting data. *Expert Syst.* **2005**, *22*, 89–93. [CrossRef]
21. Mastrangelo, S.; Verna, U.; Spirito, L. *Applicazione della metodologia User-Centred Design (UCD) a Prodotti Editoriali per Una Manualistica Usabile, Atti del IX Congresso Nazionale Società Italiana di Ergonomia*; SIE, Società Italiana di ergonomia: Rome, Italy, 2010.
22. Krueger, R.A. *Focus Groups: A Practical Guide for Applied Research*; Sage: Thousand Oaks, CA, USA, 1994.
23. Nielsen, J. Card Sorting: How Many Users to Test. Available online: https://www.nngroup.com/articles/card-sorting-how-many-users-to-test (accessed on 4 August 2017).
24. European Commission (EC). *How to Write Clearly*; European Commission: Luxembourg, Luxembourg, 2012.
25. Kirwan, B.; Ainsworth, L.K. *A Guide to Task Analysis: The Task Analysis Working Group*; Taylor and Francis: Oxford, UK, 2005.
26. Stanton, N.A. Hierarchical task analysis: Developments, applications and extensions. *Appl. Ergon.* **2006**, *37*, 55–79. [CrossRef] [PubMed]
27. Adams, J. *Risk*; UCL Press: London, UK, 1995.
28. Keene, E.O.; Zimmermann, S. *Mosaic of Thought: Teaching Comprehension in A Reader's Workshop*; Heinemann: Portsmouth, NH, USA, 1997.

agriculture

MDPI

Article

Adoption of Web-Based Spatial Tools by Agricultural Producers: Conversations with Seven Northeastern Ontario Farmers Using the *GeoVisage* Decision Support System

Daniel H. Jarvis [1],*, Mark P. Wachowiak [2], Dan F. Walters [3] and John M. Kovacs [3]

[1] Schulich School of Education, Nipissing University, North Bay, ON P1B8L7, Canada
[2] Department of Computer Science and Mathematics, Nipissing University, North Bay, ON P1B8L7, Canada;
     markw@nipissingu.ca
[3] Department of Geography, Nipissing University, North Bay, ON P1B8L7, Canada;
     danw@nipissingu.ca (D.F.W.); johnmk@nipissingu.ca (J M K)
*   Correspondence: danj@nipissingu.ca; Tel.: +1-705-474-3461 (ext. 4445)

Received: 9 June 2017; Accepted: 2 August 2017; Published: 8 August 2017

Abstract: This paper reports on the findings of a multi-site qualitative case study research project designed to document the utility and perceived usefulness of weather station and imagery data associated with the online resource *GeoVisage* among northeastern Ontario farmers. Interviews were conducted onsite at five participating farms (three dairy, one cash crop, and one public access fruit/vegetable) in 2014–2016, and these conversations were transcribed and returned to participants for member checking. Interview data was then entered into *Atlas.ti* software for the purpose of qualitative thematic analysis. Fifteen codes emerged from the data and findings center around three overarching themes: common uses of weather station data (e.g., air/soil temperature, rainfall); the use of *GeoVisage* Imagery data/tools (e.g., acreage calculations, remotely sensed imagery); and future recommendations for the online resource (e.g., communication, secure crop imagery, mobile access). Overall, weather station data and tools freely accessible through the *GeoVisage* site were viewed as representing a timely, positive, and important addition to contemporary agricultural decision-making in northeastern Ontario farming.

Keywords: GIS technology; precision agriculture; web-based access; weather; weather station; decision support

1. Introduction

For millennia, farming has been characterized by compound and unpredictable factors. Aubert et al. [1] discuss this complexity and the perennial uncertainty of crop farming as follows:

> A crop farmer needs to consider a variety of parameters such as crop yield, availability of water and nutrients, and a range of site- and soil-specific factors to optimize the plant treatment (e.g., application of fertilizer, pesticides, or irrigation). A high variability of these parameters within a single field further complicates the optimization of the plant treatment. (p. 510)

Agriculture, in terms of both the cash cropping and the livestock industry, has been under increasing pressure from both governmental agencies and the general public to change traditional farming practices to minimize adverse environmental and social effects. Agricultural decision support systems (DSSs) have been designed to help farmers implement more sustainable practices by aiding them in optimizing farming practices to maximize economic efficiency and to reduce impacts off the farm—good stewardship. Over the past 40 years, researchers have speculated that it was only a matter of time before agricultural decision support systems became an essential tool in the management of agricultural operations. Despite these systems being readily available and affordable [2–8], implementation at the farm-scale has not met expectations [9–12]. The theoretical rationale for supporting decision support technology in the 1970s and 1980s was based on cognitive science. It was thought that computation could overcome the limitations of human ability to process information and to make rational decisions based on scientific evidence. McCown [13] argues that this information-processing view of human decision-making has been replaced by "more 'ecological' theories . . . in which there is an emphasis on sense-making and experimental learning in complex, lived-in environments" (p. 190). This idea represents a fundament shift from developing decision support systems based on scientific evidence to one that also emphasizes social values and constructs. There needs to be less emphasis on the implementation problem and more attention on how to achieve mutual understanding between developers and practitioners. Less emphasis should be placed on recommendations to farmers, and "more about facilitation of decision process adaptation" (p. 181). Decisional guidance is defined by Silver [14] (p. 107) as "how a decision support system enlightens or sways its users as they structure and execute their decision making process—that is, as they choose among and use the system's functional capabilities."

In an effort to improve the adoption of decision support systems among the agricultural industry, designers are increasingly taking a more participatory approach. The benefit of a participatory approach is that the end user can assess the usefulness of the tool. Several studies suggest that the adoption of a participatory approach that integrates designers and users in the development of agricultural decision support tools will reduce the implementation gap of these decision support tools [15,16]. The key to user acceptance of this technology is the perceived usefulness of the tool and the farmer's purpose or task. Rose et al. [17] identify fifteen factors affecting use of agricultural decision support tools, including relevance to producers, reliability, peer endorsement, ease of use, and cost, among others. Ultimately, implementation is dependent upon the perception that the decisional guidance tool provides a high value of return with limited risk [13,18,19].

The development of online video tutorials, training programs, and access to knowledgeable practitioners can help improve implementation [15]. However, Jakku and Thorburn [16] suggest that social learning is a fundamental component of the participatory approach. Collaboration can enhance opportunities for innovation. The qualitative research described in this paper provided the opportunity to directly interview a number of northeastern Ontario farmers to assess whether, and in what ways, they have adopted a freely available, locally developed decision support systems involving farm-based weather stations and the related, freely accessible online resource known as *GeoVisage*. Such qualitative investigations based upon case studies are valuable for assessing the extent to which new precision agriculture technologies are being adopted, and underscore the importance of participatory approaches when designing technology tools for agricultural producers. First, we shall provide some background information regarding the northeastern Ontario weather stations and the development of the *GeoVisage* decision support system (Section 2), and then we will elaborate on the case study that was conducted with local agricultural producers and the related findings and conclusions (Sections 3–6).

2. *GeoVisage* and the Northeastern Ontario Weather Stations

GeoVisage [20] is a web-based decision support tool that represents a multi-year, multi-disciplinary project that has been funded by the Northern Ontario Heritage Fund Corporation, the Ontario Soil and Crop Improvement Association, and Nipissing University [21]. *GeoVisage* was designed for use by northeastern Ontario agricultural producers to support key decisions in the increasingly important emerging agricultural regions of Temiskaming Shores, Verner, and Cochrane. To collect data for the *GeoVisage* system, starting in May 2009, Nipissing University has installed and has maintained seven weather stations throughout northeastern Ontario (see Figure 1).

Figure 1. One of the many weather stations maintained for the *GeoVisage* system. This HOBO U30 station, located in Temiskaming Shores, has been actively collecting data since 3 July 2009.

Five additional stations have since been erected (Figure 2) and will soon be integrated into the system. It is important to note that local producers first contacted the university hoping to help them collect weather data. Prior to the weather station network, the local producers had to rely on federal government (Environment Canada) stations that, for the most part, are located at airports (e.g., North Bay (NB) & Earlton stations) which are often not representative of local conditions (e.g., NB airport is located on an escarpment away from Lake Nipissing). Most importantly, there are very few weather stations in northeastern Ontario; for example, prior to 2009, the West Nipissing agricultural district (including Verner and Sturgeon Falls) had to rely on the Sudbury and North Bay Environment Canada weather stations which are approximately 80 km west and 50 km east of their district, respectively (see Figure 2).

Each station collects data on its microclimates, including such vital real-time weather information as air and soil temperature, relative humidity, wind speed, leaf wetness, and photosynthetically active radiation. The *GeoVisage* system also calculates growing-degree-days and crop-heating-unit values to assist farmers in planning [22]. These data are made available via an interactive website. The system provides four main services: real-time data acquisition and display, in-depth visualization of comparative and historical data, imagery obtained from a variety of modalities, and sharing of geo-referenced digital photographs.

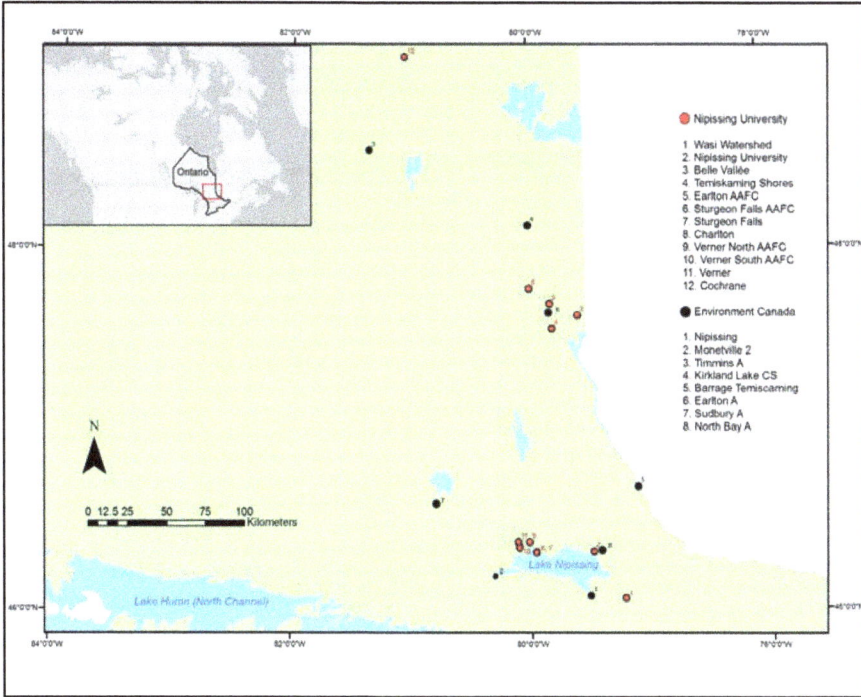

Figure 2. Location of current weather stations maintained by Nipissing University (NU) and by Environment Canada. NU stations numbered 1, 3, 4, 7, 8, 11, and 12 are currently fully integrated into *GeoVisage*.

2.1. Real-Time Data Acquisition and Display

GeoVisage features a real-time data tool that displays graphs of current sensor data from several weather stations. With the Weather Station Data tool, all the data collected by the weather stations from the earliest collected by Nipissing weather stations (some stations as early as 2009) to near present can be viewed. The real-time data component is the standard interface provided by HOBOlink [23], part of the HOBO service. The weather stations use HOBO U30 or RX3000 data loggers (Onset®, Bourne, MA, USA) to acquire and wirelessly transmit data at regular five-minute intervals. Both atmospheric and soil properties are measured, with most stations recording gust speed, wind speed, leaf wetness, soil moisture, photosynthetically active radiation (PAR), soil temperature, air pressure, air temperature, rainfall, relative humidity, dew point, and solar radiation.

2.2. Weather Station Data Visualization

The real-time data offer a quick view of recent conditions, but lacks the exploratory features needed for more detailed or in-depth analysis. Consequently, *GeoVisage* provides other features to allow properties from different weather stations or time periods to be visualized and compared. The Weather Station Data Visualization tool offers a much larger selection of features than the real-time data, but is updated less frequently. The tool has interactive time selectors that allow different properties to be compared and assessed across different weather stations and time periods. Basic statistics are also provided in a separate table below the plots. An example time series comparing two stations and two properties is shown in Figure 3.

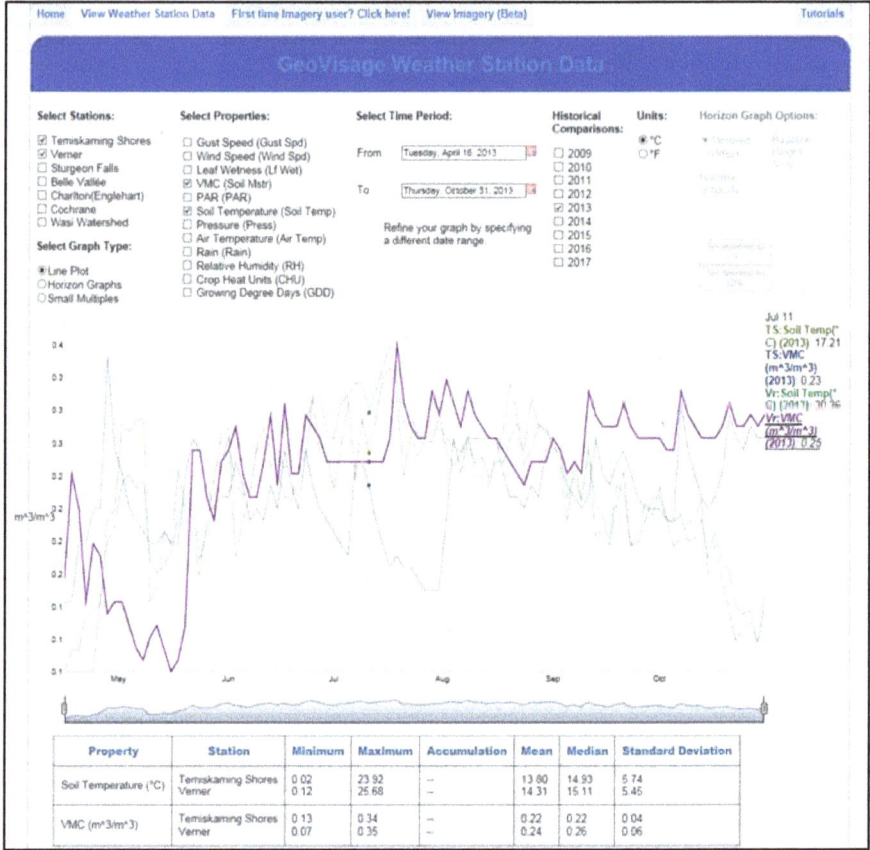

Figure 3. Interface for weather station data. Time series for two environmental properties (soil temperature and volumetric moisture content), comparing two sites (Verner and the more northern Temiskaming Shores region) over a growing season. The time selector below the line plot allows time zooming. Hovering over a time series displays the sensor readings for a specific time and date. The statistics panel is located below the selector.

Visual analytics capabilities, currently in testing with limited availability, have recently been added as a research tool and to facilitate community-based participatory research. Visual analytics features include small multiples and horizon plots for displaying a large amount of time series data on a limited screen space and more advanced time-based correlation and multi-resolution visualizations.

2.3. Imagery

Additional features of the *GeoVisage* resource include an imagery application, based on the Java-based NASA World Wind GIS framework, to display remote sensing images, soil maps, yield maps, and field imagery captured with an unmanned aerial system (UAS, or "drone" technology). *GeoVisage* employs various types of imagery, such as remote sensing images, soil maps, yield maps, normalized difference vegetation index (NDVI) images (Figure 4), and field imagery captured with either of the two UAS deployed by the university (Figure 5).

Figure 4. One of many current grey-scale Normalized Difference Vegetation Index (NDVI) images of Temiskaming Shores shown over the World Wind standard image globe (in colour). Bright tones are indicative of high biomass, or healthy crops, whereas dark tones indicate low biomass or no vegetative cover (e.g., lakes and rivers appearing as black).

The remotely sensed imagery includes both UAS and satellite based sensor data. Since the UAS imagery is collected close to the surface (e.g., ~90 m) the spatial resolution is extremely high, in the order of a few centimeters per pixel. However, UAS imagery is typically based on hundreds of individual photos that are stitched together using specialized software. Although these images are of high spatial resolution, they cover relatively small areas, often just one field, and require considerable time to collect in the field and process in the lab. In contrast, most of the historical satellite imagery are collected from LandSat sensors which have a 30 m pixel resolution, or if pansharpened, 15 m. However, these images are collected regularly, every 16 days, and cover an immense area of 185 by 185 km (i.e., hundreds of fields).

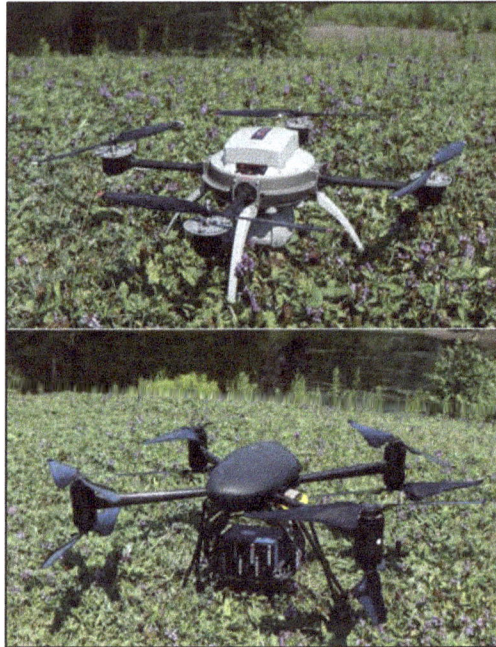

Figure 5. Two unmanned arial system (UAS) quadrocopters deployed by Nipissing University researchers for agricultural monitoring. The Aeryon Scout (Aeryon Labs Inc., Waterloo, ON, Canada) carries (top photo) an infrared ADC lite camera (Tetracam, Chatsworth, CA, USA), whereas the Dragan Flyer X-8 (Draganfly Innovations Inc., Saskatoon, SK, Canada) carries a Mini-MCA camera (Tetracam, Chatsworth, CA, USA).

Measurement tools, provided with the World Wind framework and customized for the application, are provided to facilitate quantitative assessment. An example of imagery overlaid onto terrain as a semi-transparent layer, as well as a field measurement, is shown in Figure 6.

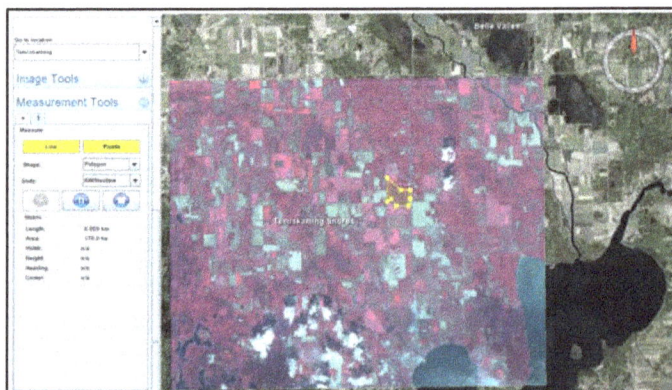

Figure 6. Example imagery overlaid atop terrain provided by the NASA World Wind framework. Tools for the imaging interface are shown at the left. The transparency tool, shown here, allows blending of several images and the underlying terrain.

2.4. Geo-Referenced Digital Photographs

GeoVisage allows producers to upload images of their crops, pests, etc., to facilitate information sharing. Users also have the capability to display digital photos to better identify crop conditions and to alert the community of possible pest infestations. These photographs are geo-referenced to a specific location or to a general region, and are then made available to all *GeoVisage* users. The feature directly supports the community-based participatory research and "citizen science" aspects of the *GeoVisage* initiative. An example of this feature is shown in Figure 7.

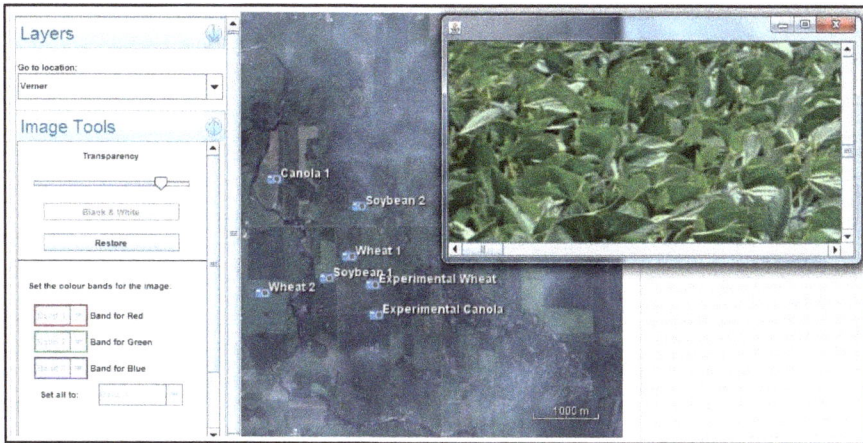

Figure 7. Example of the digital photo-sharing feature. The soybean image is geo-referenced to the "Soybean 2" label on the map.

2.5. The Importance of Human Computer Interaction

To encourage adoption by agricultural producers, ease-of-use and intuitiveness were primary considerations in designing the *GeoVisage* system interface [24]. For instance, the NASA World Wind visualization framework [25] provides producers with the usual zooming, panning, and navigation tools to quickly identify and focus on specific locations, such as their own farms, and is based on standard paradigms characteristic of many GIS products and services. In another example, producers are provided with a variety of colour-maps for displaying time series data to enhance exploratory capabilities, and to take into account users' preferences. Intuitive location, viewing, basic image processing, and measuring tools are also provided. The "broad" (i.e., complex, or "busy") interface paradigm was adopted so that producers have access to a wide variety of tools simultaneously. Because of the importance of training to the adoption of technology tools [11], instructional videos and tutorials (Figure 8) are also available through the system interface.

Finally, but very importantly, major benchmarks in the development of the interface were discussed and assessed with the producers for whom *GeoVisage* was designed. In the design and development of the system, groups of producers have visited with developers for training, and to provide valuable feedback on the intuitiveness of using various features. The developers also attended farm shows, association crop tours, and other events to communicate with a wider audience of users (see Figure 9).

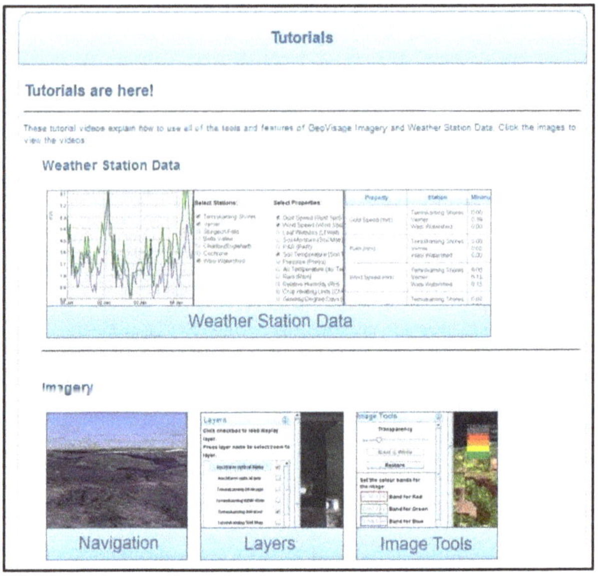

Figure 8. *GeoVisage* online video tutorials screen-capture.

Figure 9. Nipissing weather station as part of the Ontario Soil and Crop Improvement Association/Temiskaming Crops Coalition booth at the 2009 International Plowing Match. The event was held for the first time in Earlton with an estimated 75,000 visitors over 5 days.

2.6. Precision Agriculture and Spatial Decision Support Tools

Precision agriculture may be loosely defined as "The application of geospatial techniques and sensors (e.g., geographic information systems, remote sensing, GPS) to identify variations in the field and to deal with them using alternative strategies" [26]. Although there are many tools for supporting precision agriculture, web services for spatial decision support in particular are growing in popularity because spatial information is easily understood, and because of the importance of sharing agricultural and environmental information. Remote sensing imagery, including images captured by UAS and satellite-based sensors, is an integral component of this new web-based paradigm, characterized by a growing number of research contributions. For instance, *CropScape*, an initiative of the National Agricultural Statistics Service in the United States [27], is an interactive geospatial web-based system for disseminating, querying, visualizing, and analyzing crop and land cover classifications obtained from remote sensing data. Since it became available as a website in 2011, *CropScape* has been visited by over 81,000 users (reported as of 2013) [28]. In another example, a web service has been designed for exploiting the availability of global navigation satellite system reflectometry (GNSS-R) signals for assessing soil moisture [29]. Agricultural monitoring is also facilitated by open geospatial web services that integrate heterogeneous information from multiple wireless and wired sensor networks [30].

As mentioned above, *GeoVisage* provides visualization services, which are key components in enhancing agricultural decision support. *GeoVisage* and other web services that support data from GPS-based farming vehicles and remote sensing technologies need to visualize a large amount of data that, because of its volume, is difficult to interpret. Visualization is the first part of an extensive data mining process for enhancing interpretation and understanding [31]. Examples of this synergy include a recent study which described a prototype system that supports agricultural decision-making that is enhanced through cartographic visualizations and which integrate current local environmental and agro-monitoring data with a GIS system [32], as well as a web-based system targeted to policymakers and other stakeholders with the goal of assessing the consequences of crop changes on multiple ecosystems [33].

In addition to the technical aspects of web-based tools for agricultural decision support, it is important to understand whether, how, and to what degree these technologies have been adopted by producers [26]. The current study analyzes the factors involved in adoption of *GeoVisage* through interviews with local producers in two northeastern Ontario agricultural areas: Temiskaming Shores and Verner.

3. Methods

The use of interviews and case studies to investigate the adoption of precision agriculture has been successful in pinpointing specific technological needs (e.g., yield monitoring and mapping services) [34]. The current study extends this approach to an investigation of the adoption of the web-based dissemination of location-specific weather data and imagery, and is therefore relevant to other researchers developing web-based precision farming and analytics tools.

With a view to better understanding how northeastern Ontario farmers were using the *GeoVisage* online resource and weather station data, a multi-site qualitative case study was conducted in 2014–2016. Following the approval by the Nipissing University Research Ethics Board, the study involved site visits to two farms in the Temiskaming Shores area in summer 2014, and site visits to three farms in the Verner area in spring 2016 (see Table 1).

Table 1. Participating farm details.

Farmer(s) (Pseudonyms)	Farm Type	Farm Acreage/Size	Weather Station Access
Aidan	Dairy + Cash Crops	84 head milking (tie-stall with milkers and pipeline); 1500 acres	3 weather stations used (Belle Vallee, Charlton, Temiskaming Shores (located on his farm))
Ben	Sheep + Cash Crops	250 sheep; 1235 acres	3 weather stations used (Belle Vallee, Charlton, Temiskaming Shores)
Carl, Callie, Chris	Dairy + Cash Crops	80 head milking (two fully automated milking stations); 1000 acres	2 weather stations used (Verner (located on their farm), Sturgeon Falls)
Dean	Fruits/Vegetables + Cash Crops	public berry/vegetable picking, rides; 650 acres	2 weather stations used (Verner, Sturgeon Falls (located on his farm))
Evan	Dairy + Cash Crops	80 head milking (tie-stall with milkers and pipeline); 1250 acres	2 weather stations used (Verner, Sturgeon Falls)

Three of the farms involved in the study were dairy operations with approximately 80 head of cattle, two of which used traditional tie-stall milking machine technology, and the third featuring an advanced system with two fully automated milking stations. A fourth farm involved a flock of 250 sheep and was primarily focused on the growing and selling of a variety of cash crops. The fifth site represented a popular, family-run public access farm featuring seasonal berry picking, vegetables, wagon rides, gift shop, and a small eatery. Three of the five farms had agreed to host a weather station on their property; all five farms were provided access to weather station data via *GeoVisage*.

Interviews were conducted with four individual farmers, and in the fifth case a group interview was conducted with a farmer and his parents, all three of whom were significantly involved in the home farm history and daily operation. Interviews were then transcribed verbatim, followed by member checking, during which stage all participants were able to review transcripts and offer insights into minor corrections/clarifications. The seven farmers were each given pseudonyms (Aidan, Ben, Carl/Callie/Chris, Dean, Evan) to provide participant confidentiality.

The interview questions were semi-structured (i.e., open-ended in nature) and designed according to case study standards [35,36]. Participants were thereby encouraged to communicate their individual perceptions relating to their adoption of the *GeoVisage* resource and features (see Appendix A). Interviews were transcribed and checked for accuracy, and were also returned to participants for "member checking," allowing them to ensure accuracy of content and to suggest any requested revisions.

Transcripts were loaded into a qualitative software program known as *Atlas.ti* (Scientific Software Development GmbH, Berlin, Germany), which is a computer-assisted qualitative data analysis (CAQDAS) software package for analyzing text, audio, and video-based research data [37]. For this case study, the researchers employed *Atlas.ti* because it allowed for the organization of large amounts of text-based interview data; because it allowed the researchers to manually define, apply, merge, and modify thematic codes based on the transcript data; and because it allowed the researchers to establish and export to word processing software the emergent main themes and related quotations to expedite the writing process. The researchers employed a common thematic analysis method that involved familiarization with data, generating initial codes, searching for emergent themes among codes, reviewing themes, defining and naming themes, and producing the final paper.

4. Results

From the interview data, fifteen separate codes were identified and subsequently grouped into three broad themes for discussion purposes. In what follows, we shall examine the perceptions of participating farmers relating to commonly used weather station measures, other uses of *GeoVisage* website tools/data, and future recommendations for the decision support tool. Note that some of the *GeoVisage* features elaborated upon in Section 2 above are not represented in the qualitative results and discussion sections, as the participants, when asked about their own experiences with, and perceptions of, the *GeoVisage* tool, did not mention these particular features.

4.1. Commonly Used Weather Station Measures

Farmers participating in the study indicated the common use of four key types of weather station measures: air temperature; rainfall, leaf wetness, and relative humidity; soil temperature; and wind speed and direction. Relevant excerpts from the interview conversations are presented to provide further insight into how and why these measures are often used in everyday farming decisions.

Not surprisingly, there are essentially three aspects of general weather patterns that are of interest to farmers: what is happening "today" with the weather (present); what has happened historically in terms of the archived comparison data (past); and, perhaps most importantly, what is being predicted in terms of "tomorrow" and the remainder of a given week or month (future). In speaking with participants in this study, it became clear that understanding all three of the above chronological aspects of weather station and website data is vitally important for their everyday work and decision-making processes.

All of the participants check weather forecasts on at least a daily basis, if not several times per day depending on what season they are in, and on what decisions have to be made concerning crops. A number of popular weather forecast websites were reported as being used for this purpose such as AccuWeather, the Weather Network, and Environment Canada's Canadian Weather Radar site.

4.1.1. Air Temperature

Ben, one of two brothers who have taken over the operation of the family dairy farm business, indicated that he has commonly used *GeoVisage* for checking both temperature and rainfall measures.

> **Ben:** So, through the growing season we would be on that site, I would say every day. We would definitely want to know the amount of heat that's coming, like what our highs and lows were for the day, and how much moisture we actually did capture ... I am checking the amount of rainfall and temperatures through the day, and we make decisions on impending weather conditions—whether or not to plant, whether or not to spray, whether to work all night if there's an impending week of rain coming. These are more precise measurements that we would rely on, as opposed to just checking Environment Canada.

Dean, who maintains the publically accessible berry-picking farm, discusses how air temperature is one of his most important considerations in order to avoid frost damage to the berries after planting.

> **Dean:** We have alarms, but if there was frost on the strawberries, and you didn't get at them at night to put the sprinklers on, then they get damaged, so you assess the damage, and might have to cut back on your sales that day ... The water's warmer than the air, so just the effect of adding the warmer water does bring up the temperature, but if it forms ice, the creation of ice being formed actually creates heat. It's weird. So, as long as you don't let that ice become dry on the flowers—on the blossoms, then they'll be saved ... So, we keep it wet constantly. We have to make sure we come back to the same field within 15 min. So, we rotate fields ... It's all night, yes, driving around, walking around—we have to shut off valves, open valves, check a lot of things ... They're all set up around the fields, but the pump can only handle so much at a time, so it's a matter of switching from field to field.

An additional temperature sensor was added to the weather station on Dean's farm to measure air temperature above the height of the strawberries. Dean had also described the frost as a river of cold air moving over the canopy, and that a temperature sensor at the height of the berries is a better estimate of frost risk.

A dairy farmer located near Dean's farm noted the importance of knowing when to expect frost for other types of crops:

> **Evan:** Frost is still fairly important after we plant, to see how cold it is—this year we had a scare after the Canola was up. We had a couple of nights that we had frost, and Canola can be susceptible if it freezes too hard, and soybeans the same thing . . . At certain times of the year, yes, we check carefully. Come later on in the fall, it doesn't quite matter as much.

Clearly, with so much riding on decisions relating to frost, freezing, and crop failure, weather station data pertaining to air temperature ranks among the most significant use of weather station data.

4.1.2. Rainfall, Leaf Wetness, and Relative Humidity

Related closely to air temperature in terms of features of the weather station data and the *GeoVisage* website that were reported as being particularly useful for farmers was rainfall, or precipitation measurement. Dean contrasts traditional methods of tracking rainfall with more modern measures:

> **Dean:** All farmers are interested in weather and what's happening at their own farm because even before the weather stations were built, most farmers actually had a rain gauge on their farm, which we did ourselves too. But this one, we don't have to do anything—it's done for us . . . It's very useful to know how much rain has fallen in the last 24 h—for decisions if we should irrigate or not, for example . . . The one measure I look at the most, yes, is rainfall.

Evan, located within the same region in Ontario, explained how rainfall affects field conditions:

> **Evan:** Everybody used to have rain gauges on their farms to measure what you would get. Now I think a lot of people look this up online. It also gives you an idea of how much moisture is in the ground . . . If you got two inches of rain, you know you won't be going into the field for a couple of days . . . The long range doesn't really interest me that much, but the prior 2–3 days, that's useful information.

Chris and his parents, Carl and Callie, who all take part in their family dairy business, expressed the importance of consulting weather forecast websites and radar data in terms of their decision-making:

> **Chris:** For forecasting, it's sort of halfway between the Weather Network and Environment Canada. Weather Network's too optimistic. Environment Canada's too pessimistic. It's usually about halfway in between.

> **Callie:** The forecasting is really interesting because if you can see the radar, you can make the judgment call as to what's going to happen . . . We always get the radar out.

Aidan, a dairy farmer (along with his brother) who had a Nipissing University weather station located on his own farm, noted that he regularly checked the website regarding precipitation measures:

> **Aidan:** If it rained, I would definitely check it the next day, but at least once a week—just to see how much rain we got. So, you know, if I woke up in the morning and there had been any precipitation at all, I would click on it just to see what amounts we got. Because everyone has a plastic range gauge outside of their house, but how accurate are they, compared, you know, to a scientific rain weather station? It could vary greatly.

Ben reported regularly checking the data from the weather station located on nearby Aidan's farm, as well as data from two other local weather stations that were all made available through the online *GeoVisage* resource: "Because there is quite a variance—in May we had a two-and-a-half inch rain capture here . . . It was good to check Belle Vallee and Charlton as well—those three, and to see, you know, that there is a variance in the amount of moisture." Precipitation measurements would often vary throughout the region, and hence multiple data points, taken together, were viewed as providing a more accurate picture of overall rainfall patterns.

Closely related to rainfall precipitation were the separate measures of leaf wetness and relative humidity. Although neither of these factors were discussed at length by any of the farmer participants, a few comments were shared in this area. For example, Chris noted that both leaf wetness and relative humidity are important factors when deciding when to cut and bail hay crops. Furthermore, Aidan explained how both of these measures also affect decisions surrounding the harvesting of wheat crops:

> **Aidan:** When we get heavy dews during the day, in the morning, there is a critical time for wheat, and its right at heading time—so, when the wheat elongates out of the stem, that's what they call a critical time for disease prevention . . . If you think it's dry out, you don't really know the leaf wetness of the crop, and you don't know the humidity. So, two critical things are the amount of moisture of the leaf, as well as the humidity in the air, because these both breed disease.

Another commonly referenced weather station measure, regardless of the type of farm, was that of soil temperature.

4.1.3. Soil Temperature

The temperature of the soil can directly affect the survival of newly planted seeds, as well as overall sustainability depending on the nature of the crop and the soil bed in which it is planted. Aidan contrasted traditional and more current soybean planting decisions in light of the available data.

> **Aidan:** When planting beans, it's more critical that the soil temperature be high enough for the beans. They say that when you plant beans in a cold ground it actually shocks the beans, and it will delay the growth. That has really been a useful tool . . . We used to go out and plant when the ground was dry—we could get on the field without getting stuck. We planted more by the calendar, like it was the first of May, so you would go out and plant. Now we can plant by the temperature of the soil—if the soil is actually ready. It might be dry, but if it's been a few cold nights and the day temperature hasn't warmed the soil up, then, you know, we will hold off.

Evan echoed this general rule of thumb in terms of planting temperature, noting that "for cereals it's not as important, but for soybeans, they want the temperature to be at least 10 degrees—they recommend ten degrees for the first 24 h." Aidan also mentioned the snow insulation factor, which can affect certain crops and pests:

> **Aidan:** It is surprising, the soil temperature, depending on the amount of snow we get—because you don't realize the insulating factor of snow. The soil temperature will only be minus two or three degrees, so with these new pesticides, this new pest we have, the Swede midge in the Canola crops—although it was minus forty, they assumed it was going to kill the bugs in the ground. But we had such a large amount of snow it actually worked as an insulator, and so the ground didn't get that cold . . . We got a spring thaw, so that created a sheet of ice, and it cut the oxygen off . . . but it recovered from it. So, with tools like that you would go on and see the soil temperature, you know—is it below freezing still, or is the ground starting to warm up?

We have thus far looked at measures of air temperature, precipitation/humidity, and soil temperature. A fourth weather station data measure commonly referred to by farmer participants was that of wind.

4.1.4. Wind Speed and Direction

Wind speed and direction are two factors that affect several aspects of modern farming. When asked if he used pesticide sprays on his fruit and vegetable farm, Dean noted that "pesticide spray is needed, yes, especially for the cash crops. It's good to check and see what the wind speed is—if there's too much gusting . . . It depends on what you're applying—some products are at the very most 15 km per hour, and some have more chance of drifting than others."

Aidan further explained that "you could waste a lot of product, or get yourself in a bit of trouble, if the wind speeds are too high . . . Like just the drift—you don't realize, if you're in a sheltered area, just how windy it could be out there." Similarly, Carl maintained the same rule-of-thumb wind speed threshold: "Yes, I use 15 km per hour. If it's over that, I kind of hold off . . . If you're growing cereals and your neighbours have Canola or something on another side of a ditch—if the wind's blowing the wrong way, you can actually do damage to other fields."

The frequency with which a farmer checks different weather station data is also of interest. With soil temperature, for example, Evan explained that hourly or daily checking often suffices; whereas with wind speed, a farmer may more often wish to be checking this measure every 15 min given how volatile the weather patterns can be in this regard. Carl spoke about the importance of relying on the weather station data, rather than just on sometimes inaccurate visual or tactile impressions:

> **Carl:** So, there's a limit really, of 15 km per hour—beyond that, yes, it's just too windy. So, you're standing outside and a gust of wind blows by, and you think it's too windy—so you go check. But it's only reading 12 km/h. Maybe that gust was 20 km/h, so it's actually okay, it's fine, it should be all right, and so I'll go and spray. But without this reading, I might have said, 'Forget about it for today.' . . . That's why wind direction is important, especially if I'm spraying in a field next to another person's house. I don't want to spray them if I can help it.

4.2. GeoVisage Imagery Data

During the course of the five farm-based interviews, participants discussed the *GeoVisage* imagery data and tools that were also being developed for, and accessible through, the *GeoVisage* website. Using World Wind capabilities, a measurement tool was developed which allowed users to measure acreage. Further, UAS (i.e., Aeryon Scout, DraganFlyer X8) with cameras were used to capture rich crop imagery of several participating farms.

The UAS technology was demonstrated during an annual agricultural event in the Temiskaming Shores area during the weekend of the first interviews in 2014. Clearly, there was already much interest being generated around this technology and the potential utility of the high-resolution imagery being produced. *GeoVisage* offered both a land measurement tool and the UAS imagery of participating farms. Farmers referenced both of these features during interviews, some having used them more than others. However, the post-processing of UAS imagery and integration into farm-scale technology remains a constraining factor in widespread adoption by small-scale producers [38].

4.2.1. Acreage Measurement

Aidan described two critical uses of such a measurement tool in farming:

> **Aidan:** We are now in the year of high rent, you know. We are up into the $160-per-acre-per-year range for Temiskaming, which is unheard of. So, if you can scientifically measure out the acres that you are cropping, and then only pay for those acres. If the farmer you are renting it from claims it's a 160-acre farm, but really it's only a 140 workable acres, that's 20 extra acres that's not being used. But also for crop insurance. Agricorp wants to know exactly how many acres you are cropping. Markers on drills and acre meters on drills are only so accurate.

Accurately calculating crop acreage also relates to projected crop yield, and hence has sale implications:

> **Dean:** I thought it might be a tool in the future to say, you know, "Close to how many tons of soybeans can you pull out of that certain field?" And so that could help you gauge how to sell it—if you know you have so many tons, you can sell them in advance because you don't want to sell too much. And if you sell on the stock market you owe the stock market regardless of your crop yield—you have to fill in what you agreed to sell.

Two other significant areas of land use decision making that relate to acreage measurement are the ability to plan for adequate tile drainage, and the annual purchase and application of crop fertilizers.

4.2.2. Tile Drainage and Fertilization

Another aspect of remotely sensed imagery relates to irrigation and tile drainage decisions. Farmers can view their land using various filters within *GeoVisage* Imagery, and low/wet spots can be identified. For instance, particularly wet fields may not be appropriate for winter wheat because of the episodes of freezing in this region.

> **Aidan**: People think that they are on flat ground … If you know the lay of your land, when you are out there seeding or combining—especially combining—if you have a low spot or a high spot on the farm and you didn't realize it was that much higher or lower, and you start to realize that's why the crop's not growing there very well, or that's why it gets flooded out.

Within the same township, Ben, who specializes in cash cropping using high tech equipment, agrees with the importance of remotely sensed imagery in potentially influencing decisions around drainage and yield.

> **Ben:** One of the things might be capturing moisture levels remotely. Whenever we have a combine yield monitor running and we see variability in the field, 80% of the time it will be too much moisture in the soil. This will probably be the biggest benefit financially to the farmer. So, whether that means more tile drainage needed, or that something is wrong with the existing tile, you have got to answer those questions. We had a wet spring and farmers went out and compacted their soil. You will see lot of yellow fields, where farmers were impatient and went out and spread manure and compacted the soil with their machines … So, if you could have those elevations in there, all of those things might aid you in planning efficient water movement.

Not only does imagery provide farmers with watershed and field drainage information, it can also generate imagery that can allow farmers to analyze crop performance relating to fertilization decisions. Dean explained how he had found this aspect to be the most promising for his purposes.

> **Dean:** Well, the first time I used the imagery it was more for prediction. I wanted to see the difference between applying different amounts of fertilizer … We could see the difference made between the different application rates of fertilizer when we were done harvesting.

Thus far we have discussed how farmers had been using the *GeoVisage* online resource for weather station data access, land measurement, and remotely sensed imagery information. In the next section we will now turn to future recommendations of the participating farmers regarding the *GeoVisage* resource. According to our group of northeastern Ontario agriculture producers, how might this tool be improved?

5. Discussion

Participants had used the *GeoVisage* online resource in varying degrees, and for a variety of purposes as detailed above. Three of the five farms involved in the study had allowed weather station towers to be constructed on their respective farm properties. Long-term working relationships had clearly been established between the researchers and the participating farms. When asked about feedback and recommendations, interviewees were open, positive, and respectful, offering a number of insightful suggestions dealing with new or adapted *GeoVisage* features and ongoing support.

5.1. Internet Access and Communication

One of the difficult issues surrounding the development of *GeoVisage* was that, since a variety of organizations had been funding the initiative over time, there would sometimes be periods wherein weather station maintenance and/or online resource development would be temporarily put on hold pending new funding and/or changes in development and support staff. This situation occasionally led to what was perceived as a lack of communication between the farms and the university.

Because of the computational demands of imaging applications in general, and the bandwidth requirements of web-based visualizations in particular, there were also some Internet access and browser issues that limited the ability of some participants to access certain project features and tools. For example, when asked about high-speed Internet access, which is preferred for the proper functioning of certain *GeoVisage* features, Evan noted "it's a private service provider—they call it high speed but really it's not . . . So, I mean some days it's great and other days it's not as good, but I think it should be able to handle it."

Some of the issues related to properly accessing *GeoVisage* had to do with the particular Internet browser being used (e.g., Chrome, Firefox, Internet Explorer, Safari). Due to several technical factors, including the utilization of Java technology in the World Wind framework, some browsers required specific security settings to make *GeoVisage* imagery features accessible. In response to these various scenarios, the research team created a series of online video tutorials with graduate assistants to help farmers download the required Java updates and to modify security settings in order to access the various tools. Aidan found these browser set-up tutorials, which could be reviewed as needed, to be very helpful: "I was lost until you had her speak through it on the video. Then you can pretty well keep up with her, and then you can pause it, and find out where or what she is talking about, and then just continue on." However, some participants still found the process to be challenging due to the technical requirements of the imagery tool.

Ben indicated during his interview that he would have liked more frequent communication between the university and his farm in terms of the new features being added and improved over time: "as long as we have a contact, I think that's the biggest thing. If somebody was maybe appointed, so when more farmers see it, and they have a go-to person, or even if there was a chat line perhaps?" Evan mentioned the idea of further education around the resource: "I think it would be good to promote it a little bit more to get farmers interested in using it regularly. Maybe even a little course, just to know how to use all of the features."

Overall, however, all five participants clearly thought very highly of the university research team and were excited to be actively taking part in this project over time. What follows are related comments made by participating farmers from within the two different communities involved in the research study.

> **Aidan:** If it's easy to use, that'll be the big thing—everybody will use something that is easy to use . . . It's a great tool. Nipissing has sort of stuck with us here in the north, and I appreciate it . . . It'll always be new to us—it's all cutting edge information, right? But I know when they come to the farm show they always get a good response at their booth. There is always somebody talking about something, and I just hope that they are getting out of it what we are getting out of it. We don't show it sometimes, but we do appreciate what they are doing, especially for now, and beyond—because farming is only going to get more and more high tech.

Having built strong working relationships with farmers in these two northeastern Ontario communities over multiple years and involving a number of different project initiatives, the researchers had opportunities for open communication and to acquire useful feedback.

5.2. Crop Imagery

Along with his sons, Ben runs a large cash crop farming operation that utilizes very complex, technology-enhanced machinery and processes. Not only does he sell his crops through a local co-operative, he also contracts out to international businesses in terms of planting, monitoring, and harvesting select crops for the international market. Part of this process involves carefully negotiating these contracts and maintaining open communication with his various employers. As he explained during the interview, these companies often require ongoing, documented photographic data of the crop being grown for them. Hence, Ben sees the addition of such a feature in *GeoVisage* as a useful tool.

> **Ben:** Some of the things I could see fitting in this are crop progress reports to end-users. So, for instance, we have a buckwheat project with Japanese buyers, and they want to be very involved in this project. They like to see imagery, they want field notes from individual producers, and how that specific field is progressing. So, the ability to actually import pics, and to add notes to those pics, to report on the progress—everything from disease pressure to crop conditions—that we could import imagery ourselves, and export that file to these end-buyers so that they can view all of their fields, and all of their sites, because they are dealing with multiple farmers in northeastern Ontario.

What was being recommended here was a potential secure photo/data sharing system that could perhaps incorporate not only digital still photography taken by local farmers and shared with employers through the *GeoVisage* resource, but also UAS imagery of field growth conditions and crop progress.

5.3. Mobile Access

Perhaps the most commonly mentioned item by all participants during the interviews, regarding future recommendations for the online *GeoVisage* resource, was that of making it more accessible with a user-friendly mobile application version for smartphones. It should be noted *GeoVisage* was originally developed as a decision-support and research tool, and not primarily as a real-time monitoring system; consequently, mobile utility is somewhat limited. Farmers in the study all used smartphones in their daily work for communication and information purposes, occasionally even checking them during the interview proper regarding incoming calls or messages.

One key use of smartphones in contemporary farming is that of crop rotation and tracking in order to help replenish soil and to avoid crop diseases. Ben explained the nature of his crop rotations.

> **Ben:** To grow crops you need a rotation. Oats, wheat, and barley are a grass plant, susceptible to the same diseases and the same kind of sprays. You need a broad leaf plant, like a soybean, or Canola because they require a different spray, and so you break disease pressure. If you just grow a grass plant year after year, any of those fungus diseases, if they attack one grass plant, they will probably attack again and again. But, if you start rotating a broad leaf canola, soybean, or flax, for example, you would get a mix of weeds and disease, and you could then throw something new at it all the time.

Similarly on the fruit/vegetable farm, crop rotation is a key feature of Dean's annual planting decisions:

> **Dean:** We try to never grow the same thing in the same spot, year after year ... The berries, we'll let them produce for three years, but after that, I won't go into strawberries at the same spot for at least five years ... We don't fumigate any of our crops, so in order to avoid

a lot of diseases we have to do crop rotation . . . Well, you plant something else—every crop has its own specific problems, specific funguses, so by rotating it, you break the cycle and disease doesn't have a chance to build up . . . For berries, I'll plant maybe corn afterwards, and then I'll put maybe pumpkins the year after that.

In order to keep careful track of the annual planting decisions, maps are used, both print and digital.

Dean: We have a laminated map in every tractor and all of the fields are numbered. But all of the actual inputs, everything is logged into an app that I have called *Field Tracker*—Carl's son created it. So, I can touch any year, any field, and I can know all the inputs from start to finish, what was done in that field . . . This use to be done on paper, but now it's done on the go. As I'm planting, I'm recording—the date, the seed name, and even the seed variety, the machine setting, and what fertilizer I've applied—everything . . . It's all cloud-based on the Internet, so I could break my phone, or lose it, and I'll never lose this data.

Clearly the farmers in this group are all using their smartphones regularly for activities such as emailing, crop tracking, weather forecast checking, price comparison and purchasing, digital photographs/video of crop growth, and general communication with their business partners, family, and other individuals/companies, etc. Developing a *GeoVisage* app that would work well on a number of different mainstream devices (e.g., iPhone, Windows phones, Android, Blackberry) would no doubt incentivize farmers to use it even more frequently. That being said, it was pointed out by several participants that certain features on the *GeoVisage* online resource (e.g., comparative weather data with graphs or charts) are easier to read and interpret on a large computer monitor, rather than viewing the data on a smartphone or even a tablet or iPad.

According to Dean, when it comes to daily email activity, he probably does "60% out of the pocket (via smartphone) throughout the day, compared to maybe 40% at the desktop" back at the farm office. Chris shared similar thoughts about the benefits of mobile access: "as soon as it's desktop-based, it becomes less helpful, because if you're spraying, you're not at home in an office, necessarily. You're at the site. You've decided what spray you're going to use, and then to decide if it's time to spray—you're not going to drive home to find out."

Finally, Evan made the point that it would be nice to not only have a *GeoVisage* mobile app for smartphone, but that it also would be convenient to include in such an app a direct link to, or an embedded measure from, one or more weather forecasting networks which farmers already check regularly throughout the day. He noted specifically that "if there was a place to go to the Weather Network right there, that would also be great. I know that the Dairy Farmers of Ontario just added this last year to their website—you can click onto the Weather Network . . . It actually locates where you are, using GPS, so you just click on it and it gives you exactly what's going on with the weather there." Four new stations serving the farmers in northeastern Ontario provide access via direct link with a mobile app. The data are not yet integrated into *GeoVisage*. However, it addresses some of the suggestions from users about increasing accessibility in the field.

6. Conclusions

As stated at the beginning of this paper, the main research goal was to gather information about the utility and perceived usefulness of the *GeoVisage* online resource by participating farmers. As the results have shown, the individuals at all five farms (three dairy, one cash crop, and one fruit/vegetable) took advantage of the online *GeoVisage* weather station data in different ways and in varying degrees; and some had also taken time to explore the newer *GeoVisage* imagery tools and video tutorials.

Several recommendations regarding existing and additional features (e.g., secure crop image sharing, mobile app access) were shared within the interviews. Overall, the *GeoVisage* project was perceived by all participants as representing a very useful set of tools for their complex everyday work. As Dean noted, "they're helpful at the moment for making decisions, and they also have much potential for future usage … All farmers are interested in weather and what's happening at their own farm."

Community-based participatory research endeavours, such as *GeoVisage*, require engagement with end-users, and such interaction with users is rapidly becoming indispensable for the successful adoption of precision agriculture and agricultural decision support tools. As an example of the community-based participatory aspects of this project wherein feedback from users is vital for continued development, from the results of the interviews of current *GeoVisage* users presented in this paper, future research and development of *GeoVisage* will focus on secure imagery interfaces that require less configuration, more support for mobile app visualization and communication, and a greater range of data graphing and visualizations options integrated into the real-time component of the system.

As researchers continue to work closely with northeastern Ontario agricultural producers in further developing this particular type of spatial decision support tool within a positive and participatory framework, they are hopeful that increased adoption and improved effectiveness will characterize this already promising and clearly beneficial resource for agronomists, fruit/vegetable growers, and livestock farmers alike.

In this paper, we have presented empirical evidence that supports McCown's research showing that collaborative decision support system development will lead to the likely adoption and use of the system. More broadly, for other agricultural decision support tool developers, we have found that the regular contact, communication, and feedback demonstrated in this collaborative research initiative are essential to maintain trust and to build lasting relationships, which ultimately contribute to prolonged and meaningful use of the technology and to continued improvements.

Acknowledgments: This research was made possible by a grant provided to D.W., J.M.K. and M.P.W. from the Northern Ontario Heritage Fund Corporation (Project #920161), and through Natural Sciences and Engineering Research Council of Canada funding to J.M.K. (Grant #RGPIN-2014-06188) and to M.P.W. (Grant #386586-2011). We also thank Renata Wachowiak-Smolíková for helpful suggestions and proofreading.

Author Contributions: All four co-authors conceived and designed the study. M.P.W., D.W. and J.M.K. have developed the *GeoVisage* software over time, and have also established ongoing professional relationships with local farmers in a number of projects. D.H.J. conducted the qualitative research interviews with the participating farmers. D.H.J., M.P.W., D.W. and J.M.K. all contributed to the paper.

Conflicts of Interest: The authors declare no conflict of interest.

Appendix A. Interview Schedule (Questions)

1. Please explain your involvement with the weather station and *GeoVisage* online support tool project to date? How and when did you become involved with this project?

2. *GeoVisage* is accessible through the Internet. Is high-speed Internet adequately available within your local area (home/barn office)? Would most local farmers be able to access Internet?

3. The *GeoVisage* website has been updated several times over the past few years. To date, could you describe how often, and in what ways, have you used the *GeoVisage* website as a farmer?

4. Do you have any suggestions regarding the improvement of the existing features, or possibly any suggestions regarding the addition of other features to the *GeoVisage* website?

5. Overall, in your opinion, how helpful is this website resource for your work as a farmer?

6. The aerial imagery is the newest addition to the website resource. Is this something that you might be interested in becoming involved with? How would this be useful data for your farming?

7. Do you have any further comments, or other questions for the research team at this time?

References

1. Aubert, B.A.; Schroeder, A.; Grimaudo, J. IT as enabler of sustainable farming: An empirical analysis of farmers' adoption decision of precision agriculture technology. *Decis. Support Syst.* **2012**, *54*, 510–520. [CrossRef]
2. Barr, S.; Sharda, R. Effectiveness of decision support systems: Learning or reliance effects? *Decis. Support Syst.* **1997**, *21*, 133–146. [CrossRef]
3. Bochtis, D.D.; Sorensen, C.G.; Green, O. A DSS for planning of soil-sensitive field operations. *Decis. Support Syst.* **2012**, *53*, 66–75. [CrossRef]
4. Burgos-Artizzu, X.P.; Ribeiro, A.; Guijarro, M.; Pajares, G. Real-time image processing for crop/weed discrimination in maize fields. *Comput. Electron. Agric.* **2011**, *75*, 337–346. [CrossRef]
5. Guillard, V.; Buche, P.; Destercke, S.; Tamani, N.; Croitoru, M.; Menut, L.; Guillaume, C.; Gontard, N. A decision support system to design modified atmosphere packaging for fresh produce based on a bipolar flexible querying approach. *Comput. Electron. Agric.* **2015**, *111*, 131–139. [CrossRef]
6. McCown, R.L.; Hochman, Z.; Carberry, P.S. Probing the enigma of decision support system for farmers: Learning from experience and from theory. *Agric. Syst.* **2002**, *74*, 1–10. [CrossRef]
7. Nute, D.; Rosenburg, G.; Nath, S.; Verma, B.; Rauscher, H.M.; Twery, M.J.; Grove, M. Goals and goal orientation in decision support systems for ecosystem management. *Comput. Electron. Agric.* **2000**, *27*, 355–375. [CrossRef]
8. Walters, D.F.; Smolikova-Wachowiak, R.; Wachowiak, M.; Shrubsole, D.; Malczewski, J. Ontario's Nutrient Calculator: Overview and focus on sensitivity analysis. *J. Agric. Sci.* **2013**, *5*, 189–200. [CrossRef]
9. Cerf, M.; Jeuffroy, M.-H.; Prost, L.; Meynard, J.-M. Participatory design of agricultural decision support tools: Taking account of the use situations. *Agric. Sustain. Dev.* **2012**, *32*, 899–910. [CrossRef]
10. Rossi, V.; Salinari, F.; Poni, S.; Caffi, T.; Bettati, T. Addressing the implementation problem in agricultural decision support systems: The example of vite.net. *Comput. Electron. Agric.* **2014**, *100*, 88–99. [CrossRef]
11. Seelan, S.K.; Laguette, S.; Casady, G.M.; Seielstad, G.A. Remote sensing applications for precision agriculture: A learning community approach. *Remote Sens. Environ.* **2003**, *88*, 157–169. [CrossRef]
12. Shibl, R.; Lawley, M.; Debuse, J. Factors influencing decision support system acceptance. *Decis. Support Syst.* **2013**, *54*, 953–961. [CrossRef]
13. McCown, R.L. Changing system's for supporting farmers' decisions: Problems, paradigms and prospects. *Agric. Syst.* **2002**, *74*, 179–220. [CrossRef]
14. Silver, M.S. *Systems that Support Decision Makers: Description and Analysis*; John Wiley Sons: New York, NY, USA, 1991.
15. Eastwood, C.; Klerkx, L.; Nettle, R. Dynamics and distribution of public and private research and extension roles for technological innovation and diffusion: Case studies of the implementation and adaptation of precision farming technologies. *J. Rural Stud.* **2017**, *49*, 1–12. [CrossRef]
16. Jakku, E.; Thorburn, P.J. A conceptual framework for guiding the participatory development of agricultural decision support systems. *Agric. Syst.* **2010**, *103*, 675–682. [CrossRef]
17. Rose, D.C.; Sutherland, W.J.; Parker, C.; Lobley, M.; Winter, M.; Morris, C.; Twining, S.; Ffoulkes, C.; Amano, T.; Dicks, L.V. Decision support tools for agriculture: Towards effective design and delivery. *Agric. Syst.* **2016**, *149*, 165–174. [CrossRef]
18. Kuhlmann, F.; Brodersen, C. Information technology and farm management: Developments and perspectives. *Comput. Electron. Agric.* **2001**, *30*, 71–83. [CrossRef]
19. Mackrell, D.; Kerr, D.; von Hellens, L. A qualitative case study of the adoption and use of an agricultural decision support system in the Australian cotton industry: The socio-technical view. *Decis. Support Syst.* **2009**, *47*, 143–153. [CrossRef]
20. GeoVisage. Available online: http://geovisage.nipissingu.ca (accessed on 7 August 2017).
21. Bond, A. Farmers get ahead of Mother Nature. Available online: http://yourontarioresearch.ca/2016/04/getting-ahead-mother-nature/ (accessed on 7 August 2017).
22. Nipissing News. GeoVisage Tool Helping Farmers Grow, Now Online. Available online: http://www.nipissingu.ca/about-us/newsroom/Pages/GeoVisage-tool-helping-farmers-grow,-now-online.aspx (accessed on 7 August 2017).
23. HOBOlink. Available online: https://www.hobolink.com/ (accessed on 7 August 2017).

24. Nittel, S.; Bodum, L.; Clarke, K.C.; Gould, M.; Raposo, P.; Sharma, J.; Vasardani, M. Emerging Technological Trends likely to Affect GIScience in the Next 20 Years. In *Advancing Geographic Information Science: The Past and Next Twenty Years*; Onsrud, H., Kuhn, W., Eds.; Global Spatial Data Infrastructure Association (GSDI), 2015. Available online: http://gsdiassociation.org/index.php/49-capacity-building/publications/343-advancing-geographic-information-science-the-past-and-next-twenty-years.html (accessed on 7 August 2017).
25. Hogan, P. NASA World Wind: A planetary visualization tool. In Proceedings of the ACM SIGGRAPH 2005 Educators Program, Los Angeles, CA, USA, 31 July–4 August 2005. [CrossRef]
26. Zhang, C.; Kovacs, J.M. The application of small unmanned aerial systems for precision agriculture: A review. *Precis. Agric.* **2012**, *13*, 693–712. [CrossRef]
27. Han, W.; Yang, Z.; Di, L.; Mueller, R. CropScape: A web service based application for exploring and disseminating US conterminous geospatial cropland data products for decision support. *Comput. Electron. Agric.* **2012**, *84*, 111–123. [CrossRef]
28. Mueller, R.; Harris, M. Reported uses of CropScape and the national cropland data layer program. In Proceedings of the International Conference on Agricultural Statistics VI, Rio de Janeiro, Brazil, 23–25 October 2013.
29. Du, W.; Chen, N.; Yan, S. Online soil moisture retrieval and sharing using geospatial web-enabled BDS-R service. *Comput. Electron. Agric.* **2016**, *121*, 354–367. [CrossRef]
30. Chen, N.; Zhang, X.; Wang, C. Integrated open geospatial web service enabled cyber-physical information infrastructure for precision agriculture monitoring. *Comput. Electron. Agric.* **2015**, *111*, 78–91. [CrossRef]
31. Georg, R.; Kruse, R.; Schneider, M.; Wagner, P. Visualization of agriculture data using self-organizing maps. In *Applications and Innovations in Intelligent Systems XVI*; Allen, T., Ellis, R., Petridis, M., Eds.; Springer: New York, NY, USA, 2009; pp. 47–60.
32. Kubicek, P.; Kozel, J.; Stampach, R.; Lukas, V. Prototyping the visualization of geographic and sensor data for agriculture. *Comput. Electron. Agric.* **2013**, *97*, 83–91. [CrossRef]
33. Tayyebi, A.; Meehan, T.D.; Dischler, J.; Radloff, G.; Ferris, M.; Gratton, C. SmartScape™: A web-based decision support system for assessing the tradeoffs among multiple ecosystem services under crop-change scenarios. *Comput. Electron. Agric.* **2016**, *121*, 108–121. [CrossRef]
34. Batte, M.T.; Arnholt, M.W. Precision farming adoption and use in Ohio: Case studies of six leading-edge adopters. *Comput. Electron. Agric.* **2003**, *38*, 125–139. [CrossRef]
35. Denzin, N.K. *The Sage Handbook of Qualitative Research*, 3rd ed.; Lincoln, Y.S., Ed.; SAGE: Thousand Oaks, CA, USA, 2005.
36. Yin, R.K. *Case Study Research: Design and Methods*, 4th ed.; SAGE: Thousand Oaks, CA, USA, 2009.
37. Atlas.ti: Qualitative Data Analysis. Available online: http://atlasti.com/ (accessed on 7 August 2017).
38. Zhang, C.; Walters, D.; Kovacs, J. Applications of low altitude remote sensing in agriculture upon farmers' requests: A case study in Northeastern Ontario, Canada. *PLoS ONE* **2014**, *9*, e112894. [CrossRef] [PubMed]

agriculture

MDPI

Technical Note

Innovative Solution for Reducing the Run-Down Time of the Chipper Disc Using a Brake Clamp Device

Andrea Colantoni [1,*], Francesco Mazzocchi [1], Vincenzo Laurendi [2], Stefano Grigolato [3], Francesca Monarca [1], Danilo Monarca [1] and Massimo Cecchini [1]

[1] Department of Agricultural and Forestry Sciences (DAFNE), Tuscia University, 01100 Viterbo, Italy; mazzocchi@unitus.it (F.M.); francescamonarca@hotmail.it (F.M.); monarca@unitus.it (D.M.); cecchini@unitus.it (M.C.)
[2] INAIL National Institute for Insurance against Accidents at Work, Via di Fontana Candida 1, 00078 Monte Porzio Catone (RM), Italy; v.laurendi@inail.it
[3] Department of Land, Environment, Agriculture and Forestry, Università degli Studi di Padova, Viale dell'Università 16, 35020 Legnaro, Italy; stefano.grigolato@unipd.it
* Correspondence: colantoni@unitus.it; Tel.: +039-076-135-7356

Received: 18 July 2017; Accepted: 17 August 2017; Published: 20 August 2017

Abstract: Wood-chippers are widely used machines in the forestry, urban and agricultural sectors. The use of these machines implies various risks for workers, primarily the risk of contact with moving and cutting parts. These machine parts have a high moment of inertia that can lead to entrainment with the cutting components. This risk is particularly high in the case of manually fed chippers. Following cases of injury with wood-chippers and the improvement of the technical standard (ComitéEuropéen de Normalisation-European Norm) EN 13525: 2005 + A2: 2009, this technical note presents the prototype of an innovative system to reduce risks related to the involved moving parts, based on the "brake caliper" system and electromagnetic clutch for the declutching of the power take-off (PTO). The prototype has demonstrated its potential for reducing the run-down time of the chipper disc (95%) and for reducing the worker's risk of entanglement and entrainment in the machine's feed mouth.

Keywords: wood chipper; brake clamp; work safety; forestry

1. Introduction

The research activities reported in this article were conducted in the framework of the "Protection of machinery operators against crush, entanglement, shearing" (PROMOSIC) project funded by the Italian National Institute for Accidents at Work Insurance (INAIL). The project covered a number of safety issues associated with most widely used agricultural machines and the article describes a prototype solution for forestry chippers with manual feeds.

Data on the occurrence of accidents provide an objective index of the danger of machinery as well as a valid reason for identifying the most critical features of these machines. At present, there is no database covering Italy and all Europe for reporting significant accident data associated with chippers, so analyses of accident indexes compiled in other countries were helpful. In North America, 2042 non-lethal accidents involving chippers and 31 deaths in the decades of 1992–2002 were reported.

An in-depth analysis of the fatal accidents disclosed that 42% involved work in gardening, 16% were related to ground-keeping work, and the remainder involved workers in the forestry and agricultural sectors [1]. These analyses also turned up the point that 68% of the accidents were caused by workers' direct contact with the mechanical components of the machinery in operation, and contact with mechanical components not during chipping operations accounted for 29% of the accidents due to the open protective case of the drum or disk chipper still in motion. The largest number of fatal accidents was concentrated each year in the period between July and August. Again, in North America, it was found that the social cost of fatal accidents with chippers came to US$28.5 million in 2003. Analyses of the non-lethal accidents showed that the majority of cases involved workers aged 25–34 and that 60% of the accidents caused immediate injury or amputation of parts of the upper-body limbs. For 25% of these injuries, the victims were unable return to work for periods of up to 30 work days [2]. Further studies indicated that 16% of these accident victims had less than three months of experience in that particular job and 18% of them had worked from three to 11 months on their jobs [3–5]. Safety in the use of machinery is, and has been in the past, a strongly pertinent problem at the European and specifically at the Italian level. The specifics of chippers, however, are actually only applied at the European Union level, with the technical (European Standard EN) regulation on safety specifically targeted to manufacturers. The technical standard was set out in the Machinery Directive EN 13525:2005 + A2:2009 on forestry machinery, mobile chippers and safety. Following a formal French objection due to accidents in 2011 and 2012, this standard with C1-type harmonization has been superseded [1]. The choice of work to perform with a chipper depends mainly on request and continuity of use (type of work and quantity of chips), characteristics of materials to feed in (the origin), and the work system (productivity and work on site). The chipping machines used in forestry, agriculture and on urban greenery are usually mobile types coupled with a tractor and are mounted, trailed or semi-mounted and for energy production by agro-forestal biomass [6,7]. They can be mounted on a truck with an independent engine, or driven by the tractor or self-propelled. Chippers in the low power category (<20 kW) can be equipped for transport and an internal combustion engine or electric motor. The project PROMOSIC has been involved with the Italian manufacturer Peruzzo Ltd. (Curtarolo, Italy) which made a number of portable chipper models available for preliminary trials for evaluating the risk for operators during manual feeding and the time needed for stopping the chipper components, disks or knives in specific cases [8]. The leading safety feature in the use of chippers is, in fact, the danger of coming into contact with the internal flywheel. This component continues to rotate by inertia even if the safety bar is inserted. According to EN 13525: 2005 + A2:2009, the safety bar installed in the bottom of the machine and on the sides of the feed chute is mandatory. The safety bar acts to block the feed rotors, stopping them completely in a quarter of a second, but the flywheel continues to rotate by inertia for more than a minute [4,5,9–14]. The consequences are the potential for dragging the worker into the chipping chamber to the chipping components, which is the major cause of serious or fatal accidents.

1.1. Manually Fed Wood-Chippers

Manually fed wood-chippers (or self-feeding mobile wood-chippers) are commonly used in small businesses or in the domestic maintenance of rural land and small forest stands as well as urban woodlands, urban parks and gardens for small tree-trimming operations and/or for comminution of small branches [15].

Generally, a manually fed wood-chipper consists of a feeding hopper, a rotating chipper unit, a power unit and a discharge system. Branches and small logs or trees are fed manually into the feeding hopper and thus pulled into the chipping unit. In detail, self-feeding mobile wood-chippers are commonly used during tree-trimming operations and consist of a self-loaded frame, a feed chute, knives mounted on a rotating disc or drum or as an alternative to an auger, a discharge unit and a power unit (using power take-off or an independent diesel power unit or electric engine).

The biomass is first entered manually by the operator into the machine's in-feed chute and then the grabbing mechanism feeds it towards the chipper unit. The chipper disc or drum rotating generally between 1000 and 2000 rpm is able to comminute the biomass in regular small wood pieces (woodchips) [16]. Through the expulsion mechanism the woodchips are discharged on the ground or into a bin or carrier (Figure 1).

A distinction between several models in the market, whether manual or load-type machines, is found in the type of drum chipper, disc and auger chipper. In drum chippers, the cutting device consists of a steel rotor rotating around its longitudinal axis and two to 24 tangential knives or hammers can be inserted (lower quality of the obtained product). The maximum cutting diameter is one-third of the diameter of the drum (diameter from 500 to 1500 mm). In the wood-chippers, the cutting member (minimum of 800mm) is made up of a flywheel which usually consists of two to four radially oriented knives [17–19]. In this case, the maximum cutting diameter corresponds to one-quarter of the disc chipper diameter.

Figure 1. Manually fed wood-chipper used in a riverbank vegetation clearing cut.

1.2. Specific Hazards of the Manually Fed Wood-Chipper

Chipper machines can lead to different risks, in particular direct contact with the sharpened components of the machine such as knives inserted into the drum or disk. Contact with the chipper's operating components (blades, discs or knives) may result in amputation or death. Workers may also be injured by material thrown from the machine. To minimize these hazards, appropriate engineering and work practice controls, including worker training, should be guaranteed.

Working with or around a wood-chipper can be dangerous and might result in death or serious injury if proper procedures are not followed. Workers feeding material into the self-feeding wood-chippers are at risk of being pulled into the chipper if they are entangled in the branches being fed into the machine. In addition, workers are at risk of being struck by unlatched, improperly secured, damaged or improperly maintained hoods that may be thrown from the wood-chipper after contact with the rotating chipper knives.

1.3. Development of an Innovative Solution

Peruzzo Ltd. made a "TIREX" chipper model (Figure 2) available for trials which mounted the three-point hitch of a tractor in order to test a disengagement chipper brake system as an innovative safety solution. In this specific case, the chipping components were made up of four knives fitted on a rotating disk and a fixed counter-blade mounted on the bodywork of the machine. The feeding system consisted of a horizontal toothed roller and chain, both powered by hydraulics to convey materials to the fixed cutting disc knives. The feeding chute rollers can be stopped or their direction of rotation reversed simply with the use of a control lever. A standard no-stress system with an hour counter automatically regulated correct branches feeding with power supplied in relation to the power required by the materials to ease the work of the chipping components and the engine. The disengagement chipper brake system was made up of an electromagnetic clutch mounted on the chipper drive shaft; a steel caliper brake; a hydraulic electro-valve for brake control; and an emergency button with normally closed (NC; closed = short circuit = creating a path for the current) and normally open (NO; open = open circuit = not creating a path for the current) contact. When a normally open push-button is pressed, a path is provided for the current. When a normally closed push-button is pressed, the current is impeded from flowing.

Figure 2. The manually fed wood-chipper used for developing the brake clamp device system (picture by Peruzzo).

NC push-buttons are used in emergency stop buttons. They are pressed when an accident has occurred or may occur, and the machine needs to be stopped immediately due to an action which could damage someone or something. Normally closed buttons are preferred for two reasons: firstly, they do not rely on creating good contact to create a signal. They just have to open a circuit, which is much easier. An NC is more robust and therefore safer. Secondly, they react quicker. For an NO button, the signal event happens at the end of the movement (when the movable part makes contact). For an NC button, the signal event happens at the beginning of the movement (when the movable part stops making contact), through a 12 V hydraulic flow regulator actuating as a brake plug with a no-stress CPU (central processing unit). In light of the components considered for completing the system, the cost is estimated to be around €900. This cost was calculated on the building of a prototype. The cost could be substantially lowered if the system became integrated in the serial production of these machines. In conclusion, the solutions studied could be useful for reducing accidents due to entanglement and dragging with relatively small costs. Moreover, the no-stress system moderating power turns out to be needed for work with the combination of the electromagnetic clutch.

2. Materials and Methods

2.1. The Wood-Chipper Used

For developing an innovative system for lowering the cut-off time of the cutting disc, it was decided to base the study on one of the most globally used machine configurations most commonly used in green maintenance in the internal business environment.

All the tests were performed in 2017 at the "Lucio Toniolo" experimental farm of the University of Padova in cooperation with Peruzzo Ltd. The chipper chosen is a type of driven chipper powered by power take-off tractors of between 30 and 60 kW. The machine has a cutting member consisting of a flywheel with a thickness of 30 mm and diameter of 620 mm, on which two knives are inserted with a width of 200 mm and a thickness of 25 mm. The feeding system is made up of a plan, whereby two rollers are provided to convey the material to the chipping chamber. Vertical oscillation rollers are hydraulically actuated and controlled by a block control.

2.2. The Run-Down Time Evaluation

For the machine found in current market conditions, the stopping times of the flywheel have been verified, following disconnection of the power outlet, by a decoupling device of the power outlet from the universal joint (Figure 3).

Figure 3. Decoupling system installed between power outlet and cardan shaft.

The tests predicted the activation of the decoupling system, both in absence and in the presence of chipping material. Tests firstly provided for a maximum rotation of the flywheel, corresponding to 1400 rpm, then activating the decoupling system and disconnecting the power outlet from the shaft. The stopping time has been calculated as from the disengagement moment until the flywheel is fully stopped. The analysis was set by observing the spin speed of the flywheel on the display of the machine's anti-stress system and by slow motion via a dedicated camera.

2.3. Braking System Design

The use of a motorcycle brake caliper mounted in the power intake shaft (additional component), specifically a Honda CBR 600 with a vacuum mass of 180 kg has been suggested.

Below, calculations are used to determine the braking torque necessary to curb the inside of the chipper in a 4s time (ex-post condition) with respect to the current state. The calculation method can be applied indifferently to flywheels, even of different dimensions (Table 1).

Table 1. Data for calculation of braking torque for brake clamp device.

Initial Data	
Diameter	0.6 m
Radius	0.3 m
Mass	80 kg
Angular Speed	157 rad/s
Stop Time	4 s

Angular acceleration a_{ang} was calculated as:

$$a_{ang} = \frac{\omega_{final} - \omega_{initial}}{t} = \frac{0 - 157}{4} = 39.25 \, \text{rad s}^{-2} \tag{1}$$

where ω_{final} is the final angular speed and $\omega_{initial}$ is the initial angular speed (rad/s).

Then, the moment of inertia I, can be calculated (I):

$$I = \frac{m \times r^2}{2} + \frac{80 \times 0.09}{2} = 3.6 \text{ kg m}^2 \tag{2}$$

where m is the mass (kg) and r is the radius (m).

Finally, the braking torque $C_{braking}$ is calculated:

$$C_{braking} = I \times A_{ang} = 3.6 \times 39.25 = 141.3 \text{ Nm} \tag{3}$$

Based on the braking torque, a hydraulic "clamp" was dimensioned, suitable for the dimensional characteristics of the internal flywheel.

3. Results and Discussion

3.1. Flywheel Stop Times, with and without Chipping Material and Disconnection of Power Take-Off

In the test, the insertion of the decoupling system had to be combined with the anti-stress system. The anti-stress system, consisting of a control unit and a sensor for measuring the speed of the flywheel rotation, controls the feed of the chipping chamber, avoiding overload situations, through the speed control of the feed rollers (Figure 4).

Figure 4. Anti-stress system installed in the tested machine.

In case the chipping flywheel is impacted by a material overload and decreases the number of revolutions, the anti-stress system effectively blocks the entry of further material into the chipping chamber by locking the feed rollers.

The anti-stress system is activated when the flywheel rotation speed drops below a certain number of revolutions per minute (rpm). Since inserting the power take-off decoupling system leads to the breaking of the drive torque at the chipping wheel, there is an immediate effect on the rotational speed, which begins to decrease immediately. As a result, the anti-stress system is activated immediately (if present and if inserted), which effectively blocks the feed of the chipping chamber.

If the anti-stress system is off or not present, the power system will continue to work for a few seconds after the power outlet is disconnected, without actually blocking the power supply of the material to the chipping chamber. It could therefore result in the real risk of damage to the machine and the same flywheel and chipping chamber, rather than blocking any contact between the operator and flywheel (Table 2). Also, in the best of cases, after removing the cause of the block, the machine, in a subsequent work phase, can strike a new flywheel block even with little time from the start of the new chipping phase, failing to work the plant material (an example of the material that locked the flywheel in Figure 5). This may require the operator to intervene by removing the crankcase and the protections on the chipper (to remove the vegetal material from the flywheel knives), creating high-risk situations for workers. This can also occur for small diameters.

Table 2. Flywheel stop times in relation to different conditions.

Condition	Average Stop Times (s)
Absence of material	79
During drumming (with no stress system inserted)	73
During stacking (with no stress system inserted)	76

Figure 5. Detail of the blocking branch.

3.2. Design for Chipper Disengagement and Brake Design and Prototype Design and Implementation

The aim of the study is to provide a solution that allows synergic decoupling with the disc braking of a chipper, coupled with a forest tractor. Current safety devices do not allow the machine to be disengaged and brake-coupled to a tractor, as it does not guarantee the safety of operators.

The chipper consists of a rotating disk where four knives are mounted and a counter blade fixed in the frame of the machine, which can chip wood up to a diameter of 18 cm. The chip size may vary according to the needs by adjusting the knives and feed rollers of the product. The feed rollers are hydraulically driven, with a no-stress device to preserve disk blades. If necessary, the power cycle can be reversed or stopped. The discharge of the product takes place through a swivel and 360° rotating tube.

Technical Aspects of the Brake Clamp Device

The brake clamp device is composed of the following components (Figure 6):

(a) An electromagnetic clutch mounted on the tractor's cardan shaft; the wood-chipper's disc cutter is made up of a steel ring 600 mm in diameter and 30 mm thick, which mounts four knives with a total mass of ~90 kg. Supplied with a tractor of 70–80 kW, considering that the disc can rotate about 1500 rpm, we have a torque of 33 Nm. The electromagnetic clutch chosen has a diameter of 173 mm, and it works at a voltage of 12 V with a maximum power absorption of 68 W. It resists at a maximum torque of 47 Nm.

(b) The brake clamp of a motorcycle type mounted on Power Take Off (Figure 7). A "motorcycle" brake clamp was chosen, in particular a Honda CBR 600 with a 180 kg vacuum mass. Considering the driver and passenger (140 kg), the total estimated mass is about 320 kg. Divided for two motor pliers we have a load of 160 kg per gripper. The braking disc diameter of the bike is 300 mm, and at a speed of 130 km/h, the disc rotates at 2300 rpm with a peripheral speed of 36 m/s.

The disk of the chipper machine in question rotates at a speed of 1500 rpm, considering:

- mounting a Ø 300 mm brake disc will give a peripheral speed of 23.5 m/s
- the brake mass = 160 kg
- the peripheral speed drive disk Ø300 mm = 36m/s (brake disc mass = 80 kg)
- the peripheral speed of brake disc Ø300 mm = 23.5 m/s
- the chipper disc stop time (estimated) = 4 s

(c)   The brake disc Ø 300 mm in stainless steel is mounted in the inlet PTO shaft.

(d)   There are three lectro valves to control the brake clamp device.

(e)   An emergency stop.

(f)   A hydraulic flow regulator to set the brake actuation.

(g)   A 12 V power plug.

(h)   Anti-stress safety systems.

The brake clamp device patented phase has the application number 102017000052858.

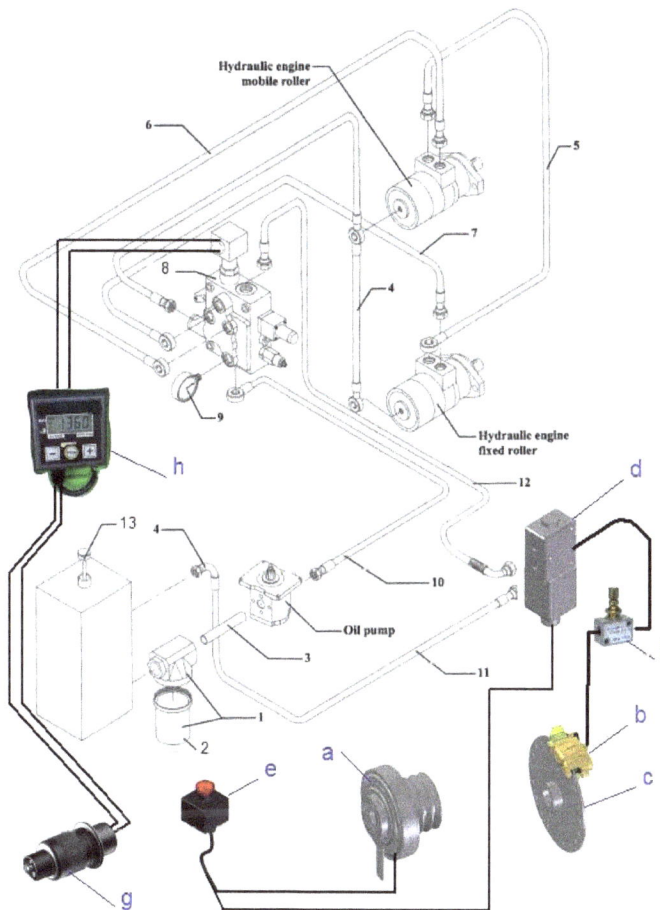

Figure 6. Components to be added to the standard machine (description in the text).

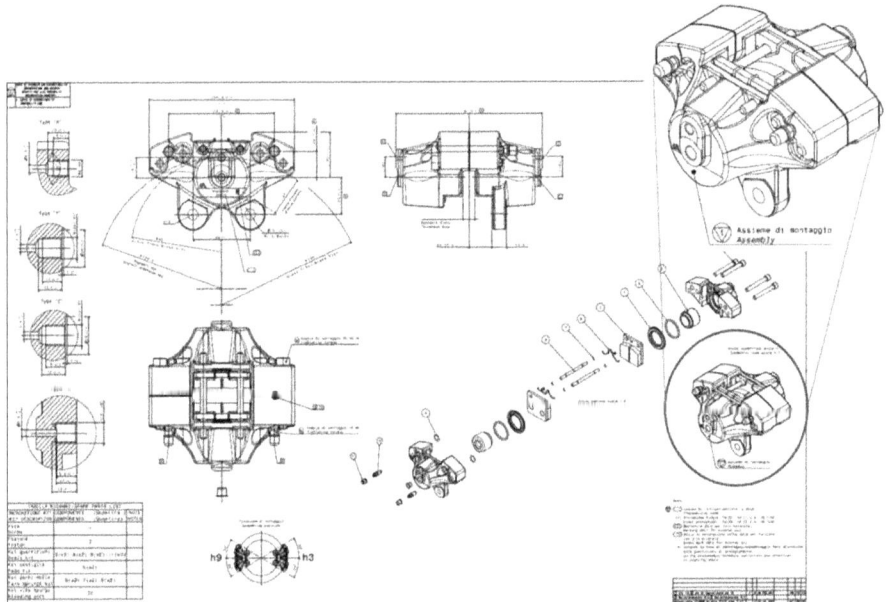

DATI TECNICI(FUNZIONAMENTO IDRAULICO)	
TECHNICAL DATA(HYDRAULIC FUNCTION)	
Diametro pistoncino idraulico *Hydraulic piston diameter*	Dp=38 mm
Area pistone idraulico *Hydraulic piston face area*	Ap= 11.34 cm²
Numero pistoncini *Number of hydraulic pistons*	Np=2
Pressione max di servizio *Max line pressure*	Pm=160 bar
Forza di serraggio totale(pres.max) *Total clamping force(max pres.)*	Fs=32340 N
Raggio efficace *Effective radius*	Re=(Gd/2-19)mm
Momento frenante(alla pres.max) *Braking torque(at max pres.)*	Mf=(Fs*m*Re)/1000 Nm
Spessore MAX del disco *MAX disc thickness*	Sd= 11 mm
Spessore MIN del disco *MIN disc thickness*	Sd= 9 mm
Diametro MAX del disco *MAX disc diameter*	Dd= 280 mm
Diametro MIN del disco *MIN disc diameter*	Dd= 180 mm
Materiale d'attrito senza amianto *Friction material asbestons-free*	
Valore coeff. d'attrito(di progetto) *Coeff.friction mat(design purposes)*	m=0,35
Area di una pastiglia *Brake pad working area*	S=19.8 cm²
Spess.max utilizzabile mater.d'attrito *Max alloveble pad wear*	Sp= 9 mm
Volume di fluido assorbito *Fluid displacement*	Va=xx cm3
Peso della pinza freno *Caliper weight*	P=3.5 Kg

Figure 7. Brake clamp technical data.

4. Conclusions

The main objective of reducing and/or eliminating the risk of operator entrapment in the mechanical part of the chipper has been met through the design of appropriate prevention and protection systems to be applied directly to the chipper. The reduction of component stop times and the consequent risk reductions were achieved with relatively simple and cost-effective technology which can be applied to most cutting machines on the market. In addition, the decoupling system, in combination with anti-stress safety systems of the machine, was effective. A further step forward will be the realization of a prototype that acts on the inertia of the flywheel, blocking it in a timely approach thanks to a braking system whose effectiveness will be quantified in further tests. From the point of view of the technical regulations, the requests from the French Authorities for the amendment of EN 13525 have been taken into account. The experience gained suggests that these changes can certainly improve the safety of the chippers.

Acknowledgments: This study was supported by the "Protection of agricultural machinery operators from crush, entanglement, shearing" (PROMOSIC) project, funded by INAIL. The authors wish to thank Marco Vieri of the University of Florence for his help in the decoupling system test, and Peruzzo Ltd.

Author Contributions: The authors contributed in the equality in the paper.

Conflicts of Interest: The authors declare no conflict of interest.

References

1. Hallock, G.G. Mutilating shredder/chipper hand injuries. *Ann. Plast. Surg.* **1994**, *33*, 8–12. [CrossRef]
2. Marsh, S.M.; Fosbroke, D.E. Trends of occupational fatalities involving machines, United States, 1992–2010. *Am. J. Ind. Med.* **2015**, *58*. [CrossRef] [PubMed]
3. Özden, S.; Nayir, I.; Göl, C.; Ediş, S.; Yilmaz, H. Health problems and conditions of the forestry workers in Turkey. *Afr. J. Agric. Res.* **2011**, *6*. [CrossRef]
4. Marucci, A.; Monarca, D.; Cecchini, M.; Colantoni, A.; Cappuccini, A. The heat stress for workers employed in laying hens houses. *J. Food Agric. Environ.* **2013**, *11*, 20–24.
5. Di Giacinto, S.; Colantoni, A.; Cecchini, M.; Monarca, D.; Moscetti, R.; Massantini, R. Dairy production in restricted environment and safety for the workers. *Industrie Alimentari* **2012**, *51*, 5–12.
6. Colantoni, A.; Allegrini, E.; Boubaker, K.; Longo, L.; Di Giacinto, S.; Biondi, P. New insights for renewable energy hybrid photovoltaic/wind installations in Tunisia through a mathematical model. *Energy Convers. Manag.* **2013**, *75*, 398–401. [CrossRef]
7. Colantoni, A.; Evic, N.; Lord, R.; Retschitzegger, S.; Proto, A.R.; Gallucci, F.; Monarca, D. Characterization of biochars produced from pyrolysis of pelletized agricultural residues. *Renew. Sustain. Energy Rev.* **2016**, *64*, 187–194. [CrossRef]
8. Pickett, W.; Hagel, L.; Dosman, J.A. Safety Features on Agricultural Machines and Farm Structures in Saskatchewan. *J. Agromed.* **2012**, *17*. [CrossRef] [PubMed]
9. Lindroos, O.; Aspman, E.W.; Lidestav, G.; Neely, G. Accidents in family forestry's firewood production. *Accid. Anal. Prev.* **2008**, *40*. [CrossRef] [PubMed]
10. Hoque, M.; Sokhansanj, S.; Naimi, L.; Bi, X.; Lim, J. Review and analysis of performance and productivity of size reduction equipment for fibrous materials. In Proceedings of the 2007 ASABE Annual International Meeting, Minneapolis, MN, USA, 17–20 July 2007; Volume 3, pp. 1–18. [CrossRef]
11. Spinelli, R.; Hartsough, B.R.; Magagnotti, N. Testing Mobile Chippers for Chip Size Distribution. *Int. J. For. Eng.* **2005**, *16*, 29–35. [CrossRef]
12. Poje, A.; Spinelli, R.; Magagnotti, N.; Mihelic, M. Exposure to noise in wood chipping operations under the conditions of agro-forestry. *Int. J. Ind. Ergon.* **2015**, *50*, 151–157. [CrossRef]
13. Magagnotti, N.; Picchi, G.; Sciarra, G.; Spinelli, R. Exposure of Mobile Chipper Operators to Diesel Exhaust. *Ann. Occup. Hyg.* **2014**, *58*, 217–226. [PubMed]
14. De Martino, G.; Massantini, R.; Botondi, R.; Mencarelli, F. Temperature affects impact injury on apricot fruit. *Postharvest Biol. Technol.* **2002**, *25*, 145–149. [CrossRef]
15. Rottensteiner, C.; Tsioras, P.; Neumayer, H.; Stampfer, K. Vibration and noise assessment of tractor trailer and truck-mounted chippers. *Silva Fennica* **2013**, *47*, 1–14. [CrossRef]
16. Spinelli, R.; Magagnotti, N.; Deboli, R.; Preti, C. Noise emissions in wood chipping yards: Options compared. *Sci. Total Environ.* **2016**, *563*, 145–151. [CrossRef] [PubMed]
17. Struttmann, T.W. Fatal and Nonfatal Occupational Injuries Involving Wood Chippers—United States, 1992–2002. *Morb. Mortal. Wkly. Rep.* **2004**, *53*, 1130–1131.

18. A.M.D. *Forestry Machinery-Wood Chippers-Safety*, 2nd ed.; BSI: London, UK, 2005; pp. 1–46, ISBN 978 0 580 62475 9.

19. Moscetti, R.; Frangipane, M.T.; Monarca, D.; Cecchini, M.; Massantini, R. Maintaining the quality of unripe, fresh hazelnuts through storage under modified atmospheres. *Postharvest Biol. Technol.* **2012**, *65*, 33–38. [CrossRef]

agriculture

Article

Energy and Carbon Impact of Precision Livestock Farming Technologies Implementation in the Milk Chain: From Dairy Farm to Cheese Factory

Giuseppe Todde *, Maria Caria , Filippo Gambella and Antonio Pazzona

Department of Agricultural Science, University of Sassari, Viale Italia 39, 07100 Sassari, Italy;
mariac@uniss.it (M.C.); gambella@uniss.it (F.G.); pazzona@uniss.it (A.P.)
* Correspondence: gtodde@uniss.it; Fax: +39-079-229-285

Received: 31 July 2017; Accepted: 19 September 2017; Published: 21 September 2017

Abstract: Precision Livestock Farming (PLF) is being developed in livestock farms to relieve the human workload and to help farmers to optimize production and management procedure. The objectives of this study were to evaluate the consequences in energy intensity and the related carbon impact, from dairy farm to cheese factory, due to the implementation of a real-time milk analysis and separation (AfiMilk MCS) in milking parlors. The research carried out involved three conventional dairy farms, the collection and delivery of milk from dairy farms to cheese factory and the processing line of a traditional soft cheese into a dairy factory. The AfiMilk MCS system installed in the milking parlors allowed to obtain a large number of information related to the quantity and quality of milk from each individual cow and to separate milk with two different composition (one with high coagulation properties and the other one with low coagulation properties), with different percentage of separation. Due to the presence of an additional milkline and the AfiMilk MCS components, the energy requirements and the related environmental impact at farm level were slightly higher, among 1.1% and 4.4%. The logistic of milk collection was also significantly reorganized in view of the collection of two separate type of milk, hence, it leads an increment of 44% of the energy requirements. The logistic of milk collection and delivery represents the process which the highest incidence in energy consumption occurred after the installation of the PLF technology. Thanks to the availability of milk with high coagulation properties, the dairy plant, produced traditional soft cheese avoiding the standardization of the formula, as a result, the energy uses decreased about 44%, while considering the whole chain, the emissions of carbon dioxide was reduced by 69%. In this study, the application of advance technologies in milking parlors modified not only the on-farm management but mainly the procedure carried out in cheese making plant. This aspect makes precision livestock farming implementation unimportant technology that may provide important benefits throughout the overall milk chain, avoiding about 2.65 MJ of primary energy every 100 kg of processed milk.

Keywords: electricity and diesel consumptions; PLF; Afimilk; coagulation properties

1. Introduction

Precision Livestock Farming (PLF) is being developed in livestock farms to relieve the human workload and to help farmers to optimize production and management procedure. PLF consists in monitoring and measuring animal data in order to model the information gathered and then use this information to evaluate the on-going processes [1]. The use of computerized sensors and online measurements are commonly adopted in modern milking parlors, in order to analyze several parameters in milk composition and to monitor the health status of cows. The bulk tank milk is than collected and delivered to cheese factory, where the milk is analyzed and used for cheese making and/or drinking milk.

The milk composition for drinking milk is well defined than that of milk intended for manufacturing cheese products. Accordingly, the protein level in drinking milk is often higher than the minimum level required by the legislations, in the other hand the cheese factory has to buy proteins for cheese making, with economic loss for the company. For this reason, the cheese factory may be interested to collect, directly from the farm, milk with different characteristics: in particular one with high coagulation properties (high level in protein and fat content) and the other one with low coagulation properties (standard milk).

The milk composition depends on breed, time in lactation and nutrition, as well as udder infections and hygiene in milking. At the moment, different animals and breeds may be selected to produce diverse products accordingly to standard desirable dairy goods [2,3]. The separation in real-time of individual cow's milk, in accordance to its optical measured coagulation properties [4–8], through the implementation of a new system (AfiMilk MCS) in milking parlors, allows to save process steps in milk chain. In fact, the two types of milk, that are collected and delivered separately with different milk tanker trucks, follow diverse production lines defined by the cheese factory (i.e., drinking milk or cheese making). The implementations of such equipment in milking parlors would enable milk producers and dairy industry, to separate and convey milk based on determined standards set by the operators, in order to increase both parties' benefits from the use of that milk [4]. Studies on AfiMilk MCS showed that the performance of this equipment installed in milking parlor provides an opportunity to divert 200 mL resolution pulses of milk into two different tanks, according to its suitability for cheese production [9].

Each AfiLab uses near-infrared spectroscopy (NIR) for on-line milk analysis [10]. The advantages of an NIR systems, over other systems, are related with a prompt and nondestructive on-line measurements [11]. In general, the AfiMilk MCS has the potential to control and define the milk quality [5].

Certainly, the implementation of such devices, would change the energy consumption and the related carbon impact in milk supply chain (from farm to factory). The main reasons are due to the introduction of a supplementary milkline, besides the new management of collection and delivery of milk with two different compositions. The optimization of milk collection routes is useful not only for reducing total energy consumptions and costs, but also to decrease the related environmental emissions [12–14].

The on-farm activities that contribute most to the electricity requirements in dairy farms were milk cooling and milk harvesting, showing how important these activities are in the farm energy management [15,16]. The consumption of diesel fuel, electricity or gas is defined as direct energy, while the term indirect energy refers to the usage of inputs depleted in one production period [17–20].

In dairy farms, technological investments are related to the herd dimension; however, when the level of mechanization is reported to the number of raised heads as indices, larger farms were more efficient and utilized less power per unit [21].

The objectives of this study were to evaluate the consequences in direct energy intensity and the related carbon impact, from dairy farm to cheese factory, due to the implementation of a real time milk analysis and separation in milking parlors.

2. Materials and Methods

The research carried out involved 3 conventional dairy farms (named A, B and C), the collection and delivery of milk from farms to cheese factory and the processing line of a traditional soft cheese into a dairy factory.

The boundaries of this study were set from dairy farm to cheese factory (first weight control, before cheese ageing) involving in particular: direct energy consumptions in dairy farms related to the milking and washing operations; direct energy use for collection and delivery of milk from dairy farm to cheese factory; electrical and thermal energy use in the production line of a traditional soft cheese, including the energy embodied in the ingredients used to standardized the milk formula.

An energy audit was carried out to collect the overall information related with energy and related carbon dioxide issues into the boundaries defined above. In this study other direct and indirect energy consumption and related emissions in dairy farms, in milk transportation and in dairy factory were not accounted since the objective of this study was to evaluate the consequence of a real-time milk analysis installation in milking parlors to avoid the standardization of the milk formula for the production of a traditional soft cheese.

The functional units were set as a 100 kg of milk processed and 1 kg of a traditional soft cheese produced.

2.1. Dairy Farms

The conventional dairy farms were located in the Oristano province, Sardinia (Italy). Farms were specialized in milk production, with a confinement management, on-farm feed production was based on grass hay and grass silage in spring and corn silage in late summer. The dairy farms were located in valley. All the equipment found in the milking parlor were inventoried, reporting the operational power and their usage time in order to calculate the annual energy consumption [22,23].

The characteristics of the investigated farms are reported as follow, farm "A" held 72 hectares of cultivated land, the farm was specialized in milk production rising 500 heads of Holstein and Brown cows and producing about 2140 tons of milk per year.

Farm "B" raised 600 heads of Holstein cows producing about 2453 tons of milk per year, the total cultivated area was about 85 of irrigated hectares.

Farm "C" was specialized in milk production (2317 t year^{-1}) raising about 480 Holstein cows in 100 hectares of land extent of which 60 hectares were irrigated.

Real Time Milk Analysis

The Afimilk MCS system (on line milk classification service; Afikim, Israel), an innovative solution to improve milk value, was installed in the milking parlors of the dairy farms involved in this project. The equipment is able to separate in real time the milk in two fractions according to predetermined coagulation properties, obtaining higher cheese-milk quality with increased cheese yield.

The milk analysis is performed by Afilab™ (S.A.E., Afikim, Israel), a spectrometer that works in real time and performs milk component measurements. Afilab is installed on the milking parlor, next to the milkmeter, in every stall and sorts milk into two milklines (Figure 1). The instrument thus becomes an integral part of the system, allowing to analyze the milk of each animal at every milking session. Afilab's analysis is based on spectroscopy in the near infrared: during milking, the cow's milk (every 200 g) is crossed by a light beam and Afilab, reading the light refraction of milk, is capable to determine the content of fat, protein, lactose and somatic cells and separate it into two target milk tanks. In this study, the AfiMilk MCS system was set to separate milk with high coagulation properties and milk with low coagulation properties with a ratio of 50%.

Figure 1. The Afilab (B), installed next to the AfiMilk (A) milkmeter at each milking unit, performs real-time analysis of milk components using spectroscopy in the near infrared. Milk with predefined coagulation properties is separated and conveyed, thanks to controlled valves (C), through two different milklines (D or E) in a predetermined cooling tank.

2.2. Dairy Factory

The cheese making plant was located in the Oristano Province (Sardinia) Italy, where an energy audit was performed in order to identify the electrical and thermal energy usage through the production line of a traditional soft cheese. In addition, the amount of ingredients used to standardize the production formula were included as primary energy and related carbon dioxide emissions.

The Traditional Soft Cheese Production Line

The raw milk is stored in cooling tanks, after that, chemical analysis take place to confirm milk suitability for processing and to assess the amount and type of ingredient (on-factory milk cream, ultrafiltered milk, etc.) needed to standardize the production formula.

Milk standardization lead to increase the indirect energy requirement of the whole process due to the incorporated energy embodied in the ingredients, in particular for milk cream was found 50.5 MJ per kg of product used and 29.2 MJ per kg of ultrafiltered milk.

After thermal treatment (74 °C per 16 s) milk is conveyed to the processing line, where starter and rennet are added when milk reach the temperature of 39 °C. Once the cheese curd is ready, it is cut into small pieces to allow whey drainage. The whey and cheese curd are placed automatically into cheese moulds and conveyed in the warm chambers, where the cheese will be held for 3 h between 37 °C and 40 °C. After drainage, the cheese is placed into the brine solution for about 3.5 h at 10 °C, when the salting procedure is done, the moulds are removed automatically and the wheels of cheese are sent to the first weight control, where the production yield is assessed. The cheese yield, specifically for the production of the traditional soft cheese, is about 12.3%. The following step is represented by the ageing at 5–7 °C for about 15 days, followed by the packaging process.

Operational parameter of cheese production process per vat: 4000 kg of processed milk; 12.3% of cheese yield; 379 wheels of traditional soft cheese; 1.3 kg per cheese wheel; 492 kg of cheese.

2.3. Milk Collection and Delivery

Adopting these PLF technologies in dairy farms lead to reorganize the logistic of milk collection. We assessed the delivery of two types of milk obtained by quality separation (milk with high coagulation properties and standard milk) from dairy farms to cheese making plant, since each type of milk needs to be collected every day and transported independently. The calculation was effectuated by a tool for the optimization of collection routes of dairy farms (MilkTour), developed by the University of Sassari.

The MilkTour software allows optimizing the collection routes of milk collection and transport phases by a client–server architecture with the clients using HTML and JS, while the server used PHP and a MySQL database. The client software for the integrated management of the collection and transport phases of milk collection used an Internet browser, from which the HTML code of the server was downloaded through an HTTP request [24]. Once the collection route has been defined, the user can calculate the distance between the different points of the map. Once the distance has been calculated, the user could then optimize the route by choosing and inserting certain parameters such us the time spent collecting milk, the start time of the route, daily collection amount and the costs (calculated by official parameters from the Ministry of Infrastructure and Transport or from the cheese factory data records). The MilkTour software allows to assess the minimum distance needed to visit all the collection points. Once the routes are optimized the software will estimate the best sequence for visiting the suppliers, the time taken, the cost of the route, the density of the collection points, and CO_2 emissions. The software takes also into account the information of suppliers (farm location and milk volume) and from cheese factory (maximum capacity of the milk truck tankers, number of trucks and time limitations). Fuel consumptions were assessed using as a benchmark minimum cost for road transport companies operating tankers, published and updated monthly on the website of the Italian Ministry of Infrastructure and Transport. The CO_2 emissions were calculated considering the fuel consumptions and the emission factor for diesel (3.15 kg CO_2–eq kg^{-1}; [25]).

The 3 dairy farms investigated in this study, belonged to a collection route involving 21 dairy farms situated in nearby areas. In order to assess the energy and carbon impact due to milk collection and delivery, in the pre and post installation phase, a simulation has been performed for the whole collection route.

3. Results and Discussion

The results for each farm are shown in Table 1, farm "A" was equipped with a herringbone type milking parlor holding 8 + 8 stalls, the single milkline measured 70 mm of diameter and was connected with two vacuum pumps of 4 kW each. Lactating cows were milked twice a day corresponding to 6 h a day of using time (milking and washing). The milking parlor was equipped with Variable Speed Drive (VSD) connected to the vacuum pumps, Water Heat Recovery (WHR) for water heating and a Milk Pre-Cooling (MPC). The milk was stocked into two milk tanks with a capacity of 5000 L each.

The milking parlor of farm "B" held 30 stalls and it was equipped with three vacuum pumps of 4 kW each one. Cows were milked three times a day corresponding to 8.5 h a day of operational use. The farm was provided with a water heating recovery system and two milk bulk tanks, containing respectively 5000 and 8000 L of milk.

Farm "C" held a milking parlor equipped with 18 stalls powered by two vacuum pumps of 4 kW each and was furnished whit VSD. The milking machine was used for 7 h a day and the milk was stocked into two cooling tanks with a capacity of 5000 L each.

The environmental and energy audit is a methodology adopted to assess energy consumptions and related emission of carbon dioxide (expressed as kg of CO_2). These requirements were estimated based on the power of the equipment used through one year. The data considered at farm level was related to the use of the milking parlours, milk cooling and washing procedure, thus, to underline the effects, on direct energy requirements, due to the installation of the AfiMilk MCS system.

Table 1. Main characteristics and appliances found in the milking parlors of the three dairy farms involved in the study.

Milking Parlor	Farm A	Farm B	Farm C
Type	Herringbone	Herringbone	Herringbone
Stalls (n)	8 + 8	15 + 15	9 + 9
Milkline (Ø ext. mm)	70	50	60
Vacuum pump (n)	2	3	2
Power (kW)	4	4	4
Using time (h/day)	6	8.5	7
Milking routine (times/day)	2	3	2
Variable Speed Drive (yes/no)	Yes	No	Yes
Water heating (type)	Electric	Diesel	Electric
Power (W)	1200	-	1200
Water volume (L/day) Ex-Ante	668	1272	708
Water volume (L/day) Ex-Post	764	1542	816
Water Heating Recovery (yes/no)	yes	Yes	No
Milk tanks (n)	2	2	2
Milk tanks power (kW)	10 + 10	10 + 16	12 + 10
Milk tanks volume (L)	5000 + 5000	5000 + 8000	5000 + 5000
Milk Pre-Cooling (yes/no)	Yes	No	No

The electricity and diesel fuel consumptions, as well as the related primary energy and carbon indicators, were reported in Table 2 per each farm during pre and post PLF technologies implementation (respectively ex-ante and ex-post phase). The ex-ante electricity requirement accounted from 51.3 to 116.8 kW·h per head (farm A and C, respectively), while diesel fuel consumption was found only in farm B using about 1.45 kg per head. The primary energy requirements and related carbon dioxide emissions showed farm B having the highest values (26.6 MJ and 1.24 kg CO_2 per 100 kg of milk). The availability of energy saving devices, installed in the milking facilities, allowed farm "A" to have lower energy utilization indices, then the other two dairy farms.

Table 2. Energy and carbon dioxide emissions summary, in the ex-ante and ex-post phase, reported for each dairy farm and the related energy and carbon indicators.

Farms	Farm A		Farm B		Farm C	
Phases	Ex-Ante	Ex-Post	Ex-Ante	Ex-Post	Ex-Ante	Ex-Post
Electricity (kW·h.head^{-1})	51.3	52.6	112.5	112.5	116.8	121.8
Diesel (kg.head^{-1})	-	-	1.45	1.70	-	-
Primary energy (MJ.head^{-1})	465	477	1087	1098	1059	1105
Emissions (kgCO$_2$.head^{-1})	21.0	21.6	50.7	51.5	47.9	49.9
Primary energy (MJ.100 kgMilk^{-1})	10.9	11.1	26.6	26.9	21.9	22.9
Emissions (kgCO$_2$.100 kgMilk^{-1})	0.491	0.504	1.241	1.260	0.993	1.036

The implementation at the milking parlour level, of the AfiMilk MCS system, slight increased the requirements of electricity and diesel fuel in each dairy farm. Expressing this growth in terms of primary energy demand per unit of milk produced, a range among 1.1% and 4.4% was observed in the investigated farms. One of the activities that showed greater energy requirements in the ex-post phase was related to washing procedure, since the hot water volumes increased significantly due to the additional milkline and components. Even though herd size was very close among farms, the availability of saving devices in the milking facilities, allowed to contain energy consumption due to the installation of the AfiMilk MCS system. The ex-post direct energy consumptions in milking activities, increased the related emission of carbon dioxide for farm A, B and C, emitting respectively 0.504, 1.260 and 1.036 kg of CO_2 per 100 kg of milk.

The energy and environmental audit carried out in the traditional soft cheese production line is reported in Table 3. The milk storage and thermal treatment operations were the most direct energy demanding activities using about 2.71 MJ per 100 kg of processed milk, which represent the 58% of total consumptions. The indirect energy requirements derived from the use of ingredient adopted to standardize the traditional soft cheese formula, which accounted to 3.64 MJ per 100 kg of milk. Considering the direct and indirect energy emissions, the standardization of the cheese formula was the most pollutant source, emitting about the 86% of the total carbon dioxide emissions in the ex-ante phase.

Table 3. Direct and indirect energy and carbon dioxide emissions summary, in the ex-ante and ex-post phase, reported for each main process carried out in the cheese factory and expressed per 100 kg of milk.

Phases	Ex-Ante			Ex-Post		
Energy and Emissions	Direct Energy (MJ)	Indirect Energy (MJ)	kg CO_2	Direct Energy (MJ)	Indirect Energy (MJ)	kg CO_2
Milk storage and Treatments	2.71	-	0.41	2.71	-	0.41
Formula standardization	-	3.64	3.24	-	0.00	0.00
Cheese making process	0.97	-	0.05	0.97	-	0.05
Stewing and Brine	1.01	-	0.05	1.01	-	0.05
Total Cheese Factory	4.69	3.64	3.75	4.69	0.00	0.51

The implementation of milking parlours with the AfiMilk MCS system, allowed to obtain two types of milk based on quality characteristics. The milk with high coagulation properties (greater in fat and protein content) was conveyed to the production line of the traditional soft cheese, while the standard milk was used for other products that do not require to be processed (i.e., drinking milk). Processing milk with high coagulation properties allowed to avoid the use of ingredients to standardized the production formula of the soft cheese and keep, at the same time, the other activities carried out during the production line, unaltered. This advantage reduced considerably the total emissions of carbon dioxide and increased the cheese yield. In particular, the use of milk with high coagulation properties for the traditional soft cheese, increased the production yield from 12.3% (ex-ante phase) to 13.1% (ex-post phase).

The MilkTour software creates the results for 21 suppliers of the original collection route (Table 4) for the ex-ante phase. The cheese factory has 4 milk tanker trucks available, two trucks of 25.4 tons and two of 29.5 tons of a maximum volume capacity. To collect milk from the 21 dairy farms in the ex-ante phase, the software calculated 4 collection routes with a total travelled distance of 53.48 km. The total fuel consumption for the collection routes accounted to 13.76 kg of diesel, with a related emission of 43.35 kg CO_2 for the overall suppliers. Large differences in fuel consumption per collection route have been observed, from 1.75 to 5.76 kg of diesel, even though the quantity of milk collected was similar among routes, the main issues that affects diverse fuels consumptions was related with the distances travelled by trucks, thus the distance of farms from the cheese factory.

Table 4. Distance (km) and CO_2 emissions (kg) of milk collection routes in the ex-ante phase with a simulation of 21 suppliers.

Collection Route	Suppliers (n)	Distance (km)	Milk (L)	Milk Truck Volume (tons)	Distance Covered * (km/L)	Fuel per Collection Route (kg)	CO_2 Per Collection Route (kg)	CO_2/t Milk (kg)
1	5	8.37	25,100	25.4	4	1.75	5.50	0.219
2	5	9.76	27,600	29.5	2.8	2.91	9.17	0.332
3	6	19.3	25,900	29.5	2.8	5.76	18.13	0.700
4	5	16.04	22,400	25.4	4	3.35	10.55	0.471
Total	21	53.48	101,000	-	-	13.76	43.35	-

* Diesel fuel consumption assessed from the milk tanker weight (Ministry of infrastructure and transport).

The reorganization of milk collection and delivery comported an increase of energy requirements and emissions of carbon dioxide, which accounted respectively to 24.53 kg of diesel fuel and 77.26 kg CO_2 for the overall collection routes (Table 5). In fact, the two types of milk were collected every day from each supplier by means of two different milk tanker trucks which lead to increase the total distance travelled from 53.48 km to 93.60 km. Based on the data obtained from the MilkTour software, the collection of milk from dairy farms in the ex-ante phase required about 0.63 MJ and emitting 0.043 kg CO_2 per 100 kg of milk collected. The optimization of milk collection, in the ex-post phase, increased the energy demand and the related carbon dioxide emissions, which accounted respectively to 1.12 MJ and 0.08 kg per 100 kg of milk.

Table 5. Distance (km) and CO_2 emissions (kg) of milk collection routes in the ex-post phase, separation rate 50%, with a simulation of 21 suppliers.

Collection Route	Suppliers (n)	Distance (km)	Milk (L)	Milk Truck Volume (tons)	Distance Covered * (km/L)	Fuel Per Collection Route (kg)	CO_2 per Collection Route (kg)	CO_2/t Milk (kg)
1	10	18.93	21,900	25.4	4	3.95	12.45	0.568
2	11	27.87	28,500	29.5	2.8	8.31	26.18	0.919
1	10	18.93	22,100	25.4	4	3.95	12.45	0.563
2	11	27.87	28,500	29.5	2.8	8.31	26.18	0.919
Total	42	93.60	101,000	-	-	24.53	77.26	-

* Diesel fuel consumption assessed from the milk tanker weight (Ministry of infrastructure and transport).

The final assessment of the energy demand and the related emissions of carbon dioxide, due to the implementation of PLF technologies in dairy farms, is shown in Table 6. In the ex-ante phase energy requirements of dairy farms represent the most demanding activities corresponding to the 69% of total consumption and followed by the dairy production plant (29%), while milk collection and delivery represents only the 2% of the total assessment. Focusing on the emissions of carbon dioxide, in the ex-ante phase, the dairy plant held the higher value with about 3.75 kg CO_2 per 100 kg of milk processed, corresponding to the 80% of total carbon emissions.

Table 6. Energy and carbon dioxide emissions summary, in the ex-ante and ex-post phase, reported for each main production step (from dairy farm to cheese factory) expressed per 100 kg of milk.

Phases	Ex-Ante		Ex-Post	
Energy and Emissions	Energy (MJ)	kg CO_2	Energy (MJ)	kg CO_2
Dairy Farm	20.16	0.93	20.66	0.95
Collection and Delivery	0.63	0.043	1.12	0.08
Cheese Factory	8.33	3.75	4.69	0.51
Total Chain	29.12	4.72	26.47	1.54

Expressing the final results of the ex-ante phase per unit of traditional soft cheese produced, 2.23 MJ and 0.36 kg of carbon dioxide per kg of cheese have been assessed.

In the ex-post phase larger energy requirements resulted in the logistic of milk collection (+44%), followed by dairy farms (+2.4%), while in the cheese factory the energy demand was reduced by the 44%, due to the ingredients saved by the use of milk with high coagulation properties.

The implementation of PLF technologies in milking parlors led a considerable reduction of total energy requirements and related carbon impact in the production of the traditional soft cheese. The magnitude of those reductions accounted to 14% of primary energy demand, corresponding to 2.03 MJ per kg of soft cheese produced. The emissions of carbon dioxide through the entire production chain, from farm to factory in the ex-post phase, were reduced by the 69% going from 0.384 to 0.118 kg of carbon dioxide per kg of traditional soft cheese.

4. Conclusions

The application of PLF technologies in dairy sector help to improve the performances of the animals raised. Monitoring in real time the health status, the quality and quantity of the productions in dairy farms allowed to know and manage important aspects of the milk production chain. This study assessed how the milk separation by means of the AfiMilk MCS system affects the energy requirements and the related carbon impact throughout the milk chain, from dairy farms to cheese factory. The equipment installed in the milking parlors allowed to obtain a large number of information related to the quantity and quality of milk from each individual cow and to separate milk with two different composition (one with high coagulation properties and the other one with low coagulation properties), with a separation ratio of 50%. Due to the presence of an additional milkline and the AfiMilk MCS components, the energy requirements and the related environmental impact at farm level were slightly larger, among 1.1% and 4.4% in respect to the ex-ante phase. One of the most process that affects the energy consumption in dairy farms was related to the increase of hot water volume used to wash the milking parlors.

The logistic of milk collection was also significantly reorganized in view of the collection of two separate type of milk, hence, it leads an increment of 44% of the energy demand. The logistic of milk collection and delivery represents the process in which the highest incidence in energy consumption occurred in the ex-post phase.

Thanks to the availability of milk with high coagulation properties, the dairy factory, produced traditional soft cheese avoiding the standardization of the milk formula, as a result, the energy uses decreased about 44%, while considering the whole chain, the emissions of carbon dioxide was reduced by 69%. In this study, the application of advance technologies in milking parlors modified not only the on-farm management but mainly the procedure carried out in cheese making plant.

The implementation of PLF technologies increased energy requirement and carbon dioxide emissions in dairy farms and in the collection of milk, however, the large amount of energy saved in the cheese factory and the increase in cheese production yield make these technologies respectful to the natural resources and to the environment, avoiding about 2.65 MJ of primary energy every 100 kg of processed milk.

Acknowledgments: This study was performed as part of the project "INNOVALATTE", supported by PSR program of Sardinian Region (Italy).

Author Contributions: A.P., G.T. and M.C. conceived and designed the experiments; G.T. and F.G. did data collection; G.T. and M.C. analyzed the data and wrote the paper; G.T., M.C., F.G. and A.P. revisioned the article for the final approval of the version to be published.

Conflicts of Interest: The authors declare no conflicts of interest.

Abbreviations

MCS	Milk Classification Service
NIR	Near Infra-Red
VSD	Variable Speed Drive
WHR	Water Heat Recovery
MPC	Milk Pre-Cooling
HTML	HyperText Markup Language
JS	JavaScript
PHP	Hypertext Preprocessor
MySQL	Open source database
HTTP	HyperText Transfer Protocol
PLF	Precision livestock farming

References

1. Berckmans, D. Precision livestock farming (PLF). *Comput. Electron. Agric.* **2008**, *62*, 22–28. [CrossRef]
2. Wedholm, A.; Larsen, L.B.; Lindmark-Mansson, H.; Karlsson, A.H.; Andrén, A. Effect of protein composition on the cheese-making properties of milk from individual dairy cows. *J. Dairy Sci.* **2006**, *89*, 3296–3305. [CrossRef]
3. De Marchi, M.; Bittante, G.; Dal Zotto, R.; Dalvit, C.; Cassandro, M. Effect of Holstein Friesian and Brown Swiss breeds on quality of milk and cheese. *J. Dairy Sci.* **2008**, *91*, 4092–4102. [CrossRef] [PubMed]
4. Leitner, G.; Lavi, Y.; Merin, U.; Lemberskiy-Kuzin, L.; Katz, G. Online evaluation of milk quality according to coagulation properties for its optimal distribution for industrial applications. *J. Dairy Sci.* **2011**, *94*, 2923–2932. [CrossRef] [PubMed]
5. Leitner, G.; Merin, U.; Lemberskiy-Kuzin, L.; Bezman, D.; Katz, G. Real-time visual/near-infrared analysis of milk-clotting parameters for industrial applications. *Animal* **2012**, *6*, 1170–1177. [CrossRef] [PubMed]
6. Katz, G.; Merin, U.; Bezman, D.; Lavie, S.; Lemberskiy-Kuzin, L.; Leitner, G. Real-time evaluation of individual cow milk for higher cheese-milk quality with increased cheese yield. *J. Dairy Sci.* **2006**, *99*, 4178–4187. [CrossRef] [PubMed]
7. Weller, J.I.; Ezra, E. Genetic and phenotypic analysis of daily Israeli Holstein milk, fat, and protein production as determined by a real-time milk analyzer. *J. Dairy Sci.* **2016**, *99*, 9782–9795. [CrossRef] [PubMed]
8. Kaniyamattam, K.; De Vries, A. Agreement between milk fat, protein, and lactose observations collected from the Dairy Herd Improvement Association (DHIA) and a real-time milk analyzer. *J. Dairy Sci.* **2016**, *97*, 2896–2908. [CrossRef] [PubMed]
9. Leitner, G.; Merin, U.; Jacoby, S.; Bezman, D.; Lemberskiy-Kuzin, L.; Katz, G. Real-time evaluation of milk quality as reflected by clotting parameters of individual cow's milk during the milking session, between day-to-day and during lactation. *Animal* **2013**, *7*, 1551–1558. [CrossRef] [PubMed]
10. Tsenkova, R.; Atanassova, S.; Toyoda, K.; Ozaki, Y.; Itoh, K.; Fearn, T. Near-infrared spectroscopy for dairy management: Measurement of unhomogenized milk composition. *J. Dairy Sci.* **1999**, *82*, 2344–2351. [CrossRef]
11. Schmilovitch, Z.; Shmuelevich, I.; Notea, A.; Maltz, E. Near infrared spectrometry of milk in its heterogeneous state. *Comput. Electron. Agric.* **2000**, *29*, 195–207. [CrossRef]
12. Lou, Z.; Li, Z.; Luo, L.; Dai, X. Study on Multi-Depot Collaborative Transportation Problem of Milk-Run Pattern. *MATEC Web Conf.* **2016**. [CrossRef]
13. Sethanan, K.; Pitakaso, R. Differential evolution algorithms for scheduling raw milk transportation. *Comput. Electron. Agric.* **2016**, *121*, 245–259. [CrossRef]
14. Paredes-Belmar, G.; Lüer-Villagra, A.; Marianov, V.; Cortés, C.E.; Bronfman, A. The milk collection problem with blending and collection points. *Comput. Electron. Agric.* **2017**, *134*, 109–123. [CrossRef]
15. Wells, C. *Total Energy Indicators of Agricultural Sustainability: Dairy Farming Case Study*; Technical Paper 2001/3; Ministry of Agriculture and Forestry: Wellington, New Zealand, 2001; pp. 1–79. ISBN 0-478-07968-0. ISSN 1171-4662.
16. Murgia, L.; Todde, G.; Caria, M.; Pazzona, A. A partial life cycle assessment approach to evaluate the energy intensity and related greenhouse gas emission in dairy farms. *J. Agric. Eng.* **2013**, *XLIV*. [CrossRef]

17. Lockeretz, W. *Agriculture and Energy. Washington University Through*; Academic Press: New York, NY, USA, 1997.
18. Kraatz, S. Energy intensity in livestock operations—Modeling of dairy farming systems in Germany. *Agric. Syst.* **2012**, *110*, 90–106. [CrossRef]
19. Sefeedpari, P.; Rafiee, S.; Akram, A.; Pishgar Komleh, S.H. Modeling output energy based on fossil fuels and electricity energy consumption on dairy farms of Iran: Application of adaptive neural-fuzzy inference system technique. *Comput. Electron. Agric.* **2014**, *109*, 80–85. [CrossRef]
20. Pagani, M.; Vittuari, M.; Johnson, T.G.; De Menna, F. An assessment of the energy footprint of dairy farms in Missouri and Emilia-Romagna. *Agric. Syst.* **2016**, *145*, 116–126. [CrossRef]
21. Todde, G.; Murgia, L.; Caria, M.; Pazzona, A. A multivariate statistical analysis approach to characterize mechanization, structural and energy profile in Italian dairy farms. *Energy Rep.* **2016**, *2*, 129–134. [CrossRef]
22. Todde, G.; Murgia, L.; Caria, M.; Pazzona, A. Dairy Energy Prediction (DEP) model: A tool for predicting energy use and related emissions and costs in dairy farms. *Comput. Electron. Agric.* **2017**, *135*, 216–221. [CrossRef]
23. Edens, W.C.; Pordesimo, L.O.; Wilhelm, L.R.; Burns, R.T. Energy Use Analysis of Major Milking Center Components at a Dairy Experiment Station. *Appl. Eng. Agric.* **2003**, *19*, 711–716. [CrossRef]
24. Caria, M.; Murgia, L.; Todde, G.; Chessa, G.; Pazzona, A. A model to improve the logistic of milk collection and delivery to cheese production factories. In Proceedings of the International Mid-Term Conference. Italian society of Agricultural Engineering, Naples, Italy, 22–23 June 2015.
25. ENEA. *Inventario Annuale Delle Emissioni di Gas Serra su Scala Regionale, Le Emissioni di Anidride Carbonica dal Sistema Energetico*; Rapporto; ENEA: Kista, Sweden, 2010.

agriculture

MDPI

Review

Whole-Body Vibration in Farming: Background Document for Creating a Simplified Procedure to Determine Agricultural Tractor Vibration Comfort

Maurizio Cutini * , Massimo Brambilla and Carlo Bisaglia

CREA Research Centre for Engineering and Agro-Food Processing, via Milano 43, 24047 Treviglio (BG), Italy; massimo.brambilla@crea.gov.it (M.B.); carlo.bisaglia@crea.gov.it (C.B.)
* Correspondence: maurizio.cutini@crea.gov.it; Tel.: +39-0363-49603

Received: 19 July 2017; Accepted: 26 September 2017; Published: 29 September 2017

Abstract: Operator exposure to high levels of whole-body vibration (WDV) presents risks to health and safety and it is reported to worsen or even cause back injuries. Work activities resulting in operator exposure to whole-body vibration have a common onset in off-road work such as farming. Despite the wide variability of agricultural surface profiles, studies have shown that with changing soil profile and tractor speed, the accelerations resulting from ground input present similar spectral trends. While on the one hand such studies confirmed that tractor WBV emission levels are very dependent upon the nature of the operation performed, on the other, irrespective of the wide range of conditions characterizing agricultural operations, they led researchers to set up a possible and realistic simplification and standardization of tractor driver comfort testing activities. The studies presented herewith indicate the usefulness, and the possibility, of developing simplified procedures to determine agricultural tractor vibration comfort. The results obtained could be used effectively to compare tractors of the same category or a given tractor when equipped with different seats, suspension, tyres, etc.

Keywords: lower back pain; safety; comfort; test track; whole-body vibration

1. Introduction

Agricultural tasks rank among the most hazardous occupations—according to work injury statistics, the related fatality rate is six times higher than that of all other industrial activities together and there is concern about the growing number of leisure-related farm injuries [1,2].

Occupational exposure has been recognized as one of the most important contributors to the onset of chronic diseases the outcomes of which, even if not resulting in premature mortality, can result in substantial disability, thus representing an extremely important cost from human and socio-economic perspectives [3,4]. Workers' safety issues range from proper accident prevention (resulting from the improvements to devices) to attention to the operator's welfare and comfort; this is related to microclimate as well as to exposure to physical and chemical agents [5,6]. Agricultural operators can also be exposed at the same time to different risk factors: for example, when running agricultural machines, workers experience physical (noise, vibrations), chemical (dust and chemical agents in the air, smoke) and biological (spores, micro-organisms and pollen which are conveyed with the dust) hazard exposure [7].

A risk factor requiring attention in agriculture is the exposure to vibration, both whole-body and hand harm. It has been pointed out that vibrations with a frequency lower than 2 Hz can induce minor and temporary effects like carsickness that, producing remarkable discomfort, interfere with the desired working performance while long-term exposure to vibrations ranging from 2 to 20 Hz can cause severe diseases such as spinal column degenerative pathologies [8–10]. Such harmful vibrations, when worsened by difficult working conditions (e.g., uncomfortable postures, inappropriate seats, frequent handling operations, etc.), could even lead to spinal disorders [11]. Operating hand-held olive harvesters has shown high levels of hand-arm vibrations (HAV) due to hand contact with the handle [12,13]. The prolonged exposure to these types of stresses could cause the so-called hand arm vibration syndrome (HAVS), which affects the various structures of the upper limb (musculoskeletal, nervous and vascular) [14–16]. The rise of these problems in both developed and less developed countries confirms that the ergonomic aspects of agriculture need adequate attention [17].

European Directive 2002/44/EC [18] defines "Whole-body vibration" (WBV) as "the mechanical vibration that, when transmitted to the whole-body, entails risks to the health and safety of workers, in particular lower-back morbidity and trauma of the spine." As far as seated operators are concerned, the ISO 5008/2002 [19] standard further specifies that WBV is the "vibration transmitted to the body as a whole through the buttocks of a seated operator." Exposure to high levels of whole-body vibration can cause or aggravate back injuries. Such risks are greatest when vibration magnitudes are high, exposure durations are/become long, frequent and regular and, furthermore, vibration includes severe shocks or jolts.

Ergonomic factors such as manual handling of loads as well as restricted or awkward postures [20] can be as important as exposure to WBV in causing back injury [21]: as a matter of fact, all these factors can separately result in the onset of back pain [22] and the risk turns out to be increased when, during WBV exposure, the operator is additionally exposed to one or more of these factors (environmental factors like temperature included).

People using agricultural machinery are likely to be exposed to the risk of vibration, therefore action to reduce workplace exposure to WBV is required for most operators who often run agricultural machinery. Manufacturers are continuously improving tractor comfort with active seats, the adoption of front suspension and on purpose designed cab suspension systems. Nevertheless, these efforts suffer from the lack of a focused approach to define tractor comfort. Studies on tractor dynamics indicate the technical possibility of filling this gap by developing a standard aimed at characterizing tractors in terms of comfort levels.

In this paper, after reviewing WBV risk analysis and the methods used for WBV assessment, we provide a different approach to WBV evaluation that introduces the possibility of setting up a simplified procedure to determine the vibration comfort of agricultural tractors.

The paper indicates the usefulness and the possibility of developing such a simplified procedure so that the outcome can be used: (1) to compare different tractors, provided that they belong to the same category; (2) the exposure resulting from varying the equipment mounted on a given tractor (seats, suspension, tyres, etc.). Such simplified testing procedures cannot, however, be considered suitable for determining operators' daily exposure to vibration in open field conditions.

2. The Risk of Whole Body Vibration in Farming

2.1. Whole-Body Vibration and Lower Back Pain

Lower back disorders have been significantly associated with heavy machinery operating tasks because of the biological mechanisms arising from WBV exposure and wrong postures which, in turn, are related to workplace characteristics and use, like the kind of seat, operating speed, track or tyres, cab design, the amount of time spent while seated, and the task performed [23].

In particular, results of interest can be found in the work of Lings [24] who reviewed the literature of the past seven years to find out: (i) whether there is evidence in the epidemiological literature of a causal association between WBV and lower back pain (LBP); and (ii) if there is evidence in the recent literature of a dose response relationship between whole-body vibration and LBP. In his study, twenty-four original articles concerning the association between WBV and LBP were selected for use. The six reports that best fulfilled the quality criteria were predominantly in favour of a positive association between WBV and LBP. Nevertheless, evidence in favour of a dose-response association was weak. The author concluded that there may be an association between WBV and LBP. However, it is not possible to decide whether WBV exposure per se is capable of causing LBP, or if WBV only constitutes a risk in combination with other factors, such as prolonged sitting as well as certain work postures. All the same, the current knowledge yields sufficient reasons for reducing WBV exposure to the lowest possible level.

In one of the six abovementioned studies, a survey was carried out on 1155 tractor drivers: tractor vibration and/or incorrect posture in driving activities were identified as causing more than 80% of the interviewees to suffer from lower back disorders [25].

The findings of this epidemiological study indicate that tractor driving is significantly related to an increased risk of lower back symptoms. While checking for several potential confounders by logistic modelling, total vibration dose and awkward postures at work were found to be the most predictive occupational factors for the occurrence of LBP among the tractor drivers. Quantitative regression analysis evinced a linear effect of postural load on the increased risk of LBP, while for WBV exposure this risk was found to be proportional to the power of the estimated vibration dose.

Despite this, more epidemiological and exposure data are needed in order to improve the knowledge of the dose-effect relationship between WBV exposure and lower back troubles among professional drivers.

2.2. Whole-Body Vibration and Professional Diseases

It is not simple to correlate LBP with professional lower back pain diseases (LBPD) for several reasons regarding the different approaches that different countries have to defining LBPD.

With reference to the systems, criteria and diagnostic tools used to recognize LBPD, Lötters [26] aimed to develop a model for determining the work-relatedness of lower back pain for workers with these symptoms, using both personal exposure profiles for well-established risk factors and the probability of lower back pain in the event of no exposure to such factors. To provide information on the level of work-relatedness of LBP, he developed a model based on the epidemiologic information available in the literature. Clinical decision-making modelling techniques enabled the design of a tool that could help general practitioners and occupational health physicians to assess the work-relatedness of LBP for an individual worker given the worker's exposure profile to well-established risk factors. The physical risk factors included in the model were manual handling of materials, frequent bending or twisting of the trunk, and whole-body vibration exposure.

The cut-offs that were used for high exposure were approximately: lifting weights of more than 15 kg for 10% of worktime for the manual handling of materials; 30 degrees of bending for more than 10% of worktime for frequent bending or twisting of the trunk and, with reference to WBV, 5 years of exposure to 1 ms^{-2} or an equivalent vibration dose. The model was built based on the age-dependent prevalence of lower back pain for unexposed persons: the additional presence of one or more of the risk factors under examination further raised the probability of LBP. The transformation of the model into a flow chart yielded a score ranging from +3 to +5 for manual handling of materials, frequent bending or twisting of the trunk, whole-body vibration, and job dissatisfaction. The score can increase to higher values (from +5 to +7) in the event of estimates of high exposure to the abovementioned physical risk factors. From all these possible scores, a concomitant probability for LBP could be derived: its transposition into an etiologic fraction indicates the level of work-relatedness for lower back pain.

In another analysis carried out by Laštovková [27], epidemiological studies pointed out the existence of a statistically significant correlation between LBPD and certain types of occupational burden. This important public and economic issue has been solved in different ways across Europe. Diagnostic criteria differ substantially with respect to both verification of the workload and the range of diagnoses of diseases accepted, and not all EU countries currently include LBPD caused by overload and/or WBV in their list of occupational diseases. Those who take it into account use different systems, criteria and diagnoses to recognize LBPD as an occupational disease. On the one hand, in 13 out of the 23 studied countries, LBPD caused by overload can be recognized as occupational provided that the diagnosis is sufficiently proven, exposure criteria and/or listed occupation are satisfied and the duration of exposure is confirmed (Belgium, Denmark, France, Germany, Hungary, Italy, Lithuania, Macedonia, the Netherlands, Romania, Slovakia, Sweden and Switzerland). On the other, 14 countries recognize LBPD arising from vibrations as an occupational disease. Despite this, 8 countries do not accept LBPD as an occupational disease unless it is the outcome of an injury at work.

Specific criteria to evaluate the occupational exposure of patients with LBPD have been established in Belgium, Denmark, France, Germany, Lithuania, Macedonia, the Netherlands and Slovakia. In other countries, the evaluation is made on an individual basis. Most of these countries use an individual evaluation of the patient's disorder and related work overload as assessed by medical, hygienic and ergonomics specialists. One country (Germany) uses computer models, while others (the Netherlands, Slovakia) rely on the use of mathematical models aimed at individually assessing the relationship between occupational workload and LBPD and their causality for an affected worker. The Belgian system, which was enacted in 2004 and in which LBPD is recognized as an occupational disease, both due to overload and whole-body vibrations, was inspired by the German model. In Macedonia, LBPD is only acknowledged if it is the outcome of excessive strain by muscles, tendons and their attachments.

In the Netherlands, since 2005, a rather different system based on the results of Lötters [26] has been used to evaluate the presence of a causal link between working operations and LBP in the event of non-specific lower back pain. As previously described, this probability model is designed as a three-step plan. First, it makes the right diagnosis, second, it evaluates the work-relatedness of the risk factors (exposure to lifting and carrying, bending of the trunk and whole-body vibration), while the last step consists of totting up the scores arising from the exposure to each of the three risk factors: these result in the total score for work-relatedness probability. Recently, again in the Netherlands, a criteria document was also developed which aims to assess whether lumbar herniated disc disease (lumbosacral radicular syndrome) can be classified as an occupational disease. The work-relatedness of lumbar herniated disc disease can be recognized as an occupational disease if the worker exposure is characterized by more than 10 years of physically demanding work (daily lifting and carrying of loads of at least 5 kg for, on average, 2 h or 25 times per day, including bending of the trunk more than 20° for at least one hour a day). The assumption that driving a vehicle by itself is a risk factor for lumbar herniated disc disease was not supported in this review.

In Switzerland, the etiologic contribution of occupation is estimated individually by physicians specialized in occupational medicine. For overdose-induced LBPD, the causality of the occupational workload has to be 75% or more, provided that diagnosis is confirmed by imaging methods. In the event of vibration-induced damages, an etiologic contribution of the occupation of 50% or more is sufficient.

France uses item no. 98 of the French list of occupational diseases related to LBPD (lifting heavy weights), which was introduced in 1999 [15]. To benefit from the work-relatedness presumption and allow the "automatic" recognition of LBPD as occupational in a patient that has been working in a defined occupation/branch of industry, he/she has to meet three of the following conditions: (i) being affected by sciatica/radiculalgia and having a corresponding disc hernia; (ii) having been exposed to lifting heavy weights for at least 5 years; (iii) a less than six-month interval between the last exposure and the diagnosis.

In practice, the assessment of occupational overload and its contribution to the onset of LBPD as well as its inclusion in the compensation system are important for several reasons. Firstly, it may be considered essentially preventable. Secondly, cases with a significant contribution of occupational etiology may be viewed as occupational diseases for which compensation may be claimed, as is the case in many European countries. Furthermore, including LBPD in the list of occupational diseases or another system of compensation may be viewed as a preventive measure as it increases the visibility of this problem not only for the workers, but especially for the employers [27]. These conclusions are also reported in the findings of Hulshof [28], who indicated that significant differences exist in the established and applied criteria for WBV-related injury in four EU countries (the Netherlands, Belgium,

France and Germany) where this injury is currently established as an occupational disease. Whereas "Mrs Robinson" would get recognition and compensation in one or two countries, she would be rejected in the other ones. Furthermore, the large variance in the annual incidence of this occupational disease in countries with comparable WBV exposure distribution in the working population confirms the disparity between them. This disparity, on the one hand, is partly due to differences in the occupational disease systems in general, but on the other it is also caused by the differences in the specific criteria considered for the evaluation of this occupational disease [28].

2.3. The Risks Arising from Physical Agents: Directive 2002/44

Directive 2002/44/EC (European Community) [18], on the exposure of workers to the risks arising from physical agents (vibration), seeks to introduce, at the Community level, minimum protection requirements for workers when they are exposed to risks arising from vibration in the course of their work. Directive 2002/44/EC sets out 'exposure action values' (EAVs) and 'exposure limit values' (ELVs). The EAV is the amount of daily exposure to whole-body vibration above which you are required to take action to reduce risk: it is set at a daily exposure of 0.5 ms^{-2} A(8). Whole-body vibration risks are low for exposures around the action value and only simple countermeasures are usually necessary in these circumstances.

The ELV is the maximum amount of vibration an employee may be exposed to on any single day: it is set at a daily exposure of 1.15 ms^{-2} A(8). Operators of some off-road machines and vehicles may exceed the limit value, but this depends on the task, vehicle speed, ground conditions, driver skill and duration of the operation. Moreover, the ELV also: (i) specifies employers' obligations with regard to risk assessment and determination; (ii) sets out the measures to be taken to reduce or avoid exposure and (iii) details how to provide information and training for workers. Any employer who intends to carry out work involving risks arising from exposure to vibration must implement a series of protection measures before and during the work. The Directive also requires EU Member States to put in place a suitable system for monitoring the health of workers exposed to risks arising from vibrations. When carrying out the risk assessment, employers must pay particular attention to the following:

- the level, type and duration of exposure, including any exposure to intermittent vibration or repeated shocks;
- the exposure limit values and the exposure action values;
- any effects concerning the health and safety of workers at particularly sensitive risk;
- any indirect effects on worker safety resulting from interactions between mechanical vibration and the workplace or other work equipment;
- information provided by the work equipment manufacturers in accordance with the relevant Community Directives;
- the existence of replacement equipment designed to reduce the levels of exposure to mechanical vibration;
- the extension of exposure to whole-body vibration beyond normal working hours under the employer's responsibility;
- specific working conditions such as low temperatures;
- appropriate information obtained from health surveillance, including published information, as far as possible.

Risks arising from exposure to mechanical vibration must be eliminated at their source or reduced to a minimum by taking account of technical progress and the availability of measures that enable control of the risk at source. The reduction of such risks must be based on the general principles of prevention. On the basis of the risk assessment, once the exposure action value is exceeded, employers must establish and implement a programme of technical and/or organizational measures aimed at reducing exposure to mechanical vibration and attendant risks to a minimum. In particular, they must take into account:

- other working methods that require less exposure to mechanical vibration;
- the choice of appropriate work equipment of an appropriate ergonomic design and, in line with the work to be done, producing the least possible vibration;
- the provision of auxiliary equipment that reduces the risk of injuries caused by vibration, such as seats that effectively reduce whole-body vibration and handles that reduce the vibration transmitted to the hand-arm system;
- appropriate maintenance programmes for work equipment, the workplace and workplace systems;
- the design and layout of workplaces and work stations;
- adequate information and training to instruct workers to use work equipment correctly and safely in order to reduce their exposure to mechanical vibration to a minimum;
- limitation of the duration and intensity of the exposure;
- appropriate work schedules with adequate rest periods;
- the provision of clothing to protect exposed workers from cold and damp.

2.4. Current Results of In-Field Measurements

Several studies reporting the results of measuring tractor operator WBV can be retrieved in the literature. These data are always quite similar and confirm that people using agricultural machinery are likely to undergo vibration exposure above the EAV. In some cases, action may be required to keep workers' exposure below the ELV.

2.4.1. HSE (Health and Safety Executive, UK) Information Sheet No. 20, Revision 2

Among the published works, we can note HSE information sheet No. 20, Rev. 2 [29], "Whole-body vibration in agriculture", which groups agricultural tasks according to likely exposure:

Group 1: WBV unlikely to be a risk. For these tasks, the estimated exposure is likely to be below the EAV with no significant shocks. The adoption of low-cost vibration reduction measures and the management of WBV will reduce the likelihood and the persistence of back pain. It is unusual for machinery-related tasks in agriculture to fall into this category. Even if machinery is shared among a large workforce and exposure durations are short enough to maintain exposures below the EAV, it is highly likely that there will be some exposure to significant shocks.

Group 2: You must manage exposure to WBV. This refers to tasks according to which exposures are likely to exceed the EAV on at least some days, but shocks are expected to be small. These may be:

(i) combining, hedging and ditching;
(ii) self-propelled foragers;
(iii) duties not otherwise listed requiring use of a power take-off shaft.

Here, the risk of back pain from WBV is likely to be low and back pain is more likely to be caused by other factors. Low-cost vibration reduction and management measures must be put in place (costly or difficult measures are unlikely to be reasonably practicable).

Group 3: WBV is a likely cause of back pain. In this case exposures are likely to be much higher than the EAV and/or contain large shocks. Here effective engineering and management controls must be put in place. Health monitoring is recommended to confirm that the risk from WBV is under control. These activities are:

(i) Baling, drilling, foraging, spraying, ploughing, harrowing;
(ii) Primary cultivation (up to 5.5 h);
(iii) Mowing (up to 8 h);
(iv) Tedding (up to 5 h);
(v) Transport using unsuspended tractors (up to 4.5 h);
(vi) Transport using tractors with suspended cab or chassis (up to 7 h);

(vii) Driving an ATV (all-terrain vehicle/quad bike) up to 5.5 h.

Group 4: You must restrict exposure to WBV. When dealing with these tasks the ELV must be taken as the reference and people's exposure to WBV should be restricted accordingly. Such restrictions must be applied to the following tasks:

(i)   Primary cultivation (over 5.5 h);
(ii) Mowing (over 8 h);
(iii) Tedding (over 5 h);
(iv) Transport using unsuspended tractors (over 4.5 h);
(v)   Transport using tractors with suspended cab or chassis (over 7 h);
(vi) Driving an ATV (over 5.5 h).

2.4.2. The ENAMA Technical Document

The ENAMA (Italian Board for Agricultural Mechanization) technical document [30] on the problem of agricultural machinery vibration reports the mean values of the vertical acceleration, weighted in frequency, of the measurement at the seat of 77 tractors together with the relevant time of exposure characterizing the EAV (Table 1).

Table 1. Mean values of the acceleration, weighted in frequency, of the measurement at the tractor seat with the relevant time of exposure; EAV: Exposure action values.

No.	Task	Mean Value ms^{-2}	Exposure Time EAV (h-min)
1	Forage Baling	0.50	8–00
2	Harvesting	0.45	9–53
3	Maintenance of hedgerows and ditches	0.42	11–20
4	Eradication and harvesting of beet	0.70	4–05
5	Fertilizer spreading	1.30	1–11
6	Tillage with disk harrow	1.20	1–23
7	Mowing	1.00	2–00
8	Ploughing	1.01	1–58
9	Rotary harrow	1.70	0–42
10	Loading and unloading	1.20	1–23
11	Windrowing	1.00	2–00
12	Rolling	0.60	5–33
13	Transport with trailer	0.93	2–19
14	Manure spreading	0.60	5–33
15	Spraying	1.15	1–31
16	Rear-mounted backhoe	0.74	3–39
17	Wood hauling	1.14	1–32

2.4.3. CEMA Practical User's Guide

The European Agricultural Machinery Association (CEMA) published the leaflet: "Whole-body Vibration in Agriculture, CEMA Practical User's Guide" [31]. It contains notes on good practice and can be regarded as a guideline. It is based on the HSE information sheet "WBV in Agriculture".

2.4.4. EU WBV Good Practice Guide

The EU document, "The guide to good practice on Whole-Body Vibration, non-binding guide to good practice with a view to implementation of Directive 2002/44/EC (European Community) on the minimum health and safety requirements regarding the exposure of workers to the risks arising from physical agents (vibrations)" [32] reports examples of vibration magnitudes for common machines (Figure 1).

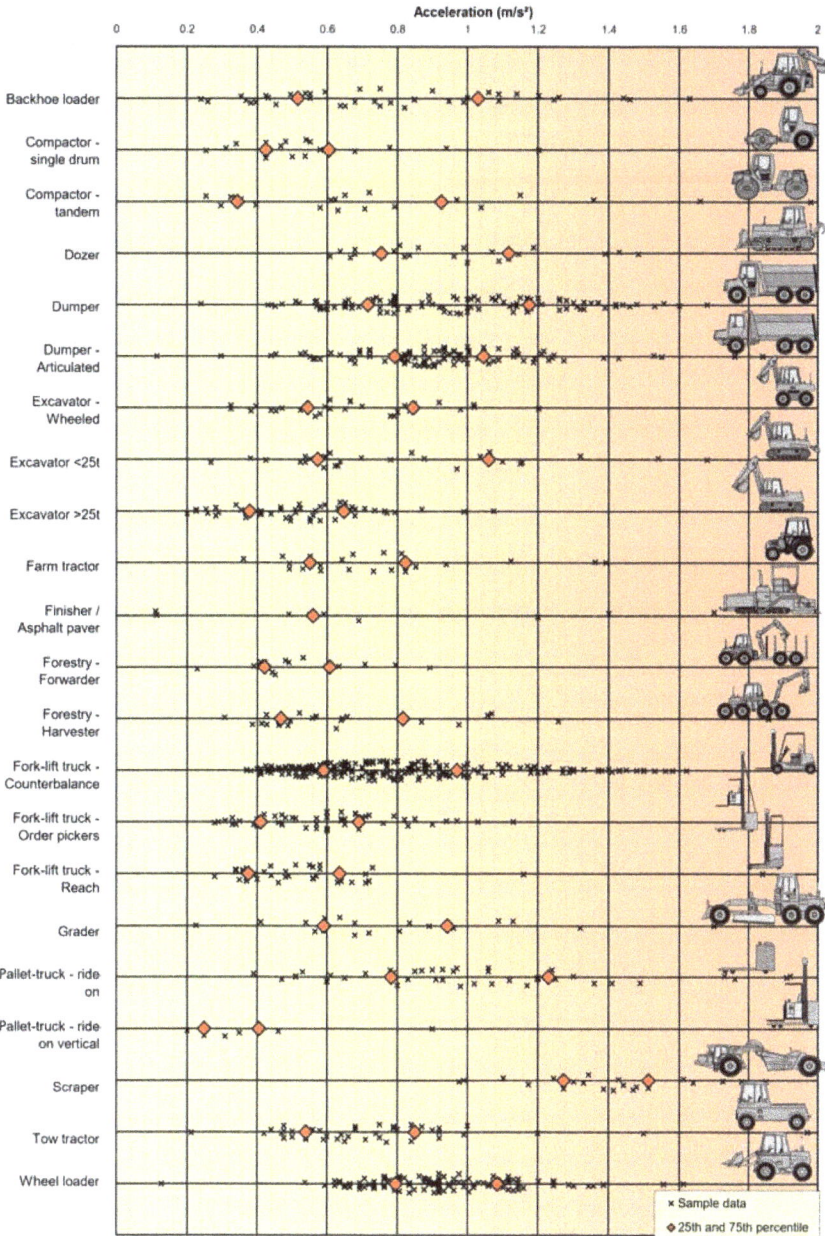

Figure 1. Examples of vibration magnitudes for common machines. Sample data based on workplace vibration measurements of the highest axis vibration values by the Institut national de recherche et de sécurité (INRS) (with the assistance of CRAM (Caisse régionale d'assurance maladie) and Prevencem), HSL (Health and Safety Laboratory) and the RMS Vibration Test Laboratory between 1997 and 2005 (EU, Guide to Good Practice on Whole-Body Vibration, 2006) [32].

The displayed data are based on workplace vibration measurements of the highest axis vibration values carried out by the INRS (Institut national de recherche et de sécurité) with the assistance of CRAM (Caisse régionale d'assurance maladie) and Prevencem, HSL (Health and Safety Laboratory) and the RMS Vibration Test Laboratory between 1997 and 2005. These data are for illustration only and may not be representative of machine use in all circumstances. The 25th and 75th percentile points show the vibration magnitude that 25% or 75% of samples are equal to or below. It is possible to see how the agricultural tractor data comply with the HSE and ENAMA document.

3. Test Methods for Measuring Vibration

Frequency represents the number of times per second the vibrating body moves back and forth. It is expressed as a value in cycles per second, more usually known as hertz (Hz).

For WBV, the frequencies thought to be important range from 0.5 Hz to 80 Hz. However, because the risk of damage is not equal at all frequencies, a frequency-weighting is used to represent the likelihood of damage arising from different frequencies. As a result, the weighted acceleration decreases when the frequency increases. A brief description of the objective measures used (AAP (Average Absorbed Power); BS (British Standard) 6841, VDI (Verein Deutscher Ingenieure) 2057, NASA (National Auronautics and Space Administration) 2299 and ISO (International Standard Organization) 2631-1997) follows [33].

3.1. Average Absorbed Power (AAP)

Average Absorbed Power (AAP) was developed by the US Army Tank-Automotive Command in 1966. Studies have shown that the human body behaves in an elastic fashion. Under vibration, the body's elasticity produces restoring forces that are related to displacement. This process continues until the energy imparted is dissipated or removed. The time rate of energy absorption is referred to as the absorbed power. The absorbed power can be computed in the frequency domain as well as in the time domain. The frequency domain ranges between 1 and 80 Hz and below 1 Hz the method is considered not successful. The AP (Absorbed Power) weighting curve strongly emphasizes the visceral resonance occurring at around 4–5 Hz, presumably because most energy is absorbed by these softer tissues.

3.2. The BS 6841 Standard

The BS 6841 standard [34] considers a frequency range of 0.5–80 Hz while introducing a new procedure based on the concept of Vibration Dose Value (VDV) instead of time dependency curves. The frequency weighting for z-axis seat vibration is modified to be in closer agreement with the results of experimental research. For each axis, a component ride value can be determined as well as an overall ride value. At first the acceleration samples are weighted using different weight functions for different directions. As for ride comfort, the root mean square (RMS) value of the weighted signal is determined as well.

3.3. The VDI 2057 Standard

In 1963 the Society of German Engineers (VDI) published the first VDI 2057 standard [35]. In principle, the VDI standard defines a calculated ride comfort index (K-factor) that is compared with a subjective table to determine the ride as subjectively experienced by humans. In 1979 the VDI standard adopted the ISO 2631-1978 tolerance curves, nevertheless it kept the K-factor for the subjective comparison of human-perceived sensations. The acceleration data is converted into the frequency domain using a fast Fourier transform (FFT) from which the RMS values are determined. This results in single values at the centre frequencies. The RMS acceleration is weighted and K-values are determined for the z-direction. The weighted signal is then plotted against limit curves. These are in principle the same as the ISO 2631-1985 standard limit curves. The frequency bandwidth ranges between 1 and 80 Hz.

3.4. The NASA 2299

This Ride Comfort standard belongs to the group of measurements that occur the least. The NASA 2299 [36] standard weights RMS acceleration values with empirically gathered weighting factors. It is defined for five degrees of freedom: vertical, lateral, longitudinal, roll and pitch. This standard does not specify accelerometer location and placement to measure the degrees of freedom and this is in contrast to the well-defined location of origins recommended by the ISO 2631 standard. One can measure as many as all five accelerations or as few as one prior to calculating Ride Comfort. The applicable frequency range for the NASA 2299 standard ranges from 0.5 to 30.0 Hz.

3.5. The ISO 2631-1997

The ISO 2631-1:1997 International Standard ("Mechanical Vibration and Shock—Evaluation of Human Exposure to Whole-body Vibration—Part 1: General Requirements") [37] defines the means for evaluating periodic, random and transient vibration with respect to human responses: health, comfort, perception and motion sickness.

The RMS method continues to be the basis method for the ISO 2631-1:1997. In principle, the methodology and calculations of the standard are the same as the BS 6841 standard. The main difference between the BS 6841 standard and this one is that the vertical weighting W_k (resulting from intensive laboratory studies) replaces W_b.

The ISO standard specifies the direction and location of the measurements, the equipment to be used, the duration of the measurements and the frequency weighting, as well as the measurement assessment methods and the evaluation of weighted root-mean-square acceleration. The effect of frequency is reflected in frequency weightings labelled W_k, W_d, W_f, W_c, W_e and W_j and different frequency weightings are required for all different axes of the body. The different sensitivity of the body to vibration in each axis is accounted for by multiplying factors. The frequency-weighted acceleration (expressed as m/s^2) is multiplied by the weighting factor before its effect is assessed. For vertical seat vibration (z-axis), the acceleration weighting W_k has the greatest sensitivity in the range 4–13 Hz. For horizontal seat vibration (x- and y-axes), the acceleration weighting W_d has the greatest sensitivity in the range 0.5–2 Hz.

In the motion sickness assessment, the weightings W_k, W_d, W_c, W_e and W_j are used in the frequency range 0.5–80 Hz, whereas the W_f weighting is used in the 0.1–0.5 Hz range.

The standard requires the evaluation to include measurement of the weighted root-mean-square (RMS) acceleration. This is expressed in metres per second squared (ms^{-2}) for translational vibration and vibration assessment, Equation (1), requires the calculation of the weighted and gained root-mean-square (RMS) acceleration (a_{vi}) along the three axes ($i = x, y, z$):

$$a_{vi} = k_{i(x, y, z)} \sqrt{\sum_{i=1}^{n} (W_i \cdot a_i)^2} \tag{1}$$

where:

- $k_{i(x,y,z)}$ is a multiplying factor (dimensionless), set at 1.4 for the x- and y-axes and at 1 for the z-axis
- W_i is a dimensionless weighting factor given by the standard itself
- a_i is the acceleration acquired at the seat (ms^{-2}).

The calculated a_{vi} were subsequently used to define WBV risk conditions according to European Directive 2002/44/CEE (EC, 2002)

3.6. The Situation in the USA

In the USA, the three most commonly referenced voluntary WBV standards are those of the American National Standards Institute (ANSI), American Conference of Governmental Industrial Hygienists (ACGIH) and ISO. In 1979 the ANSI originally published American National Standard

S3.18, which was almost identical to ISO 2631 and at a later date released the ANSI S3.18-2002 ISO 2631-1-1997, an adaptation of the most recent ISO standard. WBV exposure limits published by the ACGIH are based upon the ISO standard. Neither the National Institute for Occupational Safety and Health (NIOSH) nor the Occupational Safety and Health Administration (OSHA) has issued WBV standards [38].

4. State of the Art and Possible Improvements Aimed at Reducing WBV Exposure in Farming

The European legislation sets out the basic rules based on risk assessment and the actions to be undertaken to reduce the risk. The issue of evaluating WBV for tractor drivers involves several different themes and, consequently, requires different approaches and methods.

Considering the requirements of Directive 2002/44/EC [18], the need for WBV measurement can be summarized in the following steps:

1. *Risk assessment carried out by employers*: in this way vibration exposure is evaluated either with direct measurement in operating conditions representative of daily working, or by considering the data from the database proposed by the national authority on safety [39].
2. *Implementation of appropriate actions to reduce the risk of mechanical vibration exposure*: this can be carried out in different ways (e.g., by choosing work equipment which includes the least possible vibration), but there is currently no reference data characterizing tractors' vibrational comfort.

Table 2 reports some examples of the main topics that must be considered.

Table 2. Main topics regarding Whole-Body Vibration (WBV) risk.

Aim	Task
Risk assessment	Evaluate the level, type and duration of exposure
	Verify the exposure limit values and the exposure action values
	Adopt information provided by the work equipment manufacturers
	Verify the existence of replacement equipment designed to reduce the levels of exposure
Reduction of the risk of mechanical vibration exposure	Adopt other working methods that require less exposure to mechanical vibration
	Choose appropriate work equipment producing the least possible vibration
	Provide auxiliary equipment that reduces the risk of injuries caused by vibration, such as seats that effectively reduce whole-body vibration

It is evident that the effort made by manufacturers to develop solutions to improve tractor comfort has to be properly assessed. Indeed, for several years tyres and seats have been the main mechanical vibration mitigation devices and the studies carried out on off-road and industrial vehicles have focused on tyre properties, dumping systems, and their interaction [40–44]: the development of devices mitigating vibration transmission to drivers [45–47] is currently a challenge both for manufacturers and research institutes.

Studies to quantify WBV emission and estimated exposure levels upon a range of agricultural tractors have been carried out in controlled conditions [48,49] while performing the following tasks:

*   traversing ISO ride vibration test tracks;
*   performing selected agricultural operations;
*   performing identical tasks during 'on-farm' use;
*   transport on paved minor roads;
*   transport on focused and dedicated terrain or stony tracks.

Tractor WBV emission levels were found to be very dependent upon the nature of the operation performed. Apart from the experiences carried out on ISO 5008, other tests found in the literature that evaluate technical solutions may not result in repeatable outcomes given the changes that dedicated tracks as well as the roads used undergo in time. The approach used to elastically characterize the tractor on a standard test track and to correlate it with some in-field operations is also very interesting [50]. But, to date, the value declared by manufacturers only refers to seat performance: indeed, with reference to agricultural tractors, there is such a wide range of data on WBV that it is impossible to presume that a single piece of data can be taken as a reference or consider the so-called worst-case scenario (as happens with noise tests) a viable solution. One step forward can be represented by the approach contained in the EN (European Norm) 13059:2002+A1:2008 (E) standard [51] that provides manufacturers with a mean value in compliance with the essential safety requirements of the Machinery Directive and associated EFTA (European Free Trade Association) regulations. It allows the comparison of industrial trucks of the same category or a given truck in different configurations (equipped with different seats, tyres, etc.). However, this standard cannot be used to assess the daily exposure of the operator to vibration in field conditions. For such a goal, the development of a simplified testing methodology must set out to define:

- a standard test track;
- a comfort index;
- the machine operating conditions.

5. Background for a Simplified Procedure to Determine Agricultural Tractor Vibration Comfort

5.1. On the Definition of the Standard Test Track

In order to tackle the first issue (defining a standard test track), the effect of soil profile on the tractor dynamics shall be analysed first.

5.1.1. Effect of Soil Profile on Tractor Dynamics: Theoretical Considerations

Terrain irregularity and vehicle forward speed are the most important sources of vibrations for tractors [49,50,52]. Increasing knowledge on terrain irregularity characteristics could improve the design of solutions for greater operator comfort. At the moment, while the ISO 8608:1995 standard [53] sets out the method to report the vertical accelerations arising from surface profiles, pointing to connections between profile roughness and mechanical failure as well as operator discomfort [53–55], it does not lay down any straightforward surface profile measurement methods or instruments, confirming the difficulties in making such measurements properly [56,57]. As a matter of fact, the methods to perform surface profile measurements can be direct or indirect:

- direct surface profile measuring methods (e.g., using optical technology) do not have the required precision or repeatability as they do not account for the surface profile deformations the machine itself induces [58]. Indeed, the ISO 8608 standard recommends taking care when making off-road measurements in the event of both soft surfaces (flattened and filtered by wheels going forward) and hard ones (because of the filtering effect of the wheel envelope);
- with reference to indirect profile measurements, one protocol provides for the adoption of a two-step procedure for performing vertical acceleration acquisition: the tractor must have previously been run on different surfaces (with the machine in operating conditions) to acquire the accelerations the machine is subjected to; afterwards test bench replication (by a deconvolution method) of the previously acquired accelerations is performed until the exact reproduction of the acquired solicitations is achieved [59,60].

An experiment carried out with the latter method [61] defined the unevenness of four different surfaces (both in the time and frequency domains): despite the wide variability of agricultural surface profiles, it pointed out which features, among soil characteristics and tractor settings (i.e., ballast or tyre pressure), could significantly affect vehicle dynamics and WBV. According to this experiment, results have shown that by changing soil profile and tractor speed, the accelerations resulting from ground input present similar spectral trends, which were found to be relevant at frequencies of less than 4 Hz. This result has been also validated by research carried out by running an agricultural tractor in four setting conditions and at four different forward speeds on an ISO 5008 standard test track [62]. Further investigations [63,64] that considered different tractors and agricultural operations confirmed the possibility of defining a standard trend of soil solicitations. It follows that, as far as agricultural tractors are concerned, it is the vertical acceleration tractors undergo that excites their elastic components (i.e., tyres, cab, rubber mountings, etc.) and that, in turn, determines vehicle dynamics in terms of frequency.

To better understand this, the physics of tyre response to road unevenness must be considered, in particular the solicitation the elastic part the vehicle undergoes when the vehicle passes at constant speed over a cleat whose length is much smaller than the contact length therefore resulting in changes in the tyre rolling radius. According to Pacejka [40] and Jianmin [65], four main factors need to be taken into account:

- tyre envelopment properties (variations occurring in vertical and longitudinal forces, as well as in the angular velocity of the wheel);
- effective road plane (the effective height and slope of a short trapezoidal cleat is approximated at the axle by a half sine wave);
- effective rolling radius when rolling over a cleat (increment in normal load, local forward slope, local forward curvature);
- the fact that the measured response is purely vertical while on the test track there is a sum of vertical and longitudinal components.

As a matter of fact, the displacement recorded at the hub of the rolling tyre, both because of the rolling and radial deflection that occurs (as a consequence of the passage over the cleat), is ascribable to a sine wave (Figure 2). Hence, the response of the hub acceleration to the effect of the cleat results in a sine function as well, characterized by the same resonance frequency as the tyre.

It can be deduced that, even though forward speed and cleat height characterize the hub amplitude response, the vehicle response is determined by the elastic properties of its components, irrespective of forward speed or the randomness of the test track profile. In the same way, it is the response of the tyres and seat to the bump that mainly affects operator comfort.

It is therefore possible to suppose that specific standard testing should aim to provide an input force that, by exciting the main frequencies of the vehicle, is representative of the main frequencies affecting operator comfort during ordinary use.

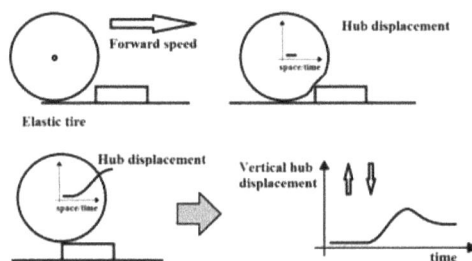

Figure 2. Diagram of the sinusoidal response of a tyre to a square input.

As a result, vertical tractor dynamics, as well as comfort and material resistance testing activities, can be greatly simplified and standardized so that elaborate test tracks can be avoided. Neither tests in the laboratory nor in the open field required the tractor to run on several test tracks, providing a variety of vertical solicitations, as it was sufficient to develop one rough test surface to be run at such a forward speed as to solicit the elastic parts of the tractor [66].

5.1.2. State of the Art of the Standard Test Tracks for WBV Assessment

With the aim of standardizing testing activity, the International Standards Organization (ISO) issued the ISO 5008:2002 standard: "Agricultural wheeled tractors and field machinery—Measurement of whole-body vibration of the operator." The purpose of this provision is to specify the instruments, measurement procedures, measurement site characteristics and frequency weighting that allow the WBV intensity of agricultural wheeled tractors and field machinery to be assessed and reported with acceptable precision. This provision defines two standard test tracks together with the relevant operating conditions. These standard test tracks are 100 m (smooth track, Figure 3) and 35 m long, composed of two different parallel, non-deformable lanes (left and right) made of wooden beams (80 mm wide and with 80 mm spacing in the smooth track, without spaces in the 35 m track) of a different standardized height to induce vibrations. EN 13059:2002 [51] sets out a method for measuring the vibration emission transmitted to the whole-body of operators of industrial trucks: it satisfies the requirements of the Machinery Directive but, as aforementioned, it cannot be used to determine the daily exposure of the operator to vibration in field conditions. With regard to industrial trucks, three predominant operating modes are considered (travelling, lifting and engine idling) and, of these, only travelling exposes the driver to significant WBV. Hence, vehicle testing is specifically based on the travelling operating mode. It consists of running the vehicle at 10 km/h on a 25 m-long track with two square obstacles whose height is related to the mean wheel diameter (Figure 4).

Figure 3. The International Standards Organization (ISO) 5008 100m smooth test track.

Figure 4. The 25 m test track defined in the EN 13059.

This approach complies with the following requirements of Directive 2002/44:

- information provided by the work equipment manufacturers in accordance with the relevant Community Directives;
- the existence of replacement equipment designed to reduce the levels of exposure to mechanical vibration;

Transferring this methodology from industrial machines to tractors still leaves some issues open, such as measurement of the exposure along three axes (longitudinal, lateral and vertical) [67].

5.1.3. Developing a Simplified Test Track

Once the effect of forward speed and cleat height on hub amplitude response has been assessed—and in light of the fact that only a small range of frequencies significantly affect operator comfort given the filtering/amplification effect that different combinations of tyres and speed have on tractor response—the square prism shape of the cleat recommended by Ente Nazionale Italiano di Unificazione (UNI) 13059 [51] cannot be deemed appropriate. On the contrary, the solicitation arising from running the vehicle on ramps seems to be the most promising. A first attempt to validate this hypothesis resulted in the development of three test tracks [68], one for each axis of solicitation (x, y, z), made of specifically designed ramps. Figure 5 shows the ramps developed for the z axis test track (1000 mm long and 50 mm high), whose solicitations on a tractor run at 8 km h^{-1} (Figure 5) were compared with those resulting from running the same tractor on field test tracks and an ISO 5008 test track.

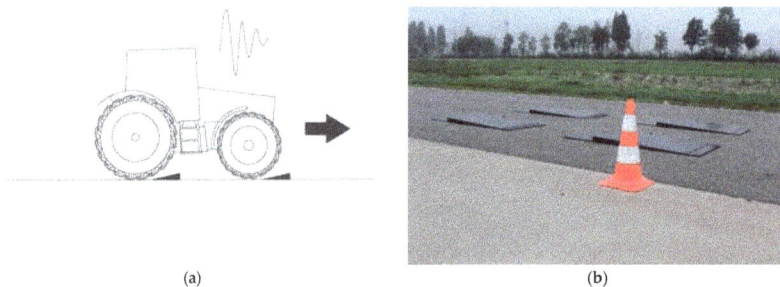

(a)   (b)

Figure 5. The simplified test layout (**a**) and the ramps specially developed (**b**) for evaluating vertical tractor comfort.

As previously mentioned, since the comfort values of each channel (x, y, z) are defined by a restricted range of frequencies, which—on all the considered tracks—depend on the elastic properties of the vehicle, the hypothesis that it is possible to develop a simplified test track was verified by pointing out that:

- each channel had a very small range of frequencies of interest;
- the frequency bandwidth mainly affecting operator comfort changes when the considered axis varies (x, y and z).

The x-axis presents two frequency ranges of interest: one from 0.6 to 1.2 Hz and another from 2 to 3.6 Hz. These two groups indicate that there are two different kinds of solicitations affecting the operator: the first is related to the elastic parts' response to impulsive solicitation (i.e., passing a ditch, moving off with a trailer, manoeuvring during harrowing), while the second one is related to the vertical movement of the vehicle (as a matter of fact, the range from 2 to 3.6 Hz corresponds to the vertical resonance frequency of the tyres): the surface roughness induces a transformation of the horizontal component of the forward speed into vertical acceleration. At the same time, the y axis also presents two ranges of interest: from 0.8 to 2 Hz and 3.2 to 3.8 Hz. Of these, the first is the most interesting as the amplitude is related to the lateral movement of the tractor. In all the conditions, the z-channel turned out to be correlated to frequencies ranging from 2.2 to 3.4 Hz, that is, the range of frequencies corresponding to the vertical resonance of the tyres.

If the combination between surface roughness and speed provides enough energy, the reaction of the elastic parts of the tractor—above all the tyres—plays a critical role in determining operator comfort.

These considerations confirmed the necessity to study the three orthogonal axes separately, opening up important issues about how these simplified test tracks could be further improved. The limit remains that it is necessary to carry out three tests, resulting in three different values that do not fit with the ultimate aim of obtaining an indication of tractor comfort with one index only.

5.2. Defining a "Comfort Index"

ISO 2631 [37] reports the most interesting parameter for representing vehicle comfort in one value: it recommends that, when assessing the effects on comfort, all the relevant vibration directions should be considered to obtain the overall total value of vibration:

$$a_w = \left(\sum_i k_i^2 a_{wi}^2 \right)^{1/2} \tag{2}$$

where:

- a_{wi} are weighted root mean square (rms) accelerations on the relevant axes;
- k_i are multiplying factors.

The complete formula requires measurement of the acceleration along the 3 axes at the seat, back, feet, roll, pitch and yaw, for a total of 12 channels. It still needs to be clarified and further investigated if this equation fits for agricultural tractors, nevertheless it can be of great interest in building a simplified approach for parameter estimation.

5.3. Defining the Machine Operating Conditions

With reference to setting up the machine operating conditions, the test setting could follow the main recommendations of the ISO 5008 and ISO 2631 standards while rethinking their application within a simplified framework.

6. Conclusions and Recommendations

To allow test laboratories to measure performance characteristics, new standards have been developed as a result of the continuous restructuring of the agricultural machinery industry and the increasing complexity of agricultural and forestry tractors.

European Parliament and Council Directive 2002/44/EC on the exposure of workers to the risks arising from physical agents introduced, at the Community level, minimum protection requirements for workers when they are exposed to risks arising from vibration in the course of their work. It specifies employers' obligations with reference to risk identification and assessment, and it sets out the measures to be taken to reduce or avoid exposure and details how to provide information and training for workers. Any employer who intends to carry out work involving risks arising from exposure to vibration must implement a series of protection measures before and during the work. The Directive also requires EU Member States to put in place a suitable system for monitoring the health of workers exposed to risks arising from vibration.

The reduction of such risks must be based on general principles of prevention. Employers must establish and implement a programme of technical and/or organizational measures intended to reduce the exposure to mechanical vibration and its attendant risks to a minimum, taking into account the choice of suitable work equipment, the provision of auxiliary equipment reducing the risk of injuries caused by vibration (such as seats), the design and the layout of workplaces.

Despite being conducted in controlled conditions (traversing ISO ride vibration test tracks; performing selected agricultural operations; performing identical tasks during "on-farm" use; transport on paved minor roads; developing focused and dedicated terrain or stony surfaces), the various studies presented herewith—carried out to quantify WBV emission and estimated exposure levels upon a range of agricultural tractors—suffer from a lack of comparability when the studied working conditions vary, with the exception of those carried out in accordance with ISO 5008.

While these studies confirmed that tractor WBV emission levels were found to be very dependent upon the nature of the operation performed, they have shown that changing soil profiles and tractor speeds give rise to similar spectral trends of the accelerations resulting from ground input. This result led researchers to investigate a possible and realistic simplification and standardization of the tractor driver comfort testing activity given the unpredictability of soil profile.

The literature presented in this review indicates the usefulness, and the possibility, of developing simplified procedures for measuring the exposure of agricultural tractor operators to WBV so that the outcome of the simplified procedure can be used to compare different tractors—provided that they belong to the same category—or a given tractor with different equipment (seats, suspension, tyres, etc.). Such simplified testing procedures cannot be considered suitable for determining operators' daily exposure to vibration in open field conditions.

Acknowledgments: The Organisation for Economic Co-operation and Development (OECD) Tractor Codes Programme commissioned this work: the opinions expressed and arguments are solely those of the authors and do not necessarily reflect the official views of the OECD or of its member countries. Authors are grateful to OECD delegates for the fruitful discussions during the OECD meetings.

Conflicts of Interest: The authors declare no conflict of interest. The founding sponsors had no role in the design of the study; in the collection, analyses, or interpretation of data; in the writing of the manuscript, or in the decision to publish the results.

Appendix A. Standards

- Directive 2002/44/EC of the European Parliament and of the Council of 25 June 2002 on the minimum health and safety requirements regarding the exposure of workers to the risks arising from physical agents (vibration). 2002.
- ISO—International Standard Organization (2002). Standard ISO 5008:2002. Agricultural wheeled tractors and field machinery—Measurement of whole-body vibration of the operator.
- BS 6841:1987 Guide to measurement and evaluation of human exposure to whole-body mechanical vibration and repeated shock. The British Standards Institution, London, UK.
- VDI 2057-1:2017-08. Human exposure to mechanical vibrations—Whole-body vibration. Beuth, Berlin, Germany.
- Leatherwood, J.D.; Barker, L.M. A User Oriented and Computerized Model for Estimating Vehicle Ride Quality. NASA Technical Paper 2299, April 1984.
- ISO—International Standard Organization (1997). Standard ISO 2631-1:1997. Mechanical vibration and shock—Evaluation of human exposure to whole-body vibration—Part 1: General requirements.
- EN—European Standard. EN 13059:2002+A1:2008. Safety of industrial trucks—Test methods for measuring vibration.
- ISO—International Standard Organization (1995). Standard ISO 8608:1995 mechanical vibration—Road surface profiles—Reporting of measured data.

References

1. European Commission (EC). *The Magnitude and Spectrum of Farm Injuries in the European Union Countries*; European Commission (EC): Athens, Greece, 2004.
2. Health and Safety Authority (HSA). *Farm Safety Action Plan 2013–2015, The Metropolitan Building 2013*; Health and Safety Authority (HSA): Dublin, Germany, 2013.

3. Litchfield, M.H. Agricultural Work Related Injury and Ill-Health and the Economic Cost. *Environ. Sci. Pollut. Res. Int.* **1999**, *6*, 175–182. [CrossRef] [PubMed]
4. Hoy, D.; Brooks, P.; Blyth, F.; Buchbinder, R. The Epidemiology of low back pain. Best 426. *Pract. Res. Clin. Rheumatol.* **2010**, *24*, 769–781. [CrossRef] [PubMed]
5. Monarca, D.; Porceddu, P.; Cecchini, M.; Babucci, V. Microclimate risk evaluation in agroindustrial work environments. *Riv. Ing. Agrar.* **2005**, *4*, 89–93.
6. Di Giacinto, S.; Colantoni, A.; Cecchini, M.; Monarca, D.; Moscetti, R.; Massantini, R. Dairy production in restricted environment and safety for the workers. *Ind. Aliment.* **2012**, *51*, 5–12.
7. Monarca, D.; Cecchini, M.; Guerrieri, M.; Santi, M.; Bedini, R.; Colantoni, A. Safety and health of workers: Exposure to dust, noise and vibrations. *Acta Hortic.* **2009**, *845*, 437–442. [CrossRef]
8. Okunribido, O.O.; Magnusson, M.; Pope, M.H. Low back pain in drivers: The relative role of whole body vibration, posture and manual materials handling. *J. Sound Vib.* **2006**, *298*, 540–555. [CrossRef]
9. Chiang, C.F.; Liang, C.C. A study on biodynamic models of seating human subjects exposed to vertical vibration. *Int. J. Ind. Ergon.* **2006**, *36*, 869–890.
10. Seidel, H.; Heide, R. Long-term effects of whole-body vibration: A critical survey of the literature. *Int. Arch. Occup. Environ. Health* **1986**, *58*, 1–26. [CrossRef] [PubMed]
11. Bonghers, P.; Boshuizen, H. Back Disorders and Whole-Body Vibrations at Work. Ph.D. Thesis, L'universite d'Amsterdam, Den Haag, The Netherland, October 1990.
12. Manetto, G.; Cerruto, E. Vibration risk evaluation in hand-held harvesters for olives. *J. Agric. Eng.* **2013**, *44*, 705–709. [CrossRef]
13. Lenzuni, P.; Deboli, R.; Preti, C.; Calvo, A. A round robin test fot the hand-transmitted vibration from an olive harvester. *Int. J. Ind. Ergon.* **2016**, *53*, 86–92. [CrossRef]
14. Bovenzi, M. Exposure-response relationship in the hand-arm vibration syndrome: An overview of current epidemiology research. *Int. Arch. Occup. Environ. Health* **1998**, *7*, 509–519. [CrossRef]
15. Bovenzi, M. Health effects of mechanical vibration. *G Ital. Med. Lav. Ergon.* **2005**, *27*, 58–64. [PubMed]
16. Punnett, L.; Wegman, D.H. Work-related musculoskeletal disorders: The epidemiologic evidence and the debate. *J. Electromyogr. Kinesiol.* **2004**, *14*, 13–23. [CrossRef] [PubMed]
17. Bishu, R.R.; Chen, Y.; Cochran, D.J.; Riley, M.W. Back injuries in farming—A pilot investigation. In *Advances in Industrial Ergonomics and Safety, Proceedings of the Annual International Industrial Ergonomics and Safety Conference, Cincinnati, OH, USA, 5–9 June 1989*; Taylor & Francis: Abingdon-on-Thames, UK, 1989; pp. 791–798.
18. European Parliament. *Directive 2002/44/EC of the European Parliament and of the Council of 25 June 2002 on the Minimum Health and Safety Requirements Regarding the Exposure of Workers to the Risks Arising from Physical Agents (Vibration)*; European Agency for Safety and Health at Work: Bilbao, Spain, 2002.
19. International Standard Organization (ISO). *Agricultural Wheeled Tractors and Field Machinery—Measurement of Whole-Body Vibration of the Operator*; Standard ISO 5008:2002; International Standard Organization (ISO): Geneva, Switzerland, 2002.
20. European Agency for Safety and Health at Work. *2005—Expert Forecast on Emerging Physical Risks Related to Occupational Safety and Health*; European Agency for Safety and Health at Work: Bilbao, Espana, 2005; ISBN 92-9191-165-8.
21. Cecchini, M.; Colantoni, A.; Monarca, D.; Longo, L.; Riccioni, S. Reducing the risk from manual handling of loads in agriculture: Proposal and assessment of easily achievable preventive measures. *Chem. Eng. Trans.* **2017**, *58*, 85–90.
22. Health and Safety Executive (HSE). *Control Back-Pain Risks from Whole-Body Vibration Advice for Employers on the Control of Vibration at Work Regulations 2005*; Health and Safety Excutive (HSE): Liverpool, UK, 2005; HSE Books; ISBN 0-7176-6119-9. Available online: www.hse.gov.uk/pubns/indg242.pdf (accessed on 9 January 2017).
23. Barriera-Viruet, H.; Genaidy, A.; Shell, R.; Salem, S.; Karwowski, W. Effect of forklift operation on lower back pain: An evidence-based approach. *Hum. Factors Ergon. Manuf.* **2008**, *18*, 125–151. [CrossRef]
24. Lings, S.; Leboeuf-Yde, C. Whole body vibration and low back pain: A sistematic, critical review of the epidemiologiacal literature 1992–1999. *Int. Arch. Occup. Environ. Health* **2000**, *73*, 290–297. [CrossRef] [PubMed]

25. Bovenzi, M.; Betta, A. Low-back disorders in agricultural tractor drivers exposed to whole-body vibration and postural stress. *Appl. Ergon.* **1994**, *25*, 231–241. [CrossRef]
26. Lötters, F.; Burdorf, A.; Kuiper, J.; Miedema, H. Model for the workrelatedness of low-back pain. *Scand. J. Work Environ. Health* **2003**, *29*, 431–440. [CrossRef] [PubMed]
27. Laštovková, A.; Nakládalová, M.; Fenclová, Z.; Urban, P.; Gaourek, P.; Lebeda, T.; Ehler, E.; Ridzoň, P.; Hlávková, J.; Boriková, A.; et al. Low-back pain disorders as occupational diseases in the Czech Republic and 22 European Countries: Comparison of national systems, related diagnoses and evaluation criteria. *Cent. Eur. J. Public Health* **2015**, *23*, 244–251. [CrossRef] [PubMed]
28. Hulshof, C.T.J.; Van der Laan, G.; Braam, I.T.J. The fate of Mrs Robinson: Criteria for recognition of whole body vibration injury as an occupational disease. *J. Sound Vib.* **2002**, *253*, 185–194. [CrossRef]
29. Health and Safety Executive (HSE). *Whole-Body Vibration in Agriculture, Agriculture Information Sheet No 20 (Revision 2)*; Health and Safety Excutive (HSE): Liverpool, UK, 2013. Available online: www.hse.gov.uk/pubns/ais20.htm (accessed on 9 January 2017).
30. Ente Nazionale Per La Meccanizzazione Agricola (ENAMA). *Produzione Documentale Tecnica Sulla Problematica Delle Vibrazioni Connessa All'uso Delle Macchine Agricole*; Ente Nazionale Per La Meccanizzazione Agricola (ENAMA): Roma, Italy, 2005.
31. European Agricultural Machinery (CEMA). Whole-body vibration in agriculture. In *Practical User's Guide*; European Agricultural Machinery (CEMA): Brussel, Belgium, 2005.
32. European Union (EU). *Guide to Good Practice on Whole-Body Vibration*; European Union (EU): Brussel, Belgium, 2006.
33. Els, P.S. The applicability of ride comfort standards to off-road vehicles. *J. Terramech.* **2005**, *42*, 47–64. [CrossRef]
34. *Guide to Measurement and Evaluation of Human Exposure to Whole-Body Mechanical Vibration and Repeated Shock*; The British Standards Institution: London, UK, 1987; BS 6841:1987.
35. *Human Exposure to Mechanical Vibrations—Whole-Body Vibration*; VDI 2057-1:2017-08; Beuth: Berlin, Germany, 2017.
36. Leatherwood, J.D.; Barker, L.M. *A User Oriented and Computerized Model for Estimating Vehicle Ride Quality*; NASA Technical Paper 2299; Scientific and Technical Information: Washington, DC, WA, USA, 1984.
37. International Standard Organization (ISO). *Mechanical Vibration and Shock—Evaluation of Human Exposure to Whole-Body Vibration—Part 1: GENERAL Requirements*; Standard ISO 2631-1:1997; International Standard Organization (ISO): Geneva, Switzerland, 1997.
38. Paschold, W.E. Whole Body Vibration Knowledge Gaps in the US. In Proceedings of the Third American Conference on Human Vibration, Iowa City, IA, USA, 1–4 June 2010. [CrossRef]
39. PAF. Available online: http://www.portaleagentifisici.it/fo_wbv_list_macchinari_avanzata.php?lg=IT&page=0 (accessed on 8 May 2017).
40. Pacejka, H.B. *Tyre and Vehicle Dynamics*, 2nd ed.; Butterworth Heinemann: Oxford, UK, 2010.
41. Taylor, R.K.; Bashford, L.L.; Schrock, M.D. Methods for measuring vertical tire stiffness. *Trans. ASABE* **2000**, *4343*, 1415–1419. [CrossRef]
42. Crolla, D.A. Off-Road Vehicle Dynamics. *Veh. Syst. Dyn.* **1981**, *10*, 253–266. [CrossRef]
43. Previati, G.; Gobbi, M.; Mastinu, G. Farm tractor models for research and development purposes. *Veh. Syst. Dyn.* **2007**, *45*, 37–60. [CrossRef]
44. Wille, R.; Bohm, F.; Duda, A. Calculation of the rolling contact between a tyre and deformable ground. *Veh. Syst. Dyn.* **2005**, *43* (Suppl. 1), 483–492. [CrossRef]
45. Kabir, M.S.N.; Ryu, M.J.; Chung, S.O.; Kim, Y.J.; Choi, C.H.; Hong, S.J.; Sung, J.H. Research Trends for Performance, Safety, and Comfort Evaluation of Agricultural Tractors: A Review. *J. Biosyst. Eng.* **2014**, *39*, 21–33. [CrossRef]
46. Pazooki, A.; Cao, D.; Rakheja, S.; Boileau, P.E. Ride dynamic evaluations and design optimization of a torsio-elastic off-road vehicle suspension. *Veh. Syst. Dyn.* **2011**, *49*, 1455–1476. [CrossRef]
47. Braghin, F.; Cheli, F.; Genoese, A.; Sabbioni, E.; Bisaglia, C.; Cutini, M. Experimental modal analysis and numerical modelling of agricultural vehicles. In Proceedings of the IMAC-XXVII A Conference and Exposition on Structural Dynamics, Orlando, FL, USA, 9–12 February 2009.
48. Oude Vrielink, H.H.E. *Comparison of High Power Agricultural Tractors: Effect on Whole Body Vibration Exposure during a Standardized Test in Practice*, Ergolab Research B.V.: Bennekom, The Nederland, 2012.

49. Scarlett, A.J.; Price, J.S.; Stayner, R.M. Whole body vibration: Evaluation of emissions and exposure levels arising from agricultural tractors. *J. Terramech.* **2007**, *44*, 65–73. [CrossRef]
50. Deboli, R.; Calvo, A.; Preti, C. Comparison between ISO 5008 and Field Whole Body Vibration Tractor Values. *J. Agric. Eng.* **2012**, *43*. [CrossRef]
51. European Committee for Standardization. *Safety of Industrial Trucks—Test Methods for Measuring Vibration*; UNI-EN 13059:2002+A1:2008; European Committee for Standardization: Milan, Italy, 2002.
52. Nguyen, V.N.; Inaba, S. Effects of tire inflation pressure and tractor velocity on dynamic wheel load and rear axle vibrations. *J. Terramech.* **2011**, *48*, 3–16. [CrossRef]
53. International Standard Organization (ISO). *Mechanical Vibration—Road Surface Profiles—Reporting of Measured Data*; Standard ISO 8608:1995; International Standard Organization (ISO): Geneva, Switzerland, 1995.
54. Roman, L.; Florea, A.; Cofaru, I.I. Software Application for assessment the reliability of suspension system at Opel cars and of road profiles. *Fascicle Manag. Technol. Eng.* **2014**, *1*, 289–294. [CrossRef]
55. Agostinacchio, M.; Ciampa, D.; Olita, S. The vibrations induced by surface irregularities in road pavements—A Matlab approach. In *European Transport. Research Review*, 2013th ed.; Springer: Berlin/Heidelberg, Germany, 2013.
56. Park, S.; Popov, A.A.; Cole, D.J. Influence of soil deformation on off-road heavy vehicle suspension vibration. *J. Terramech.* **2004**, *441*, 41–68. [CrossRef]
57. González, A.; O'brien, E.J.; Li, Y.Y.; Cashell, K. The use of vehicle acceleration measurements to estimate road roughness. *Veh. Syst. Dyn.* **2008**, *46*, 483–499. [CrossRef]
58. Fassbender, F.R.; Fervers, C.W.; Harnisch, C. Approaches to predict the vehicle dynamics on soft soil. *Veh. Syst. Dyn.* **1997**, *27*, 173–188. [CrossRef]
59. Bisaglia, C.; Cutini, M.; Gruppo, G. Assessment of vibration reproducibility on agricultural tractors by a "four poster test stand". In Proceedings of the XVI CIGR. EurAgEng 2006 64th VDI-MEG and FAO joint "World Congress—Agricultural Engineering for a Better World", Bonn, Germany, 3–7 September 2006.
60. Anthonis, J.; Vaes, D.; Engelen, K.; Ramon, H.; Swevers, J. Feedback Approach for Reproduction of Field Measurements on a Hydraulic Four Poster. *Biosyst. Eng.* **2007**, *96*, 435–445. [CrossRef]
61. Cutini, M.; Bisaglia, C.; Bertinotti, S.A. Power spectral analysis of agricultural field surface. In Proceedings of the XVII World Congress of the International Commission of Agricultural and Biosystem. Engineering, Quebec, QC, Canada, 13–17 June 2010.
62. Cutini, M.; Deboli, R.; Calvo, A.; Preti, C.; Inserillo, M.; Bisaglia, C. Spectral analysis of a standard test track profile during passage of an agricultural tractor. *J. Agric. Eng.* **2013**, *44* (Suppl. 1), 719–723. [CrossRef]
63. Cutini, M.; Bisaglia, C. Procedure and layout for the development of a fatigue test on an agricultural implement by a four poster test bench. *J. Agric. Eng.* **2013**, *44*, 402–405. [CrossRef]
64. Cutini, M.; Bisaglia, C. Experimental identification of a representative soil profile to investigate Tractor Operator's Discomfort and Material Fatigue Resistance. In Proceedings of the International Conference of Agricultural Engineering (AgEng 2014 Zurich), Zurich, Switzerland, 6–10 July 2004; p. 8.
65. Jianmin, G.; Gall, R.; Zuomin, W. Dynamic Damping and Stiffness Characteristics of the Rolling Tire. *Tire Sci. Technol.* **2001**, *29*, 258–268. [CrossRef]
66. Cutini, M.; Deboli, R.; Calvo, A.; Preti, C.; Brambilla, M.; Bisaglia, C. Ground Soil Input Characteristics Determining Agricultural Tractor Dynamics. *Appl. Eng. Agric.* **2017**, *33*. [CrossRef]
67. Gobbi, M.; Mastinu, G.; Pennati, M.; Previati, G. Farm tractor ride comfort assessment. In *The Dynamics of Vehicles on Roads and Tracks*; Resenberger, M., Plochi, M., Klaus, S., Edelmann, J., Eds.; Taylor & Francis Group: London, UK, 2016; pp. 125–136, ISBN 978-1-138-02885-2.
68. Cutini, M.; Costa, C.; Bisaglia, C. Development of a simplified method for evaluating agricultural tractor's operator whole body vibration. *J. Terramech.* **2016**, *63*, 23–32. [CrossRef]

agriculture

MDPI

Article

Monitoring and Precision Spraying for Orchid Plantation with Wireless WebCAMs

Grianggai Samseemoung [1,*], Peeyush Soni [2] and Chaiyan Sirikul [1]

[1] Department of Agricultural Engineering, Rajamangala University of Technology Thanyaburi,
    Pathumthani 12110, Thailand; chaiyan.wst@gmail.com
[2] Department of Food, Agriculture and Bioresources, Asian Institute of Technology (AIT), Pathumthani 12120,
    Thailand; soni.ait@gmail.com
*   Correspondence: grianggai.s@en.rmutt.ac.th; Tel.: +66-089-641-7532

Received: 7 September 2017; Accepted: 3 October 2017; Published: 11 October 2017

Abstract: Through processing images taken from wireless WebCAMs on the low altitude remote sensing (LARS) platform, this research monitored crop growth, pest, and disease information in a dendrobium orchid's plantation. Vegetetative indices were derived for distinguishing different stages of crop growth, and the infestation density of pests and diseases. Image data was processed through an algorithm created in MATLAB® (The MathWorks, Inc., Natick, MA, USA). Corresponding to the orchid's growth stage and its infestation density, varying levels of fertilizer and chemical injections were administered. The acquired LARS images from wireless WebCAMs were positioned using geo-referencing, and eventually processed to estimate vegetative-indices (Red = 650 nm and NIR = 800 nm band center). Good correlations and a clear cluster range were obtained in characteristic plots of the normalized difference vegetation index (NDVI) and the green normalized difference vegetation index (GNDVI) against chlorophyll content. The coefficient of determination, the chlorophyll content values (μmol m^{-2}) showed significant differences among clusters for healthy orchids (R^2 = 0.985–0.992), and for infested orchids (R^2 = 0.984–0.998). The WebCAM application, while being inexpensive, provided acceptable inputs for image processing. The LARS platform gave its best performance at an altitude of 1.2 m above canopy. The image processing software based on LARS images provided satisfactory results as compared with manual measurements.

Keywords: dendrobium orchids; pests and diseases infestation; image processing; NDVI; GNDV

1. Introduction

Dendrobium orchids are widely cultivated for both domestic and export markets. In 2015, Thailand exported around 51,811 tons of orchids to the USA, Japan and Italy [1], an increase of 5.05% from the previous year. Orchid exports are likely to grow steadily as demand continues to rise in the world market. To answer the simultaneously rising questions about orchid quality and the competitiveness of this growing market, the farmers are required to have technical expertise as well as skills in efficiently managing expensive inputs. Most growers plant orchids in containers with size 24 × 32 cm. Each container plants four orchids or the equivalent of about 75,000–94,000 plants/ha. Fertilization of orchids is done by spraying at the top of the leaves and at the roots, throughout the plant, except for the flowers. Orchids are fertilized differently at different stages of growth. At the nursery stage, fertilizer 21-21-21 should be interspersed with fertilizer 30-10-10 at the rate of 1.56–2.50 kg/200 L of water/ha every week. At the plantation stage, fertilizer 21-21-21 is interspersed with fertilizer 30-20-10 at the rate of 2.50–3.75 kg/200 L of water/ha every week. At the flowering stage, fertilizer 21-21-21 or 16-21-27 is interspersed with fertilizer 15-30-15 at the rate of 3.75–5 kg/200 L of water/ha every week. At the flower cutting stage, fertilizer 15-30-15 is interspersed with fertilizer 16-21-27 at the rate of 3.12–4.37 kg/200 L of water/ha every week [2].

The production of dendrobium orchids in greenhouses also risks disease and insect pest infestation. Most greenhouses are found to be infested with leaf spot disease or ringworm disease (Leaf Spot) caused by *Phyllostictina Pyriformis Cash and Watson*, black spot disease caused by *Alternaria alternate* and *Drechslera* spp., anthracnose disease caused by *Colletotrichum gloeosporioides (Penz.) Sacc.*, dry rot or wilt disease caused by *Fusarium oxysporum Fmoniliforme*, soft rot disease caused by *Erwinia carotovora (Jones)*, yellow leaf spots disease caused by *Pseudocercospora dendrobii* and dry rot disease caused by *Sclerotium rolfsii*. After correctly identifying the infestation, the correct amount of the chemical must be precisely sprayed at the designated areas to determine the application's efficiency as well as its production cost. Incorrect application of chemicals would not only result in economic losses, but also cause environmental damage [3,4].

An image processing technique for evaluating and recognizing crops and weeds was employed by the previous study [1,2,5]. They tended to segment crops and weeds from soil (background) in the first step. Segmentation was done by using visible color information or reflection intensity in near-infrared. Information on the variable light conditions should be taken into account to achieve good classification. In the second step, an attempt was made to classify plants as crops or weeds based on their shape, texture and color properties.

In this research, we combine variable rate spraying with the image processing technique using a wireless web camera (WebCAM) to assist in the management of dendrobium orchid plantation facilities. The process directly involves reducing production costs and increasing crop yield per plantation. The objective of this research was to design and fabricate a variable rate spraying application using the image processing techniques of wireless WebCAMs for monitoring crop growth, and infestation by pest and disease in dendrobium orchid plantations.

2. Materials and Methods

2.1. Experimental Set Up and Field Preparation

Experiments were conducted in a 72 m^2 plastic greenhouse (14.14° N, 101.48° E). There were 12 orchids m^{-2} inside the greenhouse (Figure 1). Spraying was controlled through a microcontroller by varying pumping pressure and its corresponding volume flow rate (Figure 2). Actual volumes applied were recorded.

Figure 1. Experimental set up and field preparation, (**a**) Greenhouse; (**b**) Orchid plants; (**c**) Minolta SPAD 502 Meter (Konica Minolta Sensing Inc., Osaka, Japan) measuring plant leaves; (**d**) Calibration of volume flow rate control; (**e**) SKR 1800 (Skye Instruments, Ltd., Powys, UK) illumination sensor.

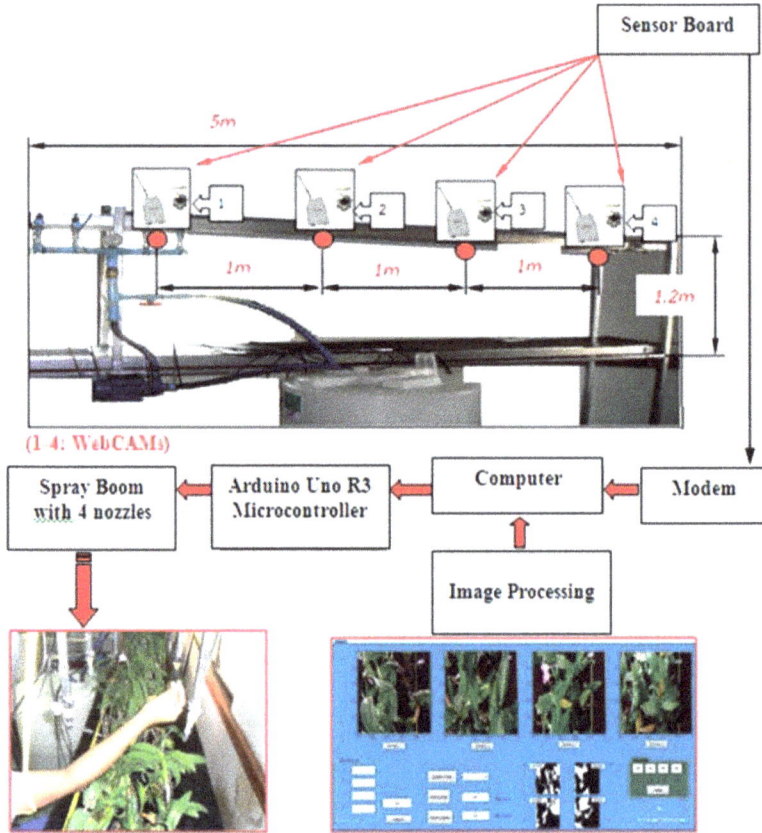

Figure 2. Variable rate spraying system with flow-based control.

2.2. Crop Growth, Pests and Diseases Infestation Monitoring and Spraying Systems

The variable rate fertilizer and chemical injection system precisely moved along the overhead rails above the orchid plantation in greenhouses that were controlled by four wireless WebCAMs, because the best resolution of each wireless WebCAMs was 1 m × 1 m with 1.2 m above the table. The rail bar system height was selected based on an appropriate coverage of the region of interest (ROI) on the table with plants. It was 2 m above the table (1 m × 5 m). The top view of the table inside the greenhouse is shown in Figure 3a. Image data was collected under cloud-free conditions between 10:30 and 12:30 h standard local time. Wireless WebCAMs (Vimicro USB2.0 UVC PC Camera, SWIFT-TECH ELECTRONICS Co. Ltd., Beijing, China) were coupled with a motor (Oriental AC Magnetic Motor 2RK6GN-AMw/2GN60K Gear Head, ORIENTAL MOTOR CO., LTD., Tokyo, Japan). A prototype of a precision sprayer was separately developed and calibrated for precise fertilizer and chemical injection. Its optimized parameters were the nozzle size and length of the boom arm, the pressure of a suitable pump kit, and an appropriate connection to the control unit for an exact injection quantity per cycle covering the ROI. Later, an algorithm to control volume per working cycle was created. This algorithm was designed on the basis of the plant-to-disease density ratio in terms of pixel areas (pixel by pixel). Finally, field tests were conducted with this automatic pesticide or chemical spraying system, and the prototype was improved until its performance was found to be satisfactory. The key variables that affect the operation of the spraying system were considered to be the relationship between the value of the disease density ratio of plants per area to injection quantity; and the relationship between the spraying frequency with the concentration of dose injected. The spraying system consisted of a series of solenoid valves (Green Water Atlantis Technology (Thailand) Co., Ltd., Nonthaburi, Thailand), an injector (Super products limited series), electric pumps (LEE SAE IMPORT (1975) LIMITED series), a microcontrollers board (Shenzhen Shanhai Technology Ltd., Shenzhen, China), and eight relay modules (Arduitronics CO., Ltd., Bangkok, Thailand) (Figure 3).

The variable rate precision sprayer's movement on rails above the plantation table in the greenhouse was controlled with wireless WebCAMs. This system was initiated based on digital image data of an orchid taken at nadir by four wireless WebCAMs installed 1 m apart. The acquired images were loaded into specially created image processing software that runs on MATLAB. The image processing software provides the pixel density expressed as the proportional growth rate of the orchids as output. This measured density was then transformed into the spraying duration of fertilizer and chemical at the upper surface of the orchid leaves. These sequences of activities were automatically controlled by the microcontroller (Arduino Uno R3; Shenzhen Shanhai Technology Ltd., Shenzhen, China) (Figure 4).

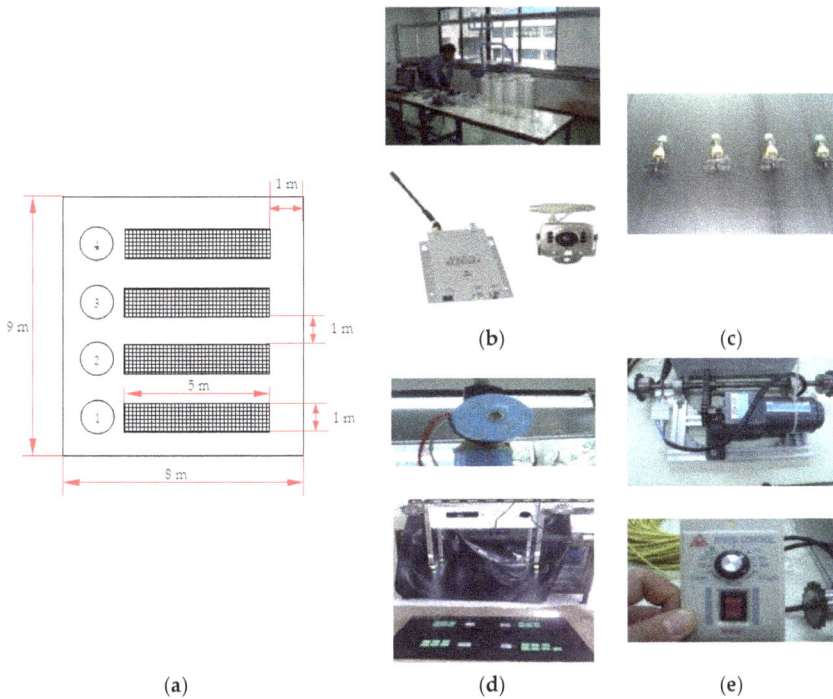

Figure 3. Specifications of the variable rate application spraying system: (**a**) Experimental layout of orchids inside the greenhouse; (**b**) A double row bar with a carrier controller and wireless WebCAM with a receiver; (**c**) Microcontroller with various sensors and nozzles; (**d**) Solenoid valve; (**e**) Gear motor and voltage adapter (220 VAC to 24 VDC, 5 A).

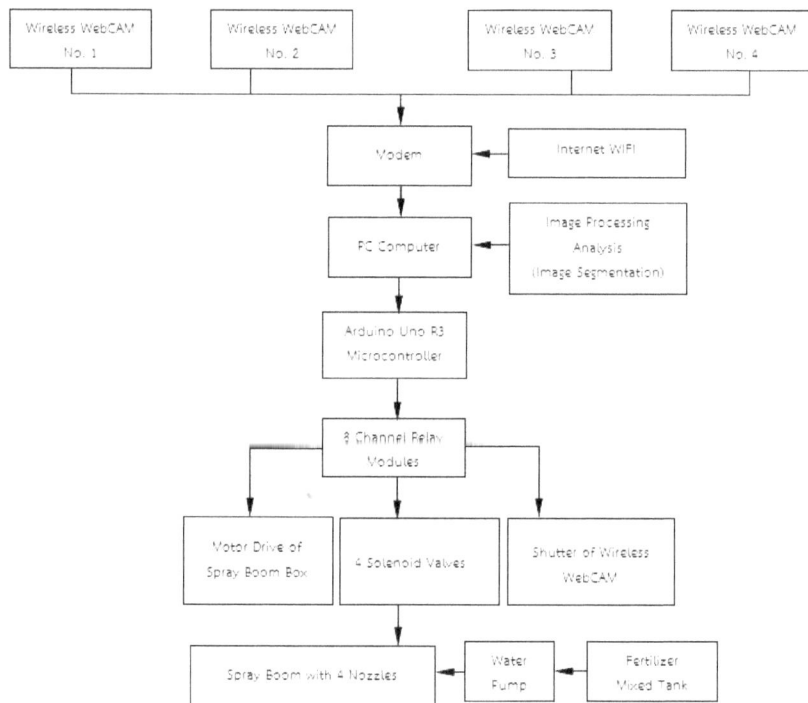

Figure 4. Variable rate spraying system with automatic direct fertilizer injection.

2.3. Low Altitude Image Data Acquisition and Processing

The developed algorithm, which separated the object from its background, was capable of system-specific command batch image processing as shown in Figure 5. An instruction set was developed to download image data from the four wireless WebCAMs that were installed 1 m apart on overhead rails above orchid plantation. A procedure was developed to determine the crop growth stage, divided into four divisions. The surface area of the green leaves was computed by the software. The pixel density of green leaves per unit of surface area was also displayed in real-time to the users. Color thresholding was done to control the time interval and the duration of the spraying, as well as the spraying target corresponding to the proportion of green pixels per area. Instructions were then sent to the solenoid valves to open/close the nozzles. The last image corresponding to the action taken was then saved to hard disk memory (gray and bimodal images) as shown in Figure 6.

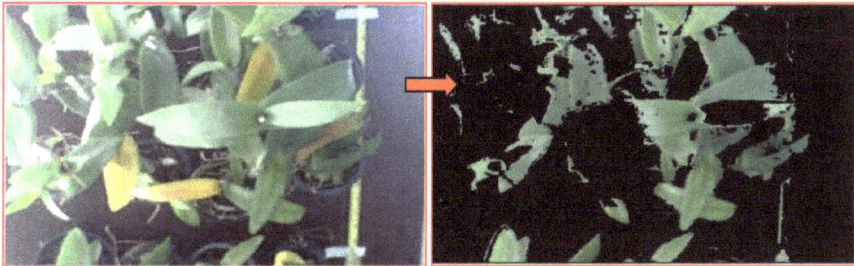

(a)

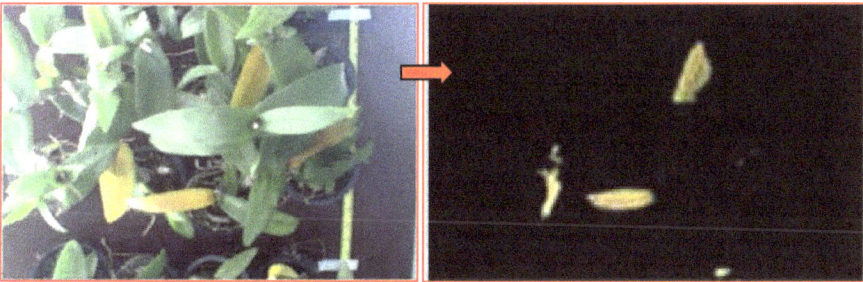

(b)

Figure 5. Image processing software with green color thresholding: (**a**) Plant growth stage determination by segmentation; (**b**) Infested area segmentation for target detection.

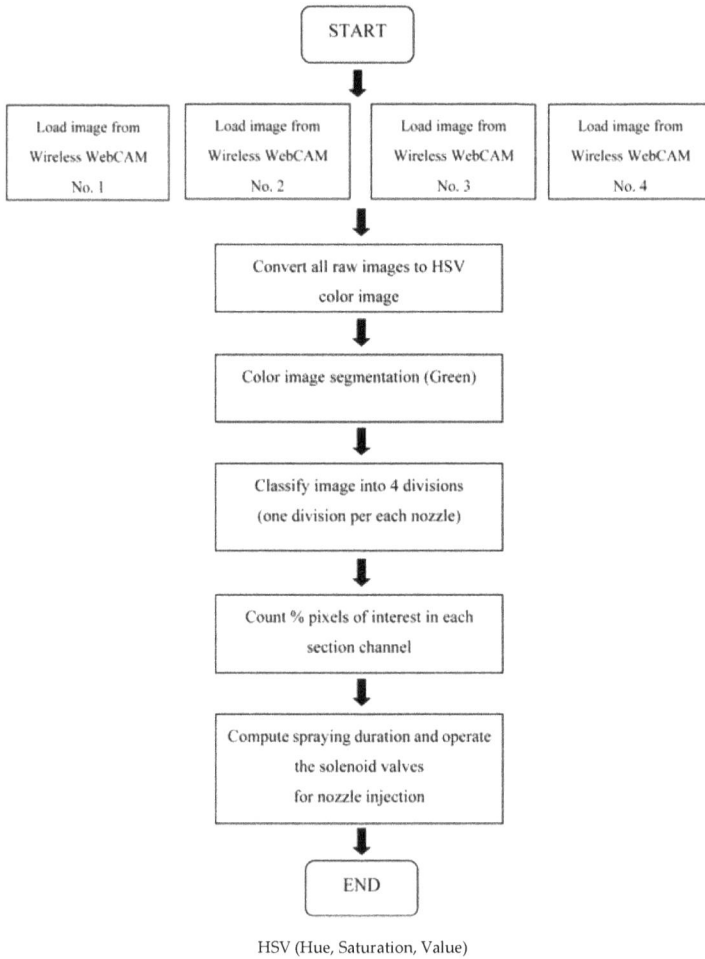

Figure 6. Functioning of the image processing software.

2.4. Image Data Calibration

For calibration, the accuracy of a particular set of commands to process the images was assessed using a standard reference frame of 65 cm × 50 cm. Different sheets were kept in the reference frame to represent the growth stage of plants, comprising a group of yellow and green cards of 2.54 cm × 2.54 cm dimensions. Trials were made with sheets of four sample color groups: 5% (24 pieces), 10% (48 pieces), 15% (72 pieces) and 20% (97 pieces) kept at different heights: 1 m, 1.2 m, 1.4 m and 1.6 m. This calibration also included variations in illumination levels with time of the day.

2.5. Statistical Analysis

All measurements were performed in triplicate. The experiment's data was analyzed by using SPSS 10.0 software (SPSS Inc., Singapore). The experiments were accomplished using a randomized complete block design (RCBD). Analysis of variance (ANOVA) was used to determine significance between treatments, and Duncan's Multiple Range test (DMRT) was used to compare means at a 95% confidence level.

2.6. Ground Truthing Measurements

Ground truthing measurements were done to understand the relationship between the stages of crop growth, infestation density of pest and disease, illumination levels and the chlorophyll content.

The SKR 1800 illumination sensor (Skye Instruments, Ltd., Powys, UK), measuring prevailing sunlight intensity, was attached to a data logger (SpectroSense-2; Skye Instruments, Powys, UK). During the experiment for calibration and ground trothing, a leaf chlorophyll meter (Minolta SPAD 502; Konica Minolta Sensing Inc., Osaka, Japan) (Table 1) was used to measure the average chlorophyll content (expressed as SPAD values) of an aged leaf, a young leaf, a young leaf infested with pests or diseases, and an aged leaf infested with pests or diseases. The units for the Minolta SPAD-502 meter can be used to express leaf chlorophyll by the following equation [4]:

$$\text{Chl } (\mu\text{mol m}^{-2}) = 10^{M^{0.265}} \tag{1}$$

where M is the leaf chlorophyll meter reading (digital number) and Chl is the chlorophyll content in μmol m^{-2}.

The green normalized differential vegetation index (GNDVI) values were estimated in this research to establish its association with different growth stages of dendrobium orchid plants. The GNDVI, based on the greenness level, representing the chlorophyll content as determined by the radiance at the leaf surface, is a significant indicator for distinguishing among young and aged, or healthy and infected orchids. The GNDVI is estimated [5] as follows:

$$\text{GNDVI} = \frac{\rho\text{NIR} - \rho\text{G}}{\rho\text{NIR} + \rho\text{G}} \tag{2}$$

where ρNIR is the reflectance value for the near infrared band and ρG is the reflectance value for the green band.

Table 1. Specifications * of the sensors used in this research.

Wireless WebCAMsVimicro USB2.0 UVC PC Camera (SWIFT-TECH ELECTRONICS Co. Ltd., Beijing, China)		Chlorophyll Meter (Minolta SPAD 502 Meter; Konica Minolta Sensing Inc., Osaka, Japan)		Illumination Sensor SKR-1800 (Spectrum Technology Inc., Powys, PA, USA)		Microcontroller Board Arduino Uno R3 (Shenzhen Shanhai Technology Ltd., Shenzhen, China)	
Feature	Value	Feature	Value	Feature	Value	Feature	Value
Image size resolution	2.0 to 6.0 Mega pixels	Type	Hand held meter	Range	Two channels each between 400–1050 nm	Type	Microcontroller ATmega328
Ground pixel size	10^X real-time digital zoom	Measuring sample	Crop leaves	Construction	Dupont "Delrin" acrylic	Operating Voltage	5 V
Spectral bands	RGB (Red, Green, Blue)	Measuring system	Optical density difference	Filters	Metal interference	Input Voltage (limits)	6–20 V
Lens type	VGA format frame rate up to 30 fps	Measuring area	2 mm × 3 mm	Detectors	GaP, GaAsP, or silicon	Digital I/O Pins	14
Mostly used	Hold true for laptop and desktop PC	Data memory	30 data points	Cable	Screened. 7-4-C military specification 3m standard length	Analog Input Pins	6
Triggering	Manual/Optional by software	Accuracy	+/− 1.0 SPAD unit reading	Temperature Range	−25 °C to +75 °C (for a fixed PVC cable)	DC Current per I/O Pins	40 mA

* As claimed by the respective manufacturer.

Tests were also conducted to observe the effect of the heights of the wireless WebCAMs on the proportion of pixel density that can be effectively detected. Furthermore, tests were conducted to determine the relationship between the height with the distribution of liquid fertilizer and chemical spraying.

For every altitude tested, the spraying quantities were varied at 20 cc, 60 cc and 100 cc. The variable rate sprayer was tested on the top of the orchid leaves surface with four levels of nozzle heights, i.e., 25 cm, 35 cm, 45 cm and 55 cm. The distribution of fertilizer and chemical spraying was observed for different combinations.

3. Results

3.1. Calibration of Variable Rate Spraying System

The image data acquisition and processing system using wireless WebCAMs was calibrated against the known values of color densities with green, yellow and brown colors, at different altitudes (1 m, 1.2 m, 1.4 m, 1.6 m). The color density of the sample sheets was varied at 5% (24 pieces), 10% (48 pieces), 15% (72 pieces) and 20% (97 pieces) for the three colors.

These color shades used for calibration are representative of the top surface area of orchids that grow in greenhouses (Figure 7). Table 2 shows the effect of height on image processing accuracy from wireless WebCAMs. An altitude of 1 m–1.2 m can be considered as having acceptably high accuracy in image processing data.

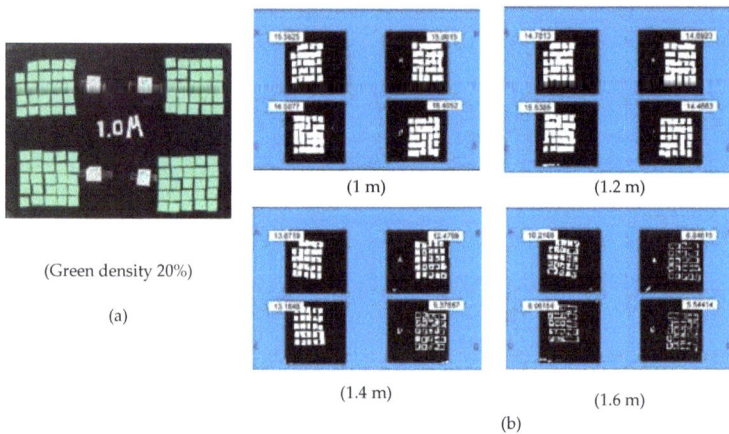

(Green density 20%)

(a)

(1 m)   (1.2 m)

(1.4 m)   (1.6 m)

(b)

Figure 7. (**a**) Image processing calibration from wireless WebCAMs; (**b**) green color density 20% image segmented at different heights.

Table 2. The effect of wireless WebCAMs altitude on the accuracy of image data acquisition with different colors and densities.

Simulated Stage of Crop Growth and Infestation	Altitude Levels (m)			
	1	1.2	1.4	1.6
1. Green color				
5%	4.340 [a]	3.995 [a]	3.756 [a]	3.410 [a]
10%	8.365 [d,e]	7.813 [c,d]	6.604 [b,c]	4.476 [a]
15%	12.400 [f]	11.284 [f]	9.515 [e]	5.721 [b]
20%	16.084 [g]	14.869 [g]	12.178 [f]	7.667 [c,d]
2. Yellow color				
5%	2.837 [a]	4.491 [b]	4.097 [b]	4.260 [b]
10%	7.816 [d]	7.970 [d]	6.724 [c]	6.750 [c]
15%	10.926 [f]	10.760 [f]	9.411 [e]	8.522 [d]
20%	14.048 [h]	12.457 [g]	11.947 [g]	10.884 [f]
3. Brown color				
5%	4.756 [b]	4.786 [b]	4.182 [b]	2.264 [a]
10%	8.241 [d,e]	9.244 [e,f]	7.402 [d]	5.228 [b,c]
15%	11.097 [g]	11.644 [g]	8.888 [e,f]	6.186 [c]
20%	14.068 [h]	14.809 [h]	11.901 [g]	9.432 [f]

Means in a column and row followed by the same letter within a color group are not significantly different at 0.05 significant levels according to Duncan's Multiple Range Test.

3.2. Estimation of Leaf Chlorophyll Content with Vegetation Index

The GNDVI value significantly correlated with the leaf chlorophyll; a higher GNDVI value corresponded to higher leaf chlorophyll. The coefficient of determination was (R^2) 0.985 for a healthy dendrobium orchid (Aged) and 0.870 for a healthy dendrobium orchid (Young) (Figure 8).

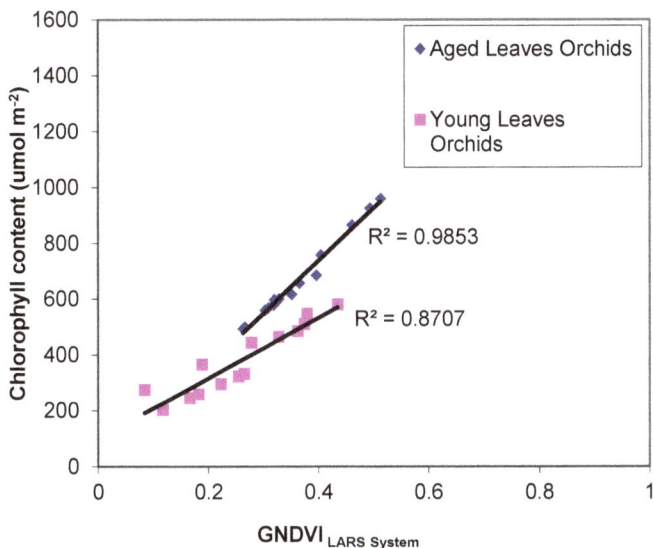

Figure 8. Leaf chlorophyll content estimation from leaf reflectance (green normalized differential vegetation index—GNDVI) for aged and young orchids (healthy).

The leaf chlorophyll values were in the range of 493–959 μmol m^{-2} for a healthy dendrobium orchid (Aged) and 275–580 μmol m^{-2} for a healthy dendrobium orchid (Young).

In relation to the proportion of color pixel density, the effect of the observation altitude was evaluated (Table 3). Results showed that 1 m–1.2 m altitude is suitable for acquiring images from WebCAMs with 5% and 10% density of orchids. Whereas, at 15% and 20% density of orchids, the height of 1.2 m is suitable.

Table 3. Effect of spraying height on spray application.

Volume Flow Rate, L/min	Spraying Height, m	Pesticide or Chemical Scatterable Distance, m
0.02 at pressure 1 bar	0.25	0.220 [a]
	0.35	0.235 [a]
	0.45	0.255 [b]
	0.55	0.280 [c]
0.06 at pressure 1 bar	0.25	0.325 [a]
	0.35	0.360 [b]
	0.45	0.375 [b]
	0.55	0.467 [c]
0.11 at pressure 1 bar	0.25	0.530 [a]
	0.35	0.635 [b]
	0.45	0.722 [c]
	0.55	0.817 [d]

Means in a column followed by the same letter within a flow rate group are not significantly different at 0.05 significant levels according to Duncan's Multiple Range Test.

The effect of spraying height on chemicals/pesticides at a scatterable distance of spraying was significant at 95% confidence level. The higher height obviously had a wider/scattered footprint of spraying. The best spraying accuracy of injection was at the height of 0.25 m, which corresponds to the prevention of fertilizer and the loss of chemicals/pesticides.

3.3. Crop Growth Status Monitoringand Application Map

Image data acquired from the wireless WebCAMs was processed to create an application map in GIS for variable rate spraying (Figure 9), in order to provide an inexpensive solution for making future plantation facility management decisions. The system provides near real-time output, thus enabling farmers to take quick actions before severe plant damage. The plot-based images acquired by four wireless WebCAMs were associated with the information from the global positioning system (GPS) receiver (24 point coordinates). In the mechanism of converting images into the geographic information system (GIS) application map, each frame of the image was combined into a small area by using location markers, i.e., coordinate (X_i, Y_i) for that area. ArcView® GIS program (Version 3.2a, Environmental Systems Research Institute, Inc., Redlands, CA, USA) was used to mask images and to trim excess areas beyond the ground margins. The altitude corrections for the images were also done beforehand. The true ground coordinates, used as the reference points, were further used to combine all the images into a matrix, as shown in Figure 9d. After all the reference images were pasted into the matrix, a combined image mosaic was obtained and then successfully converted into GIS application map layers.

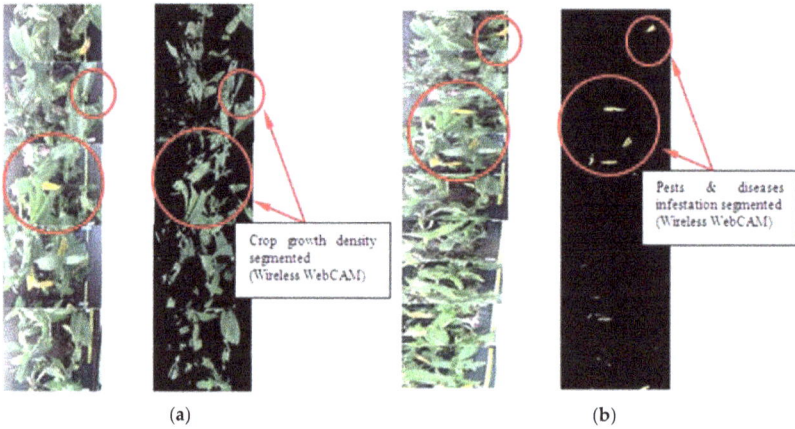

(a)   (b)

Figure 9. *Cont.*

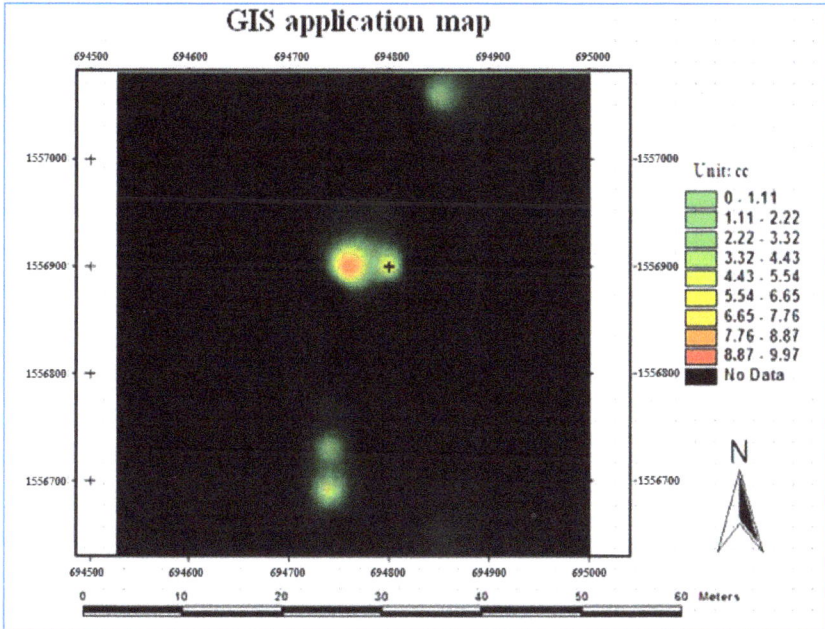

(c)

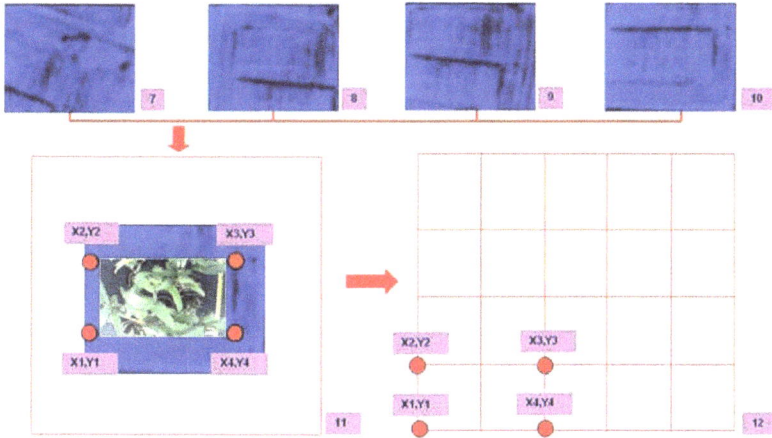

(d)

Figure 9. GIS application maps for crop status monitoring and spraying (**a**) The crop growth density segmented from wireless WebCAMs, (**b**) The infestation of pest and diseases segmented from wireless WebCAMs, (**c**) GIS application map (units are cc.), (**d**) The combination of plot-based images of wireless WebCAMs.

4. Discussion

The data presented in Tables 1–3 was derived from the results of the research to acquire and process image data using wireless WebCAMs based on laboratory and field tests [3,6–9]. These results were primarily linked to improved image processing accuracy (see Materials and methods section).

If the data is completed correctly, the developed system can be specified with an altitude of 1 m–1.2 m being considered as having acceptably high accuracy in image processing data [7], and for acquiring images from wireless WebCAMs with 5% and 10% density of orchids. A higher GNDVI value corresponded to a higher leaf chlorophyll [8], and the coefficient of determination was 0.985 (R^2) for a healthy dendrobium orchid (Aged) and 0.870 for a healthy dendrobium orchid (Young); the leaf chlorophyll values were in the range of 493–959 µmol m^{-2} for a healthy dendrobium orchid (Aged) and 275–580 µmol m^{-2} for a healthy dendrobium orchid (Young). The image data acquired from the wireless WebCAMs was processed to create an application map in GIS for variable rate spraying [1,2,4,5] in terms of the big data knowledge for agriculture.

Young healthy dendrobium orchids showed lower integrity with the R^2 value than aged healthy dendrobium orchids; although significant error could be due to smaller canopies resulting in an uncovered cultivated surface. The wireless WebCAMs images provide near-real-time and sufficiently precise results in order to develop vegetation indices and to discriminate between infected crops, regardless of their growth stage when compared to other techniques. This methodology would be useful to medium-to-large scale dendrobium orchid growers and it showed the potential scope for application to other crops [2].

5. Conclusions

The variable rate sprayer using wireless WebCAMs was developed and tested for the precise control of pesticides and chemicals in the greenhouse. The wireless WebCAMs have their best accuracy at a height of 1.2 m for digital image processing. The chlorophyll content values (µmol m^{-2}) according to *t*-test showed notable differences among clusters for healthy orchids (R^2 = 0.985–0.992) and for infested orchids (R^2 = 0.984–0.998). The image processing software based on LARS images provided satisfactory results compared with a manual measurement. The system quality was acceptable when established by the software and compared with the calibration data. The calibration data was generated at the different altitude levels of 1 m, 1.2 m, 1.4 m and 1.6 m, and the density of sample sheets of 5%, 10%, 15% and 20%, respectively. The accuracy of the image processing was found to be most effective at a height of 1.2 m.

This research could be used to form a database for the further adoption of technology with respect to the variable rate spraying application in dendrobium orchid plantations. Image processing techniques are used to increase the precision of controlling the rates of pesticide or chemicals in greenhouses.

Acknowledgments: The research team would like to thank the RMUTT annual government statement of expenditure in 2015 from Rajamangala University of Technology Thanyaburi, PathumThani, who supported the budget funds, equipment and personnel in place to prepare the test in this research; comments in the report are reflected in the research grants and Rajamangala University of Technology Thanyaburi is not always in agreement.

Author Contributions: Grianggai Samseemoung designed and conceived the experiment. Peeyush Soni carried out the field work. Chaiyan Sirikul analyzed the data. Grianggai Samseemoung, Peeyush Soni and Chaiyan Sirikul drafted the article thanks to additional funding from Rajamangala University of Technology. Thanyaburi was the project leader.

Conflicts of Interest: The authors declare no conflict of interest.

Abbreviations

Hectare	ha
Kilogram	kg
Kilowatt	kW
Kilowatt-hour	kWh
Liter	L
Meter	m
Revolutions per minute	rpm
Square	sq.
Volt	V
Watt	W

References

1. Samseemoung, G.; Soni, P.; Jayasuriya, H.P.W.; Salokhe, V.M. Oil palm pest infestation monitoring and evaluation by helicopter-mounted, low altitude remote sensing platform. *J. Appl. Remote Sens.* **2011**, *5*, 053540. [CrossRef]
2. Samseemoung, G.; Soni, P.; Jayasuriya, H.P.W.; Salokhe, V.M. Application of low altitude remote sensing (LARS) platform for monitoring crop growth and weed infestation in a soybean plantation. *Precis. Agric.* **2012**, *13*, 611–627. [CrossRef]
3. Putra, B.T.W.; Soni, P. Evaluating NIR-Red and NIR-Red edge external filters with digital cameras for assessing vegetation indices under different illumination. *Infrared Phys. Technol.* **2017**, *81*, 148–156. [CrossRef]
4. Markwell, J.; Osterman, J.; Mitchell, J. Calibration of the Minolta SPAD-502 leaf chlorophyll meter. *Photosynth* **1995**, *46*, 467–472. [CrossRef] [PubMed]
5. Stafford, J.V.; Benloch, J.V. Machine assisted detection of weeds and weed patches. In Proceedings of the First European Conference on Precision Agriculture; BIOS Scientific Publishers Limited: Oxford, UK, 1997; pp. 511–518.
6. Dasari, M.; Friedman, L.; Jesberger, J.; Stuve, T.A.; Finding, R.L.; Swales, T.P.; Schulz, S.C. A magnetic resonance imaging study of thalamic area in adolescent patients with either schizophrenia or bipolar disorder as compared to healthy controls. *Psychiatry Res. Neuronimaging* **1999**, *91*, 155–162. [CrossRef]
7. Putra, B.T.W.; Soni, P. Enhanced Broadband Greenness in Assessing Chlorophyll a and b, Carotenoid, and Nitrogen in Robusta Coffee Plantations using a Digital Camera. *Precis. Agric.* 2017. [CrossRef]
8. Chaisattapagon, C.; Zhang, N. Effective criteria for weed identification in wheat fields using machine vision. *Trans. ASAE* **1995**, *38*, 965–974. [CrossRef]
9. Franz, H.; Armanini, M.P. Characterization of a multi-component receptor for GDNF. *Nature* **1996**. [CrossRef]

Article

Analysis of Possible Noise Reduction Arrangements inside Olive Oil Mills: A Case Study

Simone Pascuzzi [†] and Francesco Santoro [*,†]

Department of Agricultural and Environmental Science (DiSAAT), University of Bari Aldo Moro,
Via Amendola 165/A—70126 Bari, Italy; simone.pascuzzi@uniba.it
* Correspondence: francesco.santoro@uniba.it; Tel.: +39-080-544-2474; Fax: +39-080-544-3080
† The Authors equally contributed to the present study.

Received: 9 August 2017; Accepted: 11 October 2017; Published: 16 October 2017

Abstract: Apulia (Southern Italy) is the leading Italian region for the production of olive oil (115×10^6 kg of oil/year), and the olive oil chain is really important from a business point of view. Currently the extraction of olive oil is essentially performed by using a mechanical pressing process (traditional olive oil mills), or by the centrifugation process (modern olive oil mills). The aim of this paper is to evaluate in detail the noise levels within a typical olive oil mill located in the northern part of the Apulia region during olive oil extraction. The feasibility of this study focusing on the assessment of workers' exposure to noise was tested in compliance with the Italian-European Regulations and US standards and criteria. Several measurements of the noise emission produced by each machine belonging to the productive cycle were carried out during olive oil production. The results obtained were then used to evaluate possible improvements to carry out in order to achieve better working conditions. An effective reduction in noise could probably be achieved through a combination of different solutions, which obviously have to be assessed not only from a technical point of view but also an economic one. A significant reduction in noise levels could be achieved by increasing the area of the room allotted to the olive oil extraction cycle by removing all the unnecessary partition walls that might be present.

Keywords: olive oil mills; noise pollution; noise reduction; workers' health and safety

1. Introduction

The remarkable development in agricultural mechanization in recent years has assured a significant increase in productivity. However, it has led to the requirement for higher attention to detail in all the areas concerning occupational health and safety [1–3], as well as in the exposure of workers' to potentially dangerous physical agents [4–7], activities [8,9] or environments [10]. Amongst these last hazards, of relevant importance is the exposure of workers to noise within plants during food processing [11–13], as is also the case inside olive oil mills during the olive oil extraction process, where noise levels are usually high and should be carefully measured.

Apulia (Southern Italy) is the leading Italian region for the production of olive oil, with a cultivated area of 378,770 hectares, from which about 7.5×10^8 kg of olives and 115×10^6 kg of olive oil is produced, accounting for 38.6% of the Italian total [14]. The extraction of olive oil is essentially currently performed by a mechanical pressing process inside traditional olive oil mills, which operate using a discontinuous working cycle, or by the centrifugation process inside modern mills with the use of a continuous cycle of extraction [15,16] (Figure 1). Both these methods require that the olives, washed and with the leaves removed, are previously crushed and reduced to a paste known as "oily juice", from which it is then easier to extract the olive oil. Mechanical crushers connected to mixers perform this step within a continuous working cycle, and the resultant "oily juice" is sent to the centrifuge (decanter), which separates it into its three components: pomace; vegetation water, and;

olive oil. Instead, inside traditional olive oil mills, stone wheels (roller crusher) crush the olives and the obtained paste is spread on synthetic fiber disks, known as pulp mats, which are stacked and brought to the mechanical press in order to separate the vegetation water and olive oil from the pomace. Finally, in both plant typologies, olive oil and vegetation water are sent to centrifugal separators to extract the olive oil [17].

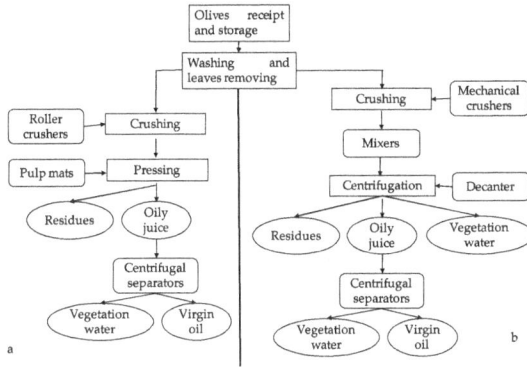

Figure 1. Schematic representation of olive oil extraction lines: (**a**) "discontinuous" (pressure extraction); (**b**) "continuous" cycle (extraction by centrifugation).

Olive crushing is a seasonal activity that normally begins at the end of October and finishes at the end of December. During this period, the work is carried out 24 h a day, seven days a week, and the workers are usually employed with both undefined and fixed-term contracts. The olive oil chain is very important in Apulia from a financial point of view and, even more so in this sector, one of the fundamental objectives of modern mechanized production is safety at work [18]. Regarding this aspect, the manual handling of loads, the use of electricity and noise exposure are the main kinds of risks occurring during working phases. Furthermore, the millers must take the necessary measures to reduce the risk linked to accidental contact with the rotating organs of the working machines, which are potentially accessible not only by the operators of the machine themselves but also by the olive producers who traditionally enter the operating environment of the olive oil mill in order to follow the processing of their own olives [17]. Many Apulian olive oil mills are still located in a single space created in a preexisting or ad hoc structure, even if the evolution of food safety standards requires a physical separation of the operations related to the olives from those related to processing them into olive oil. Therefore, inside the most developed Apulian olive oil mills the leaf removal and the washing phases are carried out in one room, whereas all the processing phases are done in another room [13].

The noise level is usually high during the activity inside the mills, and it is often the case that it is made worse for a wide range of reasons, such as inadequate plant design or improvident labor organization. Therefore, even if the problem of occupational noise exposure for olive oil mill workers is well known, scientific research could still be useful to propose suitable innovative technical solutions consistent with the financial aspects of this typology of factories. Different arrangements and technical solutions are currently used to reduce the noise inside the olive oil mill during the olive oil extraction campaign. These include a change to the layout of the machines, their replacement with newer ones which have lower acoustic impact, or specific actions such as the encapsulation of the noise source inside a box or the inclusion of sound shield or barriers [19]. It is possible to take action to change the characteristics of the acoustics in the working environment through the use of sound-absorbing materials or by means of sound-absorbing panels if the walls produce acoustic reflected waves which, added to the direct acoustic wave coming from the source, produce increased noise. Finally, a different organization of the work aimed at reducing the workers' noise exposure time is needed if other technical solutions are not good enough or, as a last line of defense, the use of personal protective equipment such as individual hearing protection devices (muffs, earplugs, headphones) should be considered in order to reduce workers' exposure to noise to below the limits [19].

The regulations in Europe and the US governing the levels of noise permitted in the workplace are someway different, but all of them define the limit for the maximum sound level for a period of eight hours of exposure ($L_{EX,8h}$ or TWA) between 85 and 90 dB(A), according to the guidelines of the International Standard ISO 1999:2013 [20]. In the US the standard OSHA (Occupational Safety and Health Administration) 1910.95 establishes values for noise exposure in the workplace and sets for anyone who operates eight hours per day the maximum permissible exposure limit (PEL) to 90 dB(A), with an exchange rate of 5 dB(A) where an increase of 5 dB(A) halves the permitted exposure time [21]. Conversely, the National Institute for Occupational Safety and Health (NIOSH) advises that the equivalent noise level at which a worker is exposed to should be limited to 85 dB(A) for an eight-hour day to minimize the risk of hearing damage [22]. In Europe, the risks linked to noise exposure at work are defined in EU Directive 2003/10/EC, which establishes the maximum limit as 87 dB(A) for an eight-hour day, even if France, Sweden, Norway and Spain allow 85 dB(A) with an exchange rate of 3 dB(A) [23].

The Italian Occupational Safety and Health legislation [24], in agreement with the EU Directive [18] and ISO standard [20], establishes that both the worker's exposure time and instantaneous peak exposure must be considered, defining both the peak sound pressure level ($L_{p,Cpeak}$), that is, the highest instantaneous sound pressure weighted through the "C" ponderation curve, and the daily A-weighted noise exposure level, $L_{EX,8h}$, that is, the average value, time-weighted, of all the noise levels at work concerning an eight-hour working day. The Italian law established exposure limit values that are: $L_{EX,8h}$ = 87 dB and $L_{p,Cpeak}$ = 140 dB or 200 Pa, respectively [24].

Taking in mind these considerations, the aim of this paper is to deeply evaluate the noise levels inside a typical olive oil mill located in the north of the Apulia region (in Italy) during the olive oil extraction process. This study focuses on the assessment of workers' exposure to noise and was tested in compliance with the Italian legislation, the EU directive and the US standards and criteria. Several measurements concerning the noise emission produced by each machine belonging to the production cycle were carried out during the olive oil production activity. The obtained data and their elaboration were then used to evaluate possible improvements to be made in order to create better working conditions.

2. Materials and Methods

2.1. Tested Olive Oil Mill

The olive oil mill under consideration, located in the Troia city area (Foggia district, Apulia region, Italy), had a so-called "mixed" or "combined" layout, commonly used in many other olive oil mills found in northern Apulia, as different production chains were integrated, and machines such as the roller crusher pertinent to the traditional discontinuous working cycle operated near to the decanter contrariwise in relation to the modern continuous working one (Figure 2).

In agreement with the extraction process, the obtained paste coming from the roller crushers is first sent to a little mixer, located below the roller crushers, then to a small finisher mechanical crusher and subsequently to the mixer. The paste is then transferred into a decanter and, finally, into the vertical axis centrifugal separators. The production cycle is also able to squeeze olive oil from stoned olives. In this case, the broken drupes coming from the stoner are transferred directly into the mixer and then into the decanter, which operates with the reduced speed of the screw conveyor.

With regards to the structural characteristics of the working environment, the leaf remover and its hopper are located in the olive storage room, built with a concrete floor, tough 4 m high walls and covered by a galvanized steel sheet (Figure 3). This space is separated from the olive oil extraction cycle area by a partition wall, the structural characteristics of which are the following (Figure 3): 4 m in height, anti-skid flooring, tough plastered walls, tiled up to 1.5 m above the ground, floor built with prestressed concrete joists and lug bricks, covered in concrete.

Figure 2. Machines that are pertinent to different working cycles side by side in the tested olive oil mill: the roller crusher (traditional discontinuous working cycle) and the decanter (modern continuous working cycle).

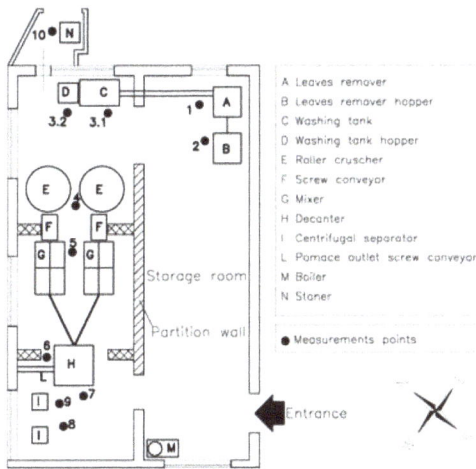

Figure 3. Layout of the tested olive oil mill with measurement points highlighted.

The stoner (Figure 3-N) is placed in a small outdoor room built with tough walls and a galvanized steel cover. Each machine is fastened to the pavement in different ways: (i) the stoner is bolted to the floor with the interposition of vibration reduction mounts; (ii) each roller crusher (Figure 3-E) is fastened to four concrete pillars, which raise the machine off the ground; (iii) the mixers are bolted directly to the floor; (iv) the decanter (Figure 3-H) is supported by a steel frame, which is in turn bolted to the floor with the interposition of vibration reduction mounts; (v) the centrifugal separators (Figure 3-I) are bolted to the floor with the interposition of vibration reduction mounts.

The olive oil mill is managed by a family-owned company and, during the olive oil extraction activities, the mill operates continuously over 24 h; in such conditions, the different workers follow the production, performing rotating shifts of 8 h each.

2.2. Noise Measurements

The noise measurements were carried out in agreement with the guidelines of the ISO standard in force [20,25]. Therefore, since the workers' tasks were limited and well defined, the noise exposure levels during the full activity of the olive oil mill under test were assessed by employing task-based measurements [20].

Taking into account the information gained from the supervisor, the work inside this olive oil mill was performed by all operators in the same way during each shift and so they were regarded as one homogeneous noise exposure group. Furthermore, the workers reported that each of them in rotation spent 1 h on job planning, briefing and breaks (quiet) with the remaining time equally spent close to the machines. Therefore the average nominal shift was distributed over the following $m = 8$ tasks: (1) job planning briefing and breaks (quiet); (2) leaf removal; (3) olive washing; (4) stone removal; (5) crushing; (6) malaxage; (7) settling; (8) centrifugation. An average duration $\overline{T_m} = 1$ h was considered for each task.

The noise contribution from work planning and breaks was of no importance to the overall noise exposure level. In fact, it was sufficient to carry out some simple noise measurements with the sound level meter, just to ensure that the sound pressure level during these working periods (tasks) had negligible influence. The average value of the actual measurements was 67.8 dB, and then an assessment for such periods was set at $L_{p,A,eqT,1} = 70$ dB.

Since the noise contribution from all the machines was highly affected by the location of the workers' ears, it was established to record the sound levels at the measuring points, placed near the following machines (Figure 3): leaf remover (1); leaf remover hopper (2); olive washer (3.1); hopper for the olive washer (3.2); roller crushers (4); mixer (5); decanter (6 and 7); centrifugal separators (8 and 9); stoner (10). The measurements were carried out at 1.5 m above the floor and at about 1 m away from the machines, as shown in Figure 3.

The A-weighted equivalent continuous sound pressure level, $L_{p,A,eqT}$ was calculated by the following equation [20]:

$$L_{p,A,eqT} = 10 lg \left[\frac{\frac{1}{T} \int_{t_1}^{t_2} p_A^2(t) dt}{p_0^2} \right] \text{dB} \tag{1}$$

where p_a is the A-weighted sound pressure during the stated time interval T, starting at t_1 and ending at t_2; p_0 is the reference pressure value (20 µPa).

On the other hand, the C-weighted peak sound pressure level, $L_{p,Cpeak}$, was calculated by the following equation [20]:

$$L_{p,Cpeak} = 10 lg \frac{p_{Cpeak}^2}{p_0^2} \text{dB} \tag{2}$$

where p_0 is the reference pressure value (20 µPa).

The measurements of $L_{p,A,eqT}$ and $L_{p,Cpeak}$ were carried out using a precision sound level meter ACOEM 01dB brand dB4 model, which complies to characteristics imposed by the standards [26,27]: Class 1 sound level meter; Class 0 octave-band and third-octave-band filter; the A weighting scale and slow response (1 s). The instrument calibration was performed before and after each measurement cycle by means of a calibrator compliant with the standard [28,29]. The sound level meter was connected to a G.R.A.S. 46 AC LEMO free-field microphone, having frequency range in the range of 3.15 Hz–40 kHz (±2 dB) and sensitivity of 12.5 mV/Pa at 250 Hz (±1 dB).

Three measurement periods were considered at each of the measuring points and these observations pointed out that the stated time interval T (measurement duration) for noise from all the machines had to be at least equal to 4 min. Therefore, in agreement with the standard, the measurement duration for all the measuring points was set to 5 min [20]. Furthermore, since the noise from quiet activities was negligible, only some brief samples of noise level were executed during these tasks.

The noise level for each of the eight considered tasks from $l = 3$ separate measurements, $L_{p,A,eqT,m}$ has been assessed through the following equation [20]:

$$L_{p,A,eqT,m} = 10 \cdot \ln\left(\frac{1}{l} \cdot \sum_{1}^{l} 10^{0.1 \cdot L_{p,AeqT,mi}}\right) \text{ dB} \tag{3}$$

where $L_{p,A,eqT,mi}$ is the A-weighted equivalent continuous sound pressure level during a task duration T_m; i the number of task samples m and l the total number of task samples m.

The noise level at the "leaf removal" task was calculated considering both the measurements recorded at the measurement points 1 and 2 (Figure 3); in the same way, the noise level at the "olive washing" task was evaluated considering both the measurements recorded at measurement points 3.1 and 3.2, the noise level at the "decanter" task considering both the measurements recorded at measurement points 6 and 7 and, finally, the noise level at the "centrifugal separation" task, considering both the measurements recorded at measurement points 8 and 9 (Figure 3). Practically, $l = 6$ was considered for the calculations concerning the "olive washing", "decanter" and "centrifugal separation" tasks.

The contribution from each of the m considered tasks to the daily A-weighted noise exposure level, $L_{EX,8h,m}$, was assessed through the following equation [20]:

$$L_{EX,8h,m} = L_{p,A,\,eqT,m} + 10lg\left(\frac{\overline{T_m}}{T_0}\right) \text{ dB} \tag{4}$$

where $\overline{T_m}$ is the average duration of the task, that is 1 h, and T_0 the duration of the nominal shift ($T_0 = 8$ h).

Finally, the evaluation of the A-weighted noise exposure level, $L_{EX,8h}$, from the noise contribution of each of the tasks was executed through the following equation [20]:

$$L_{EX,8h} = 10lg\left(\sum_{m=1}^{M=8} 10^{0.1 \times L_{EX,8h,m}}\right) \text{ dB} \tag{5}$$

where m is the task number and M the total considered number of tasks contributing to the daily noise exposure level ($M = 8$).

In agreement with the US standard and criteria, the noise levels $L_{p,A,eqT,m}$, calculated through Equation (3)for each of the eight considered tasks, were used to compute the total noise dose D over the working day, which is the amount of actual exposure relative to the allowable exposure. D equal to 100% and above represents exposures that are hazardous. According to both the OSHA standard and the NIOSH recommendations [21,22], D was evaluated through the following equation:

$$D = 100 \cdot \left(\frac{C_1}{T_{RD1}} + \frac{C_2}{T_{RD2}} + \ldots + \frac{C_8}{T_{RD8}}\right) \tag{6}$$

where C_m indicates the total time of exposure at the specific noise level m, and T_{RDm} indicates the reference duration for that level.

In agreement with OSHA Regulations 1910.95, which consider 90 dB(A) to be the maximum allowable exposure limit with an exchange rate of 5 dB(A), the reference duration $T_{RDm(OSHA)}$ (h) was assessed through the following equation [21]:

$$T_{RDm(OSHA)} = \frac{8}{2^{(L_{p,A,eqT,m}-90)/5}} \tag{7}$$

The eight-hour time-weighted average noise level (*TWA*) is then computed from the daily dose D, by means of the following formula [21]:

$$TWA_{(OSHA)} = 16.61lg\left(\frac{D}{100}\right) + 90 \text{ dB} \tag{8}$$

Conversely, according to the NIOSH recommendations, which consider 85 dB(A) the maximum permissible exposure limit with an exchange rate of 3 dB(A), the reference duration $T_{RDm(NIOSH)}$ (h) was assessed through the following equation [22]:

$$T_{RDm(NIOSH)} = \frac{8}{2^{(L_{p,A,eqT,m}-85)/3}} \tag{9}$$

In this case, the *TWA* was evaluated by means of the following formula [22]:

$$TWA_{(NIOSH)} = 10.0\lg\left(\frac{D}{100}\right) + 85 \text{ dB} \tag{10}$$

3. Results and Discussion

Figure 4 shows the A-weighted equivalent continuous sound pressure levels $L_{p,A,eqT,m}$ and the C-weighted peak sound pressure levels $L_{p,Cpeak}$ concerning the measurements carried out at each considered measured point. The range of the three measured values of $L_{p,A,eqT,m}$ never exceeded 3 dB, and then no additional measurements were made [19]. Conversely, the graph highlights that $L_{p,Cpeak}$ overcame the limit of 140 dB at no measuring point, and the lowest peak (90.1 dB) took place near the washing tank, whereas the highest one (102.3 dB) is near the mixer.

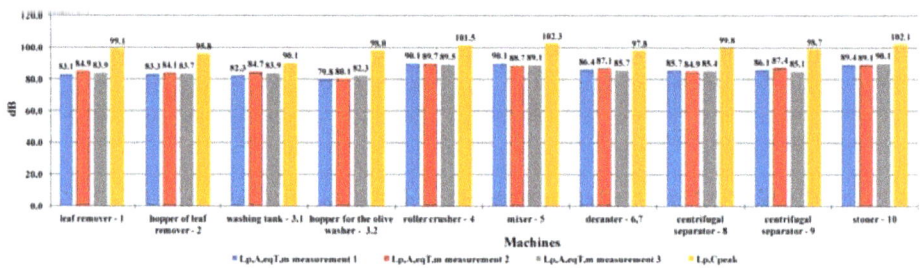

Figure 4. A-weighted equivalent continuous sound pressure levels ($L_{p,A,eqT,m}$) and C-weighted peak sound pressure levels ($L_{p,Cpeak}$) registered at the considered measuring point.

The noise level for each of the eight considered tasks $L_{p,A,eqT,m}$ are reported in Figure 5. They were calculated using Equation (3), taking into account the data pointed out in Figure 4. The chart highlights that the noisiest machines, that is, the ones with $L_{p,A,eqT,m}$ higher than 87 dB, which were the roller crushers, the mixer and the stoner, even if the last one is located in a separate room and isolated from the other machines directly involved in the production cycle (Figure 3). The results highlight the fact that workers compelled to always operate near these machines would be subjected to a daily noise exposure greater than the limit; luckily, the operation connected to these machines does not require the constant presence of employees.

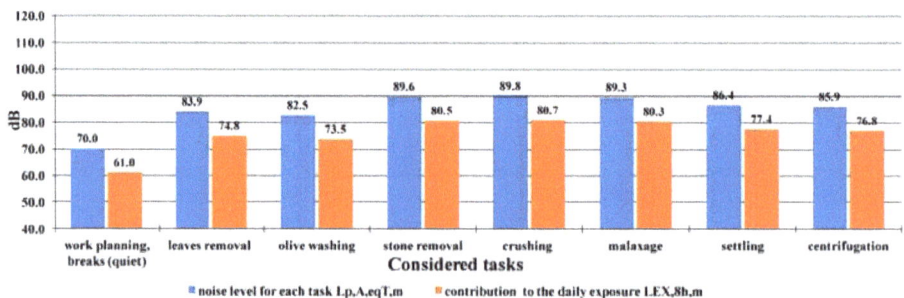

Figure 5. Noise level for $L_{p,A,eqT,m}$ and contribution to the daily A-weighted noise exposure level $L_{EX,8h,m}$ concerning the considered tasks.

Figure 5 also reports the contribution to the daily A-weighted noise exposure level, $L_{EX,8h,m}$, calculated for each activity according to Equation (4). As previously stated, the average task duration of all the considered activity was the same (1 h), and so the bar chart concerning the $L_{EX,8h,m}$ has the same shape and is only proportionally reduced with reference to the one that is relevant to $L_{p,A,eqT,m}$. In addition, this graph points out that none of the activities contribute to overcoming the limit of 87 dB.

The reference durations T_{RDm} (h), calculated according to both the OSHA Regulations 1910.95 and the NIOSH recommendations, respectively, through Equations (7) and (9), are reported in Figure 6. The chart points out the significant difference between the two approaches, mainly with reference to the noisiest machines. For example, the NIOSH recommends staying no closer to the roller crusher than T_{RDm} = 2.7 h, whereas by the OSHA standard 8.3 h are permitted (+207%), so for the stoner where T_{RDm} is 2.8 h by NIOSH, and T_{RDm} is 8.5 h by OSHA (+204%). The average task duration of all the considered activity ($\overline{T_m}$ =1 h) was in any case less than the computed values reported in Figure 6.

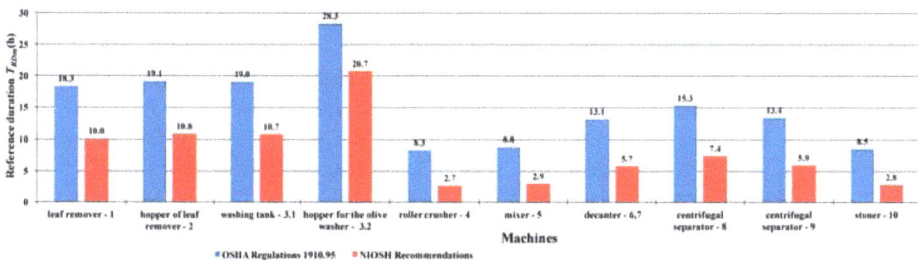

Figure 6. Reference durations T_{RDm} (h), calculated according to both the OSHA Regulations 1910.95 and the National Institute for Occupational Safety and Health (NIOSH) recommendations.

Table 1 reports the calculated values concerning the total daily dose D and the eight-hour time-weighted average noise level *TWA* with the corresponding limits according to both European and US standards. The dose D, not considered by the ISO standard, is very different if calculated according to OSHA 1910.95 or NIOSH criteria and the computed value is lower than the admissible one if evaluated through the OSHA standard. Conversely, D is much greater than the allowed one in compliance with the NIOSH criteria; moreover, the corresponding TWA, assessed through Equation (10), is obviously higher than the threshold value (Table 1). For the operators whose noise exposures equal or exceed 85 dB(A), NIOSH advises a hearing loss articulated prevention program, which contains exposure evaluation, engineering and administrative controls, suitable employment of hearing protectors, audiometric evaluation, education and motivation and recordkeeping [21].

Table 1. Average duration of the m considered tasks during the nominal shift. Time-weighted average noise level (TWA).

	Daily Noise Dose $D\%$	8 h Work Shift Noise Exposure Level $L_{EX,8h}$—TWA dB(A)	
		Computed	Limit
OSHA 1910.95	76.4	88.1	90
NIOSH criteria	188.9	87.8	85
Italian Regulations	/	86.9	87

The daily A-weighted noise exposure level, calculated through Equation (5) in agreement with the EU Directive $L_{EX,8h,m}$ = 86.9 dB, is lower than the corresponding limit, even if this value is higher than the exposure action value, which is 85 dB, so actions aimed at reducing the sound level and protecting workers' well-being have to be undertaken.

Regardless of the values pertinent to the different technical or law-making approaches, the results obtained highlight the considerable level of noise in the places where operations to process olives into olive oil are carried out, thereby emphasizing the necessity to analyze the feasibility of every possible technical solution aimed at minimizing the risk of damaging hearing and other possible types of harm such as cardiovascular diseases, fatigue, inability to concentrate and reduced motivation [30–32], and diastolic blood pressure increase [33–35].

According to the structural characteristics of the tested olive mill, together with the results of the noise level measurements, it is possible to propose some feasible arrangements aimed at reducing the noise emissions during the olive oil extraction cycle. It is useful to point out that an effective improved result in terms of noise reduction could probably be achieved through a combination of different solutions, which obviously have to be assessed not only from a technical point of view but also from an economic one. Inside the tested olive oil mill, the whole extraction cycle essentially takes places in one room, which is too small considering the number and dimensions of the machines used. There is very little free space around the machines so, in addition to hampering workers' movements, there is a higher noise level in the environment due to the occurrence of acoustic reflections. Furthermore, the structural characteristics and the dimension of the olive oil mill, together with the layout of the machines, do not allow their encapsulation, first because this solution would further decrease workers' available space. Finally, the noise level is also due to the overlap of the direct sound fields generated by machines with the indirect ones caused by the multiple reflections on the walls, so the use of noise barriers, even if placed close to the sources, may not be useful in order to obtain the expected result of reducing noise levels. The key solution in order to increase the workers' safety and their operative conditions is to eliminate the partition wall that currently separates the storage room from the room used for the olive oil extraction cycle itself (Figure 3). This new structural configuration will leave more space available to accommodate machines that will then be arranged suitably from one another. Further reductions in sound emissions can be achieved by encapsulating the noisiest machines such as the roller crusher, the mixer and the stoner. The acoustical features of the new working environment can also be varied by reducing the reflected noise waves emitted by the machines and walls by way of sound-absorbing materials or sound-absorbing panels. Personal protective equipment ultimately could be taken into account in the event that after the executed changes, noise issues still bother the workers. In any case, the plant design solutions in conjunction with suitable job organization will provide functional opportunities for operators to do their work effectively and efficiently without undue distraction or threat.

4. Conclusions

The noise levels inside a typical olive oil mill located in Southern Italy were analyzed during the operations for processing olives into olive oil. The workers' exposure to noise was assessed in compliance with the Italian-European and US Regulations. The obtained results reveal a high level of noisiness inside the working environment, and the sound measured values were very close to the corresponding threshold values covered by both the Italian-European Regulations and the US Standard, whereas they were well over the limits recommended by the US National Institute for Occupational Safety and Health (NIOSH). The executed measurements also highlight the necessity to mitigate the noise levels within the olive oil mills through solutions, doubtless cumbersome and involving technical, economic and organizational aspects. In reality, the achievement of the target could be reached by way of the synergic effect of more combined actions and consequently an integrated analysis of the problem is required. Referring to the specific tested olive oil mill, the available space where the olive oil extraction cycle takes place is not actually suitable to host all the machines used and, in addition to hampering workers' movements, widespread noise due to multiple reflections along the walls occurs. A significant reduction in the noise levels could be achieved by increasing the size of the room used for the olive oil extraction cycle by removing the partition wall that separates this room from the one next to it, which is devoted to the stockpile of the olives. It is useful to note that every technical analysis of the problem should not be kept apart from an accurate economic assessment, since the seasonal nature of work inside the olive oil mills, and the small earnings due to the high competitiveness, could be the reason for the inability to put into place any of the technical evaluated measures. These are measures which, although effective, could be too expensive from a financial point of view, even though it should be considered that the economic costs involved in the improvement of workers' safety and operative conditions are doubtless smaller than those (social as well as economic) linked to accidents at work: investing in prevention is always a good strategy.

Author Contributions: Both Authors conceived, designed and performed the experiments, did data collection, analysis and interpretation, wrote, revised and proofread the paper equally.

Conflicts of Interest: The authors declare no conflict of interest.

Abbreviations

TWA	Time Weighted Average *(Noise Levels)*
PEL	Permissible Exposure Limit
$L_{p,A,eqT}$	A-weighted equivalent continuous sound pressure level
$L_{p,A,eqT,m}$	A-weighted equivalent continuous sound pressure level during a task
$L_{p,Cpeak}$	Peak sound pressure level
$L_{EX,8h}$	Daily A-weighted noise exposure level
$L_{EX,8h,m}$	Daily A-weighted noise exposure level during a task
D	Total daily noise dose
$T_{RDm(OSHA)}$	Reference duration time according OSHA
$TWA_{(OSHA)}$	Time Weighted Average *(Noise Levels)* according OSHA
$T_{RDm(NIOSH)}$	Reference duration time according NIOSH
$TWA_{(NIOSH)}$	Time Weighted Average *(Noise Levels)* according NIOSH

References

1. Pascuzzi, S. A multibody approach applied to the study of driver injuries due to a narrow-track wheeled tractor rollover. *J. Agric. Eng.* **2015**, *46*, 105–114. [CrossRef]
2. Pascuzzi, S. The effects of the forward speed and air volume of an air-assisted sprayer on spray deposition in "tendone" trained vineyards. *J. Agric. Eng.* **2013**, *3*, 125–132. [CrossRef]
3. Baldoin, C.; Balsari, P.; Cerruto, E.; Pascuzzi, S.; Raffaelli, M. Improvement in pesticide application on greenhouse crops: Results of a survey about greenhouse structures in Italy. *Acta Hortic.* **2008**, *801*, 609–614. [CrossRef]
4. Pascuzzi, S.; Santoro, F. Evaluation of farmers' OSH hazards in operation nearby mobile telephone radio base stations. In Proceedings of the 16th International Scientific Conference on Engineering for rural development, Jelgava, Latvia, 24–26 May 2017; Latvia University of Agriculture-Faculty of Engineering: Jelgava, Latvia, 2017; pp. 748–755. [CrossRef]
5. Pascuzzi, S.; Santoro, F. Exposure of farm workers to electromagnetic radiation from cellular network radio base stations situated on rural agricultural land. *Int. J. Occup. Saf. Ergon.* **2015**, *21*, 351–358. [CrossRef] [PubMed]
6. Pascuzzi, S.; Blanco, I.; Anifantis, A.S.; Scarascia Mugnozza, G. Hazard assessment and technical actions due to the production of pressured hydrogen within a pilot photovoltaic-electrolyzer-fuel cell power system for agricultural equipment. *J. Agric. Eng.* **2016**, *47*, 89–93. [CrossRef]
7. Marucci, A.; Pagniello, B.; Monarca, D.; Colantoni, A.; Biondi, P. Heat stress suffered by workers employed in vegetable grafting in greenhouses. *J. Food Agric. Environ.* **2012**, *10*, 1117–1121.
8. Boubaker, K.; Colantoni, A.; Allegrini, E.; Longo, L.; Di Giacinto, S.; Monarca, D.; Cecchini, M. A model for musculoskeletal disorder-related fatigue in upper limb manipulation during industrial vegetables sorting. *Int. J. Ind. Ergon.* **2014**, *44*, 601–605. [CrossRef]
9. Colantoni, A.; Marucci, A.; Monarca, D.; Pagniello, B.; Cecchini, M.; Bedini, R. The risk of musculoskeletal disorders due to repetitive movements of upper limbs for workers employed to vegetable grafting. *J. Food Agric. Environ.* **2012**, *103*, 14–18.

10. Di Giacinto, S.; Colantoni, A.; Cecchini, M.; Monarca, D.; Moscetti, R.; Massantini, R. Dairy production in restricted environment and safety for the workers. *Ind. Aliment.* **2012**, *530*, 5–12.

11. Manetto, G.; Cerruto, E.; Pascuzzi, S.; Santoro, F. Improvements in citrus packing lines to reduce the mechanical damage to fruit. *Chem. Eng. Trans.* **2017**, *58*, 391–396. [CrossRef]

12. Bianchi, B.; Tamborrino, A.; Santoro, F. Assessment of the energy and separation efficiency of the decanter centrifuge with regulation capability of oil water ring in the industrial process line using a continuous method. *J. Agric. Eng.* **2013**, *44*, 278–282. [CrossRef]

13. Cecchini, M.; Contini, M.; Massantini, R.; Monarca, D.; Moscetti, R. Effects of controlled atmospheres and low temperature on storability of chestnuts manually and mechanically harvested. *Postharvest Biol. Technol.* **2011**, *61*, 131–136. [CrossRef]

14. Italian National Institute of Statistics (ISTAT). Area (Hectares) and Production (Quintals) of Olives 2016. Available online: http://agri.istat.it/sag_is_pdwout/jsp/NewDownload.jsp?id=15A\T1\textbar{}21A\T1\textbar{}32A&anid=2016 (accessed on 5 July 2017).

15. Leone, A.; Romaniello, R.; Tamborrino, A.; Xu, X.Q.; Juliano, P. Microwave and megasonics combined technology for a continuous olive oil process with enhanced extractability. *Innov. Food Sci. Emerg. Technol.* **2017**, *42*, 56–63. [CrossRef]

16. Vivaldi, G.A.; Strippoli, G.; Pascuzzi, S.; Stellacci, A.M.; Camposeo, S. Olive genotypes cultivated in an adult high-density orchard respond differently to canopy restraining by mechanical and manual pruning. *Sci. Hortic.* **2015**, *192*, 391–399. [CrossRef]

17. Clodoveo, M.L.; Camposeo, S.; de Gennaro, B.; Pascuzzi, S.; Roselli, L. In the ancient world virgin olive oil has been called "liquid gold" by Homer and the "great healer" by Hippocrates. Why is this mythic image forgotten? *Food Res. Int.* **2014**, *62*, 1062–1068. [CrossRef]

18. Manetto, G.; Cerruto, E. Vibration risk evaluation in hand-held harvesters for olives. *J. Agric. Eng.* **2013**, *44*, 705–709. [CrossRef]

19. Cirillo, E. *Applied Acoustics*; McGraw-Hill: Trento, Italy, 1997; 216p. (In Italian)

20. International Organization for Standardization. *Acoustics—Estimation of Noise-Induced Hearing Loss*; ISO 1999:2013; International Organization for Standardization: Geneva, Switzerland, 2013.

21. U.S. Occupational Safety and Health Administration. OSHA 1910.95. Available online: https://www.osha.gov/pls/oshaweb/owadisp.show_document?p_table=standards&p_id=9735 (accessed on 22 September 2017).

22. U.S. Department of Health and Human Services. Department of Health and Human Services. Revised Criteria 1998. Available online: https://www.cdc.gov/niosh/docs/98-126/pdfs/98-126.pdf (accessed on 23 August 2017).

23. EU. *Directive 2003/10/EC on the Minimum Health and Safety Requirements Regarding the Exposure of Workers to the Risks Arising from Physical Agents (Noise)*; European Parliament and the Council: Brussels, Belgium, 2003.

24. Official Gazette of the Italian Republic. *Safety and Health in Workplaces Act of 2008*; Italian Law Decree No. 81 (9 April 2008); Official Gazette of the Italian Republic: Rome, Italy, 2008. (In Italian)

25. International Organization for Standardization (ISO). *Acoustics—Determination of Occupational Noise Exposure—Engineering Method*; ISO 9612:2009; International Organization for Standardization: Geneva, Switzerland, 2009.

26. UNI. *Acoustics—Determination of Occupational Noise Exposure*; UNI 9432:2011; Italian Organization for Standardization: Milan, Italy, 2011. (In Italian)

27. International Electrotechnical Commission (IEC). *Electroacoustic—Sound Level Meters—Part 1: Specification*; IEC 61672-1:2013; International Electrotechnical Commission: Geneva, Switzerland, 2013.

28. International Electrotechnical Commission (IEC). *Electroacoustic—Octave-Band and Fractional-Octave-Band Filters—Part 1: Specification*; IEC 61260-1:2014; International Electrotechnical Commission: Geneva, Switzerland, 2014.

29. International Electrotechnical Commission (IEC). *Electroacoustic—Sound Calibration*; IEC 60942:2003; International Electrotechnical Commission: Geneva, Switzerland, 2003.

30. Lercher, P.; Hortnagl, J.; Kofler, W. Work Noise Annoyance and Blood Pressure: Combined Effects with Stressful Working Conditions. *Int. Arch. Occup. Environ. Health* **1993**, *65*, 23–28. [CrossRef] [PubMed]

31. Loewen, L.; Suedfeld, P. Cognitive and arousal effects of masking office noise. *Environ. Behav.* **1992**, *24*, 381–395. [CrossRef]

32. Evans, G.W. Environmental stress and health. In *Handbook of Health Psychology*; Baum, A., Revenson, T., Singer, J., Eds.; Wiley: New York, NY, USA, 2001; Volume 1, pp. 571–610.

33. Melamed, S.; Fried, Y.; Froom, P. The interactive effect of chronic exposure to noise and job. *J. Occup. Health Psychol.* **2001**, *6*, 182–195. [CrossRef] [PubMed]

34. Van Kempen, E.E.; Kruize, H.; Boshuizen, H.C.; Ameling, C.B.; Staatsen, B.A.M.; de Hollander, A.E.M. The association between noise exposure and blood pressure and ischemic heart disease: A meta-analysis. *Environ. Health Perspect.* **2002**, *110*, 307–317. [CrossRef] [PubMed]

35. Willich, S.N.; Wegscheider, K.; Stallmann, M.; Keil, T. Noise burden and the risk of myocardial infarction. *Eur. Heart J.* **2006**, *27*, 276–282. [CrossRef] [PubMed]

Article

Development of a Variable Rate Chemical Sprayer for Monitoring Diseases and Pests Infestation in Coconut Plantations

Grianggai Samseemoung [1,*], Peeyush Soni [2] and Pimsiri Suwan [1]

[1] Department of Agricultural Engineering, Rajamangala University of Technology Thanyaburi, Pathumthani 12110, Thailand; pimsiree_s@exchange.rmutt.ac.th
[2] Department of Food, Agriculture and Bioresources, Asian Institute of Technology (AIT), Pathumthani 12120, Thailand; soni.ait@gmail.com
* Correspondence: grianggai.s@en.rmutt.ac.th; Tel.: +66-089-641-7532

Received: 7 September 2017; Accepted: 18 October 2017; Published: 22 October 2017

Abstract: An image processing-based variable rate chemical sprayer for disease and pest-infested coconut plantations was designed and evaluated. The manual application of chemicals is considered risky and hazardous to workers, and provides low precision. The designed sprayer consisted of a sprayer frame, motors, a power system, a chemical tank and pump, a crane, a nozzle with a remote monitoring system, and motion and crane controlling systems. As the target was confirmed, the nozzle was moved towards the target area (tree canopy) using the remote monitoring system. The pump then sprayed chemicals to the target at a specified rate. The results suggested optimal design values for 5–9 m tall coconut trees, including the distance between nozzle and target (1 m), pressure (1.5 bar), spraying rate (2.712 L/min), the highest movement speed (1.5 km/h), fuel consumption (0.58 L/h), and working capacity (0.056 ha/h). The sprayer reduced labor requirements, prevented chemical hazards to workers, and increased coconut pest controlling efficiency.

Keywords: variable rate chemical spraying system; digital image processing; GNDVI; coconut plantation

---

1. Introduction

Coconut is one of the highest economic value agricultural products in Thailand. Coconut is used to cook main dishes, as well as various kinds of dessert in Thai cooking. In 2015, 0.216 M ha of area was under coconut cultivation, which produced 1.06 M tones of coconuts. According to the National Statistical Office, 8273.2 g or 18 coconuts are consumed per person per year in Thailand. With a population of 55 million, 990 M coconuts or 65% of total production were consumed domestically. The rest, 489 M coconuts (35%), went to industry or were exported. Regarding export, Thailand stood seventh in world rankings, and the fifth in a statistical database provides comprehensive data for countries in the Asia-Pacific region for a range of indicators (ASEAN rankings). Total export volume in ASEAN was 4.61 M tones, comprising 42% from Indonesia, 17% from Malaysia, 16% from Philippines, 2% from Vietnam, and 1% from Thailand [1].

Coconut pest infestation is a serious challenge to its growers; for instance, coconut shoots are bitten by two-colored coconut leaf beetles (Brontispa longissima) and coconut black-headed caterpillars (Opisina arenosella), and drilled by coconut rhinoceros beetles (Oryctes rhinoceros). These pests cause serious damage to coconut trees, which may eventually lead to dead trees [2]. These problems result in the decrease in coconut production quantity and severe economic damage to coconut plantations. Some farmers use parasitic wasps to control coconut pests naturally. However, this method did not work well in pest control, and was also labor intensive. Some farmers drill holes on coconut trees and fill chemicals inside, which may leave chemical residue in coconuts [3]. On the other hand, the limitations of chemical sprayers include their reach to the full height of coconut trees and the heavy weight of bamboo stick (sprayer boom).

Many researchers have integrated a real-time machine vision sensing system and an individual nozzle-controlling device with a commercial map-driven-ready herbicide sprayer to create a spraying system. The smart sprayer was tested to determine its effectiveness and performance under varying commercial field conditions. Using the on-board differential GPS, geo-referenced chemical input maps (equivalent to weed maps) were also recorded in real-time. The performance accuracy of the spot-applicable fertilizer spreader was evaluated both in laboratory simulation and real-time field tests. Simulation results reported that the accuracy of the developed system was 94.9%. Real-time field tests reported that the system produced acceptable results at ground speeds of 1.6 km h^{-1} and 3.2 km h^{-1} for the spot application of fertilizer at target areas (in plant areas only) within the selected field [4–11].

After considering the challenges of conventional chemical spraying, as well as the objectives of increasing plantation yield, reducing production cost, and minimizing the environmental and health hazards of excessive chemical usage, a variable rate chemical sprayer was developed and evaluated. The sprayer uses an image processing technique for disease and pest infestation in coconut plantations, and provides effective pest control, as shown in Figure 1.

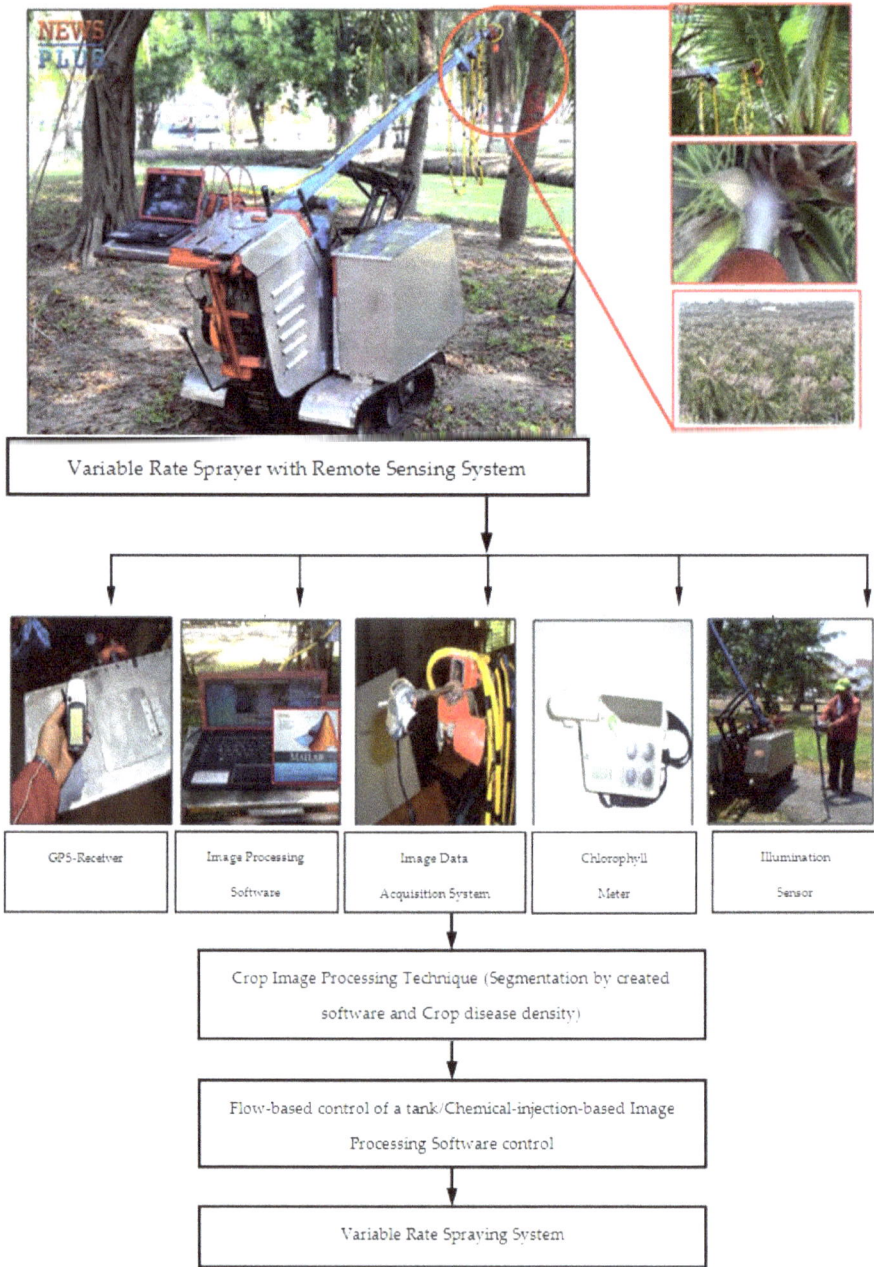

Variable Rate Sprayer with Remote Sensing System

GPS-Receiver

Image Processing Software

Image Data Acquisition System

Chlorophyll Meter

Illumination Sensor

Crop Image Processing Technique (Segmentation by created software and Crop disease density)

Flow-based control of a tank/Chemical-injection-based Image Processing Software control

Variable Rate Spraying System

(a)

Figure 1. *Cont.*

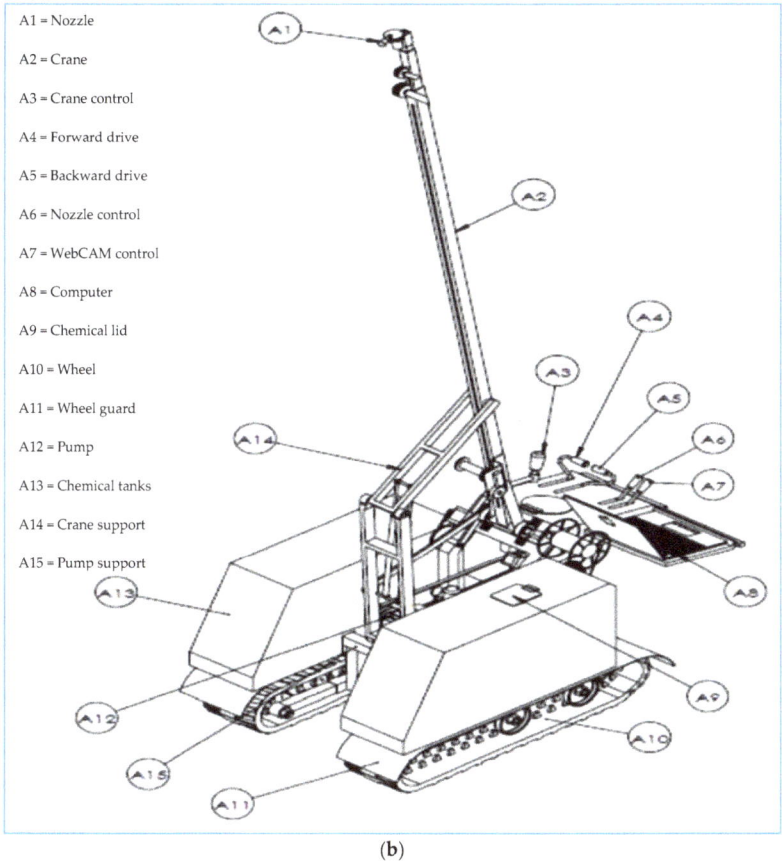

A1 = Nozzle
A2 = Crane
A3 = Crane control
A4 = Forward drive
A5 = Backward drive
A6 = Nozzle control
A7 = WebCAM control
A8 = Computer
A9 = Chemical lid
A10 = Wheel
A11 = Wheel guard
A12 = Pump
A13 = Chemical tanks
A14 = Crane support
A15 = Pump support

(b)

Figure 1. Instrumentation and overall procedural configuration; (**a**) Variable rate spraying system is a flow-based control of chemical application rate; (**b**) Isometric view of variable rate sprayer with remote sensing system.

2. Materials and Methods

2.1. Experimental Set up and Field Preparation

The experimental field was located at the Rajamangala University of Technology Thanyaburi (RMUTT), Thailand (14.03 °N, 100.61 °E). In coconut plantations, a triangle pattern is preferred, as about 15% more coconut trees can be planted compared with a square pattern. The distance between coconut trees was kept as 9 m for the chosen variety of coconut tree, and distance between rows was 7.8 m, which resulted in 138 coconut trees/ha. Coconut was transplanted in an experimental area of 40 m × 40 m. The application rate of fertilizer in the coconut plantation was 59–91 kg/ha for nitrogen, 27–40 kg/ha for phosphorous, and 85–131 kg/ha for potassium. The soil physical properties of the site are shown in Table 1.

Table 1. Physical properties of soil and fertilizer application* in the coconut plantation.

Coconut Age (year)	Rate of Fertilizer 13-13-21 or 12-12-17 (kg/m²)	Magnesium Sulfate, MgSO₄ (kg/m²)	Dolomite, CaMg (CO₃)₂ (kg/m²)
1	1	0.2	-
2	2	0.3	2
3	3	0.4	3
4 or more	4	0.5	4
Soil depth (cm)		0–20	
pH		6–7	
Soil Texture		Clay	
Sand (%)		15	
Silt (%)		30	
Clay (%)		55	
Organic matter (%)		1.54	
Particle density (g/cm³)		2.42	
Bulk density (g/cm³)		1.37	
Moisture content (%) d.b.		23	

* As claimed by respective manufacturer.

2.2. Image Data Acquisition System

The remote sensing system platform for the variable rate sprayer consisted of a WebCAM camera with two very small pieces of completely black photographic negative filter (Vimicro USB2.0 UVC PC Camera SWIFT-TECH ELECTRONICS Co. Ltd., Beijing, China) that modify an ordinary a WebCAM camera to capture images in the near infrared wavelength (>700 nm)and spraying nozzle (9000-45OS, THANAPHAN Co. Ltd., Bangkok, Thailand) with a wireless trigger control (Jelsoft Enterprises Ltd., London, England).The red/far red sensor measures radiation in μmol photons $m^{-2}s^{-1}$ in two wavebands (measured using two-channels with central bands at 660 and 730 nm; Skye Instruments, Ltd., Powys, UK) through a notebook (Pentium (R) Dual-Core CPU, T4200@2.00GHz, 1.93 GB of RAM) and software created using MATLAB® (R2013a, Windows, C++, https://www.mathworks.com/products/matlab.html, The Math Works, Inc., Natick, MA, USA)and a specifically developed data acquisition system. The specifically created software provided image orientation correction and loaded images from a WebCAM. After that, it estimated the disease density by dividing the selected image into four zones. The program converted the colored images into gray-scale images, then determined the volume flow rate of the solenoid valve for nozzle injection. The SKR 1800 illumination sensor measured prevailing sunlight intensity. The illumination sensor was attached to a data logger (Spectrum Technology Inc., Powys, PA, USA). A leaf chlorophyll meter (Minolta SPAD 502; Konica Minolta Sensing Inc., Osaka, Japan) was used to measure the average chlorophyll content (expressed as a single-photon avalanche diode (SPAD) values) of coconut tree leaves during the experiment. A feature common to disease infection in plants is the reduction in the number of chloroplasts in mesophyll. Apart from the frequent color change due to coconut shoots bitten by two-colored coconut leaf beetles (Brontispa longissima) and coconut black-headed caterpillars (Opisina arenosella), and drilled by coconut rhinoceros beetles (Oryctes rhinoceros), most of the plants showed that chlorophyll content was less in healthy plants, which had possibly been destroyed as a consequence of infection [6–8]. The units for the Minolta SPAD-502 leaf chlorophyll meter can be described by the following equation [12];

$$\text{Chl } \mu mol \ m^{-2} = 10^{M^{0.265}} \tag{1}$$

where M is the leaf chlorophyll meter reading (digital number), and Chl is the chlorophyll content in $\mu mol \ m^{-2}$.

The green normalized differential vegetation index (GNDVI) values were estimated to establish the suitability of this reflectance index for coconut plantation at different growth stages of the tree. The GNDVI, which is based on the greenness level, represented the chlorophyll content as determined by the radiance at the leaf surface. It is also a significant indicator for distinguishing young and aged healthy or infected coconut trees. The GNDVI is estimated [13] as follows;

$$GNDVI = \frac{\rho NIR - \rho G}{\rho NIR + \rho G} \qquad (2)$$

where ρNIR is the reflectance value for the near infrared band, and ρG is the reflectance value for the green band.

2.3. Design and Fabrication of Variable Rate Chemical Sprayer

2.3.1. Observation of Pests and Disease Infestation at the Coconut Tree Canopy

Information pertaining to coconut pests and disease infestation was collected at the apex of the coconut tree. The data collected were used to design a chemical sprayer with a remote monitoring system. A selected chemical (bacillus thuringiensis aizawaiwere) was tested in order to find the most appropriate dose with the highest efficiency that would also be hazard-free to the operator at the rate of 80–100 cc. in 20 liters of water. Moreover, prevailing pest control practices were studied in order to understand clearly how chemicals are used by farmers. The pests considered in the study were two-colored coconut leaf beetles, coconut black-headed caterpillars, and coconut rhinoceros beetles.

2.3.2. Fabrication of the Variable Rate Chemical Sprayer with Remote Monitoring System

The sprayer was fabricated after considering pertinent observations of trees, spraying practices, and the field. Caterpillar® track was used in the mobility of a sprayer crane. The crane must be to reach an altitude of 10 m. The nozzle was controlled manually using two operators. The sprayer consisted of seven major components, which were the sprayer frame (body), motors, the power system, a chemical tank and pump, a crane system, a nozzle with a remote monitoring system, a controlling system, and a crane controlling system.

The variable rate chemical sprayer with a remote monitoring system was fabricated at the Agricultural Machinery Engineering workshop of the Rajamangala University of Technology Thanyaburi.

The sprayer frame was fabricated from three-inch channel steel with dimensions of 1000 mm × 1500 mm × 800 mm. The sprayer was driven by a 10 hp Honda motor using belts and gears. The movement direction was controlled by a clutch system. The sprayer moved back and forth using a gear system. The crane boom was maneuvered from the motor through belts, reduction gears, and string reels. A mixture of chemicals was stored in the chemical tank, and pumped through a three-cylinder pump and nozzle.

The crane system was fabricated from 1.2 inch × 1.2 inch × 3000 mm carbon steel. The three carbon steel tubes were fixed at1.6 inch × 1.6 inch × 3000 mm, 2 inch × 2 inch × 3000 mm, and 2.2 inch × 2.2 inch × 3000 mm, which made it 10 m long at maximum stretch. Strings were attached to each tube to make it a telescopic crane boom when pulled in or out together. A WebCAM was installed at the nozzle in order to increase spraying precision and the ability to control nozzle direction.

In the design of the movement of the controller and crane, which is important in sprayer and crane operation, considered factors included crane elevation, chemical spraying, and camera angle (field of view) adjustment. The operation started when an operator controlled the sprayer to the targeted coconut tree. Crane and motors were controlled until the crane boom reached the top of the coconut tree and the infestation target was confirmed. The nozzle was pointed to the target using a remote monitoring system, and the bacillus thuringiensis aizawai, the chemical control, was pumped to the infested area at a specific rate.

2.4. Image Processing Software

An algorithm was specifically developed for image processing, particularly to separate the object of interest from its background. The developed software loaded images from the WebCAM (in near infrared-green-blue), and allowed the user to choose specific image profiles. The developed software estimated the disease density by dividing the selected image into four zones. Yellow and brown coconut shoots were detected in order to calculate density percentage per area and processed images (Figure 2). The results were displayed on the program screen. After that, the program converted colored images into gray-scale images, and determined the volume flow rate of the solenoid valve for nozzle injection. Lastly, the processed images were saved to a hard disk (Gray image, Bimodal image), as shown in Figure 3.

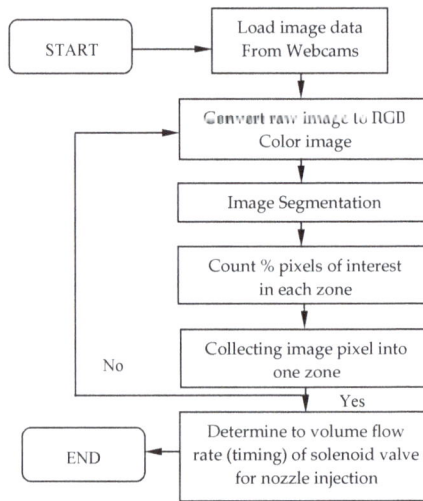

Figure 2. Image processing algorithm.

| (a) | (b) |

Figure 3. Image processing software; (a) Operating system; (b) Yellow thresholding

2.5. Field Testing and Performance Evaluation

The developed sprayer was evaluated in the field for its machine capacity. Factors considered in the assessment were the ability to work at an altitude up to 10 m, fuel consumption, and electricity consumption. The details of these indicators are presented as below.

a)   *Actual working capacity* of the sprayer (ha/h)

$$Actual\ working\ capacity = \frac{Working\ area\ (ha)}{Total\ time\ (h)} \tag{3}$$

b)   *Fuel consumption* of the sprayer (L/h)

$$Fuel\ consumption = \frac{Total\ fuel\ consumed\ (L)}{Total\ time\ (h)} \tag{4}$$

c)   *Electricity consumption* of the sprayer (kW/h)

$$Electricity\ consumption = \frac{IVt}{1000} \tag{5}$$

when

- I = Electricity current (Ampere)
- V = Electromotive force (Volt)
- t = Working time (h)

2.6. Calibration

For a suitable spraying footprint (diameter, cm) and altitude levels of the nozzle (m), the accuracy of a target covered by spray was assessed using a standard reference frame of 50 cm × 50 cm. Different samples were kept in the reference frame to represent the diseases and pests infestation area, comprising of yellow cards with 2.54 cm × 2.54 cm dimensions. Trials were made with sheets of three sample color groups: 25% (13 pieces), 50% (26 pieces), and 100% (52 pieces), which were kept at different heights from the nozzle (0.5 m, 0.7 m, and 1 m), as shown in Figure 4. This calibration also included variations in illumination levels with time of the day, as images were collected under cloud-free conditions between 10:30 and 12:30 standard local time.

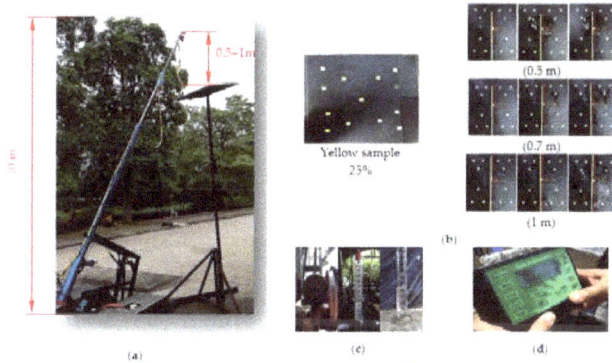

Figure 4. Experimental setup for the calibration of spraying, (**a**)Variable sprayer set up; (**b**)Test with a 25% yellow sample at different heights from the nozzle; (**c**) Average fluid flow rate determination; (**d**) SKR 1800 (Skye Instruments, Ltd., Powys, UK) illumination sensor.

2.7. Statistical Analysis

All of the measurements were performed in triplicates. The experiments were accomplished using a randomized complete block design (RCBD). The experiment data were statistically analyzed. Analysis of variance (ANOVA) was used to determine significance between treatments, and Duncan's multiple range test (DMRT) was used to compare the means at 95% confidence level [8].

3. Results

3.1. Chemical Spraying

The chemical pump was tested at three pressure levels: 1, 1.5, and 2 bar, which resulted in the spraying flow rates of 2.706, 2.712, and 3.27 L/min, respectively. The pressure value of 1.5 bar was chosen as it decreased spraying time and farmers were able to adjust the spraying angle easily (Table 2). The spraying footprint, which was measured as the diameter of the sprayed area and the distribution performance of the chemical, was tested at nozzle altitudes of 0.5, 0.7, and 1 m, and at a pressure of 1, 1.5, and 2 bar when at 25% (13 pieces), 50% (26 pieces), and 100% (52 pieces) density for the yellow color, respectively (Table 3).

Table 2. Chemical spraying rate testing.

Pressure (bar)	Fluid Flow Rate (mL/10s), $\bar{X}$	Average Fluid Flow Rate (L/min)	S.D. (L/min)
1	412.67	2.706	2.517
1.5	452.33	2.712	1.528
2	547.33	3.27	2.082

Table 3. Spraying footprint (diameter, cm) at different heights and the percentage of three sample color groups.

Altitude of Spraying Density of Yellow Color		Height of Nozzle above the Target Sample (m)		
		0.5	0.7	1
5 m				
	25%	42.00 [a]	55.00 [c]	69.33 [e]
	50%	43.00 [b]	56.67 [d]	69.67 [e,f]
	100%	43.67 [b]	56.67 [d]	70.33 [f]
7 m				
	25%	42.00 [a]	58.33 [b]	71.00 [c]
	50%	43.00 [a]	57.67 [b]	70.00 [c]
	100%	43.67 [a]	58.00 [b]	71.33 [c]
9 m				
	25%	42.00 [a]	59.67 [c]	71.00 [d]
	50%	43.00 [a]	56.67 [b]	70.67 [d]
	100%	43.67 [a]	58.67 [b,c]	72.00 [d]

* Means in a column and row followed by the same letter within a color group are not significantly different at 0.05 significant levels, according to Duncan's multiple range test.

For the calibration of the variable rate spraying system, the image data acquisition and processing system were calibrated against the known values of color densities with yellow colors at different altitudes (5 m, 7 m, and 9 m) using a wireless WebCAM. The color density of the sample sheets was varied at 25% (13 pieces), 50% (26 pieces), and 100% (52 pieces) for the yellow color.

These color panels, which were used for calibration, represented the top canopy area of the tree at the spraying footprint (Figure 4a). Table 3 shows the effect of nozzle height on the image processing accuracy of the wireless WebCAM. The altitude of the 1-m nozzle can be considered as having an acceptably high accuracy for both image processing data and the spraying area, which was 70.33 cm, 71.33 cm, and 72 cm, respectively, for 5 m, 7 m and 9 m spraying altitudes with 100% yellow color density. Figure 5 shows a sample image from the remote monitoring system.

(a)

(b)

Figure 5. Image from the remote monitoring system; (**a**) coconut pest disease infestation, (**b**) image data processing (segmentation).

3.2. Machine Performance

The results showed that the speed of the sprayer movement had a direct relationship with fuel consumption, as expected. Movement at speeds of 1, 1.5, 2, and 2.5 km/h consumed 0.46, 0.58, 0.73, 0.86 L/h of fuel, respectively. The working capacity of the sprayer was 0.048, 0.057, 0.064, and 0.076 ha/h, and electricity consumed was 0.015, 0.016, 0.018, and 0.022 kWh, respectively at the four speeds tested (Figure 6).

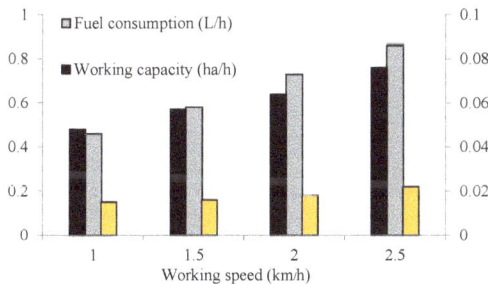

Figure 6. Variable sprayer system field performance.

4. Discussion

The stressed or diseased (infected) coconut trees are characterized with a lower level of chlorophyll than the healthy trees. Dead coconut trees had the lowest spectral reflectance. This was probably due to its lowest moisture content, which least absorbs the reflectance energy in the infrared portion of the electromagnetic spectrum, and at the same time decreases its reflectance [14].

The green normalized difference vegetation index(GNDVI) and variable rate sprayer system index also showed the coefficient of determination correlation (R^2 value of 0.687 and 0.66) slightly better when compared with the normalized difference vegetation index NDVI and variable rate sprayer system index(R^2 = 0.687 and 0.621). This may be due to a close link between the 'G' spectral band value and the greenness of the healthy young and infected mature coconut leaves, respectively (Figure 7). For healthy mature and infected young coconut trees, the GNDVI and variable rate sprayer system index also showed a higher correlation with R^2 value of 0.66 and 0.92, as compared with 0.621 and 0.71 for the NDVI and variable rate sprayer system index (Figure 8).

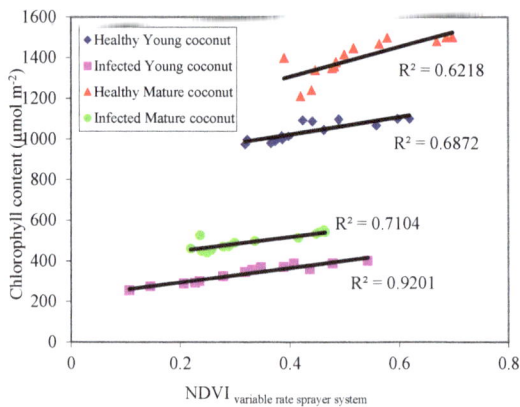

Figure 7. Estimation of leaf chlorophyll content using the normalized differential vegetation index (NDVI) and variable rate sprayer system for healthy and infected coconut trees.

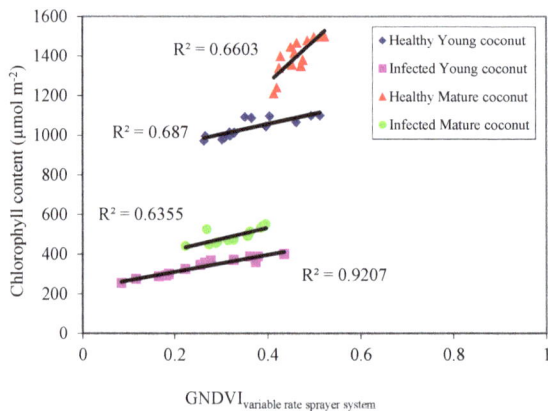

Figure 8. Estimation of leaf chlorophyll content using the green normalized differential vegetation index (GNDVI) and variable rate sprayer system for healthy and infected coconut trees.

The variable rate chemical sprayer for monitoring diseases and pests infestation in coconut plantations with various sensors was assembled and configured for an accurate reading of vegetation in coconut trees. The system was based on an Arduino board (Arduitronics CO., Ltd., Bangkok, Thailand) that is able to control the different sectors and nozzles of the system [8]. As the target was confirmed, the nozzle was moved towards the target area (tree canopy) using a remote monitoring system. The pump then sprayed the chemical to the target at a specified rate. The same system is also able to operate the position of the actuator on the air-adjusting louver system as described, and the electronic system can register data from all the systems via a serial port operating at a maximum frequency of 2 Hz [15].

5. Conclusions

A chemical sprayer with a remote monitoring system was designed and fabricated, which consisted of major parts such as a sprayer frame, motors, a power system, a chemical tank and pump, a nozzle and remote monitoring system, a movement controller, and a crane. The spraying rate was 162.72 L/h at a forward speed of 1.5 km/h, which offered fuel consumption of 0.58 L/h, and a working capacity of 0.056 ha/h. Furthermore, for the calibration of a variable rate spraying system, the image data acquisition and processing system were calibrated against the known values of color densities with yellow colors with a wireless WebCAM. The nozzle with an altitude of 1 m can be considered as having an acceptably high accuracy in image processing data and the spraying area.

The variable rate sprayer could reduce the labor force requirement, prevent chemical hazards to workers and the environment, and increase coconut pest controlling efficiency.

Acknowledgments: The authors thank Asian Institute of Technology (AIT), School of Environment, Resources and Development, Agricultural Systems and Engineering, Pathumthani, for providing experimental facilities. This work was financially supported by Rajamangala University of Technology Thanyaburi (RMUTT), Faculty of Engineering, Agricultural Engineering, Klong 6, Thanyaburi, Pathumthani, Thailand. Rajamangala University of Technology Thanyaburi is not necessary to agree with the results of this study.

Author Contributions: Grianggai Samseemoung designed and conceived the experiment. Pimsiri Suwan analyzed the data. Grianggai Samseemoung, Peeyush Soni and Pimsiri Suwan drafted the article.

Conflicts of Interest: The authors declare no conflict of interest.

Abbreviations

Full Name	Symbol
Hectare	ha
Kilogram	kg
Kilowatt	kW
Kilowatt-hour	kW h
Liter	L
Meter	m
Revolutions per minute	rpm
Square	sq.
Volt	V
Watt	W
Horse power	Hp

References

1. Office of agricultural economics: Ministry of agriculture and cooperatives. Agricultural Economic Basic Information. 2015. Available online: http://www.oae.go.th (accessed on 9 November 2015).
2. Abidin, C.M.R.Z.; Ahmad, A.H.; Salim, H.; Hamid, N.H. Population dynamics of Oryctes rhinoceros in decomposing oil palm trunks in areas practicing zero burning and partial burning. *J. Oil Palm Res.* **2014**, *26*, 140–145.

3. Bedford, G.O. Advances in the control of rhinoceros beetle, Oryctes rhinoceros in oil palm. *J. Oil Palm Res.* **2014**, *26*, 183–194.
4. Zaman, Q.U.; Esau, T.J.; Schumann, A.W.; Percival, D.C.; Chang, Y.K.; Read, S.M.; Farooque, A.A. Development of prototype automated variable rate sprayer for real-time spot-application of agrochemicals in blueberry fields. *Comput. Electron. Agric.* **2011**, *76*, 175–182. [CrossRef]
5. Stafford, J.V.; Benloch, J.V. Machine assisted detection of weeds and weed patches. In *Precision Agriculture*; Stafford, J.V., Ed.; BIOS Scientific Publishers Limited: Oxford, UK, 1997; pp. 511–518.
6. Samseemoung, G.; Jayasuriya, H.P.W.; Soni, P. Oil palm pest infestation monitoring and evaluation by helicopter-mounted, low altitude remote sensing platform. *J. Appl. Remote Sens.* **2011**, *5*, 053540. [CrossRef]
7. Samseemoung, G.; Soni, P.; Jayasuriya, H.P.W.; Salokhe, V.M. Application of low altitude remote sensing (LARS) platform for monitoring crop growth and weed infestation in a soybean plantation. *Precis. Agric.* **2012**, *13*, 611. [CrossRef]
8. Samseemoung, G.; Soni, P.; Sirikul, C. Monitoring and Precision Spraying for Orchid Plantation with Wireless WebCAMs. *Agriculture* **2017**, *7*, 87. [CrossRef]
9. Tangwongkit, R.; Salokhe, V.; Jayasuriya, H.P.W. Development of a Tractor Mounted Real-time, Variable Rate Herbicide Applicator for Sugarcane Planting *Agric. Eng. Int.* **2006**, *8*, 1 11.
10. Tian, L. Development of a sensor-based precision herbicide application system. *Comput. Electron. Agric.* **2002**, *36*, 133–149. [CrossRef]
11. Tian, L.; Steward, B.; Tang, L. Smart sprayer project: Sensor-based selective herbicide application system. *Proc. SPIE* **2000**, *4203*, 73–80.
12. Markwell, J.; Osterman, J.C.; Mitchell, J.L. Calibration of the Minolta SPAD-502 leaf chlorophyll meter. *Photosynth. Res.* **1995**, *46*, 467–472. [CrossRef] [PubMed]
13. Gitelson, A.; Kaufman, Y.J.; Merzlyak, M.N. Use of a green channel in remote sensing of global vegetation from EOS-MODIS. *Remote Sens.Environ.* **1996**, *58*, 289–298. [CrossRef]
14. Jusoff, K.; Hussein, Z.H.; SoonYew, J.; Din, M.S.H. The Life Satisfaction of academic and Non-Academic Staff in a Malaysian Higher Education Institution. *Int. Educ. Stud.* **2009**, *2*, 143–150. [CrossRef]
15. Landers, A.J.; Larzelere, W.; Muise, B. Advances in autonomous pesticide application technology for orange groves. In *Aspects of Applied Biology 114, International Advances in Pesticide Application*; Association of Applied Biologists: Wageningen, The Netherlands, 2012; pp. 91–98.

agriculture

MDPI

Article

A Study of the Lateral Stability of Self-Propelled Fruit Harvesters

Maurizio Cutini [1,*], Massimo Brambilla [1], Carlo Bisaglia [1], Stefano Melzi [2], Edoardo Sabbioni [2], Michele Vignati [2], Eugenio Cavallo [3] and Vincenzo Laurendi [4]

[1] CREA Research Centre for Engineering and Agro-Food Processingvia Milano 43, 24047 Treviglio, Italy; massimo.brambilla@crea.gov.it (M.B.); carlo.bisaglia@crea.gov.it (C.B.)

[2] Politecnico di Milano Department of Mechanical Engineering, via La Masa 1, 20156 Milan, Italy; stefano.melzi@polimi.it (S.M.); edoardo.sabbioni@polimi.it (E.S.); michele.vignati@polimi.it (M.V.)

[3] National Research Council (CNR) of Italy, Institute for Agricultural and Earthmoving Machines (IMAMOTER), Strada delle Cacce, 73, 10135 Torino, Italy; eugenio.cavallo@cnr.it

[4] INAIL, Dipartimento Innovazione Tecnologiche e Sicurezza degli Impianti, Prodotti e Insediamenti Antropici (DITSIPIA), P.le Pastore 6, 00144 Rome, Italy; v.laurendi@inail.it

* Correspondence: maurizio.cutini@crea.gov.it; Tel.: +39-0363-49603

Received: 31 August 2017; Accepted: 30 October 2017; Published: 1 November 2017

Abstract: Self-propelled fruit harvesters (SPFHs) are agricultural machines designed to facilitate fruit picking and other tasks requiring operators to stay close to the foliage or to the upper part of the canopy. They generally consist of a chassis with a variable height working platform that can be equipped with lateral extending platforms. The positioning of additional masses (operators, fruit bins) and the maximum height of the platform (up to three meters above the ground) strongly affect machine stability. Since there are no specific studies on the lateral stability of SPFHs, this study aimed to develop a specific test procedure to fill this gap. A survey of the Italian market found 20 firms manufacturing 110 different models of vehicles. Observation and monitoring of SPFHs under real operational conditions revealed the variables mostly likely to affect lateral stability: the position and mass of the operators and the fruit bin on the platform. Two SPFHs were tested in the laboratory to determine their centre of gravity and lateral stability in four different settings reproducing operational conditions. The test setting was found to affect the stability angle. Lastly, the study identified two specific settings reproducing real operational conditions most likely to affect the lateral stability of SPFHs: these should be used as standard, reproducible settings to enable a comparison of results.

Keywords: safety; tiltable platform; rollover angle; agriculture

1. Introduction

Self-propelled fruit harvesters (SPFHs) are agricultural machines designed to work on unimproved natural or disturbed terrain [1] They are intended to facilitate fruit picking and pruning, as well as any other task requiring operators to keep close to the foliage or to the upper part of the canopy. They replace ladders, so that workers are no longer required to carry them through the orchard and climb up and down; these saves time and work, thereby improving farm labour productivity [2] and operator safety [3].

Usually intended to carry at least two persons, the harvesters generally consist of a chassis, with a variable height work platform that can be equipped with lateral extending platforms enabling operators to reach fruiting branches more easily. Although SPFHs are usually powered by diesel engines, electrical engines have been introduced recently; they provide power both to the propulsion system and to the platform height and width adjusting mechanisms. The driving and operating console is located on the work platform; from this position operators can drive the machine forward or in reverse and adjust the platform height and width. During operations the machine runs on the natural ground surface, both flat and moderately inclined, and in the space between orchard rows at low operational speeds (approx. 0.4 km h^{-1}). When the machine moves away from the orchards it runs at higher speeds, usually not exceeding 15 km h^{-1}.

SPFHs can be equipped with various additional components such as self-levelling systems (to compensate for ground slope and keep the platform in a horizontal position), bin elevators and rails, different heights for front and rear or left and right portions of the platform, fruit picking assistance systems, and pneumatic or electric pruning systems.

Although the use of picking platforms has helped reduce workers' exposure to fall hazards and to risk factors associated with musculoskeletal disorders [3], operators are exposed to specific risks. For example, the French Mutualité Sociale Agricole, the second largest social security agency in France, recorded 325 accidents involving SPFHs from 2002 to 2009 and two deaths between 1995 and 2009, one following the machine's loss of stability when working on a road shoulder [4].

The positioning of additional masses (operators, fruit bins) and the maximum height of the platform (up to three meters above the ground) strongly affect machine stability. In particular, when running on unpaved ground and the lateral platform extended on one side only with the operators working on that side, the lateral displacement of the centre of gravity (CoG) may jeopardize the machine's stability even when standing still or moving slowly. The loss of machine stability cannot be effectively prevented: rollover protective structure (ROPS) would make the machine unsuitable for its intended purpose, and the possible operator error in tilt angle estimates make any recommendations about the maximum admissible operating gradient futile [5,6].

The literature contains a large number of studies on the stability of agricultural vehicles and related risks for operators [7–14], but very few address how to mitigate operator risks in relation with the loss of SPFH stability [4]. According to a survey of France the internal market for SPFHs is estimated at approximately 300–400 new machines sold per year, compared to 1500 to 2000 in Italy [4]. The figures on the French market are similar to the annual sales of large square balers, grape-pickers, and straddle tractors [4].

This the study aimed to help address SPFH safety issues by developing a procedure for assessing the lateral stability of these widely used agriculture vehicles under different working conditions.

2. Materials and Methods

The study involved a SPHF market survey, a collection of field data on real SPFH working conditions, and laboratory tests on representative SPFH components in settings reproducing real working conditions so has to define a procedure for assessing the lateral stability of these vehicles.

2.1. Market Survey

A survey of the Italian market was carried out by attending fairs and conducting telephone interviews. Technical product brochures were also collected to have an overview of companies manufacturing SPFHs and the different features of commercial harvesters. The following collected data underwent preliminary statistical processing for a functional and technical characterization of SPFHs:

- the wheel base and track width of the machine;
- the maximum platform height;
- the platform's maximum horizontal width the platform when the lateral platform (if any) are fully extended;
- the maximum weight allowed on the lateral platform.

Data were processed according to descriptive statistics using the Minitab 10.0 statistical software, State College, PA, USA [15].

2.2. Experimental Activity in Open Field Conditions

Field tests were carried out on four different SPFH models, hereafter labelled A, C, D, and E (Table 1). The vehicles were monitored while performing tasks under real operating conditions.

Table 1. Descriptive data of the tested SPFHs on operation in filed.

SPFH Label	Platform Type	Wheel Base (mm)	Width Track (mm) Front/Rear	Platform Height during the Field Tests (mm)	Operation
A	Single	1700	1670/1690	1200	Fruit picking
C	Single	2135	1600/1600	2450	Fruit picking
D	Double	2240	1780/1780	2450	Fruit picking
E	Single	2200	1680/1680	2650	Summer pruning

SPFH A was also used to simulate field operations (lateral extension of the platform, machine transfer, fruit picking). During these tests the platform was raised to the maximum height (2900 mm) whereas SPFHs C, D, and E were used for ordinary activities (fruit picking, bin loading, and unloading and summer pruning) only.

During field tests roll angles were recorded using an inertial measurement unit (IMU) and a GPS (DS-IMU1); additional data, such as georeferenced position, speed, and direction were also acquired. Other collected field data include the position and mass of the operators, bins and baskets. These data were subsequently used to define settings for laboratory tests. Field tests were carried out in apple and peach orchards, the main characteristics of which are reported in Table 2.

Table 2. Characteristics of the two fields where test have been carried out.

Species	Grown Cultivars	Pruning Style	Tree Spacing (m)	
			on the Row	between the Rows
Apple	Gala Jeromine	Taille longue	1	4
Peach	Big Top Nectacross	Spindle	1.5	4

The apple trees are trained according to the *Taille longue* pruning system (Figure 1). The trees have an axial shape and a tendency to develop free-bearing (acrotony) summit branches; the fruit-bearing branches, bent below the horizontal and never shortened, are inserted along the entire central axis [16].

Figure 1. Tree spacing plantation layout of the apple orchard.

The peach trees (Figure 2) are trained with the *Fusetto* (free spindle) system. The trees, 3.0–3.5 m tall, have a single vertical stem leader and are conical in shape; the central trunk has no permanent lower tier branches [17,18]. SPFHs are well suited for operation in the resulting hedgerows.

In all the apple and peach orchards there was hail netting in place. The nets covered the entire area above the tree tops (Figures 1 and 2).

Figure 2. The peach orchard.

In the *Big Top* peach orchard, field assessment was carried out during fruit picking, with the platform at a height of 2.45 m and extending 0.35 m on each side, while in the Nectacross peach plots, data were collected during summer pruning (Figure 3). The platform was at a height of 2.65 m and the lateral platforms were extended by operators to 200–350 mm. In both plots the ground between rows had a less than 1° slope to allow for flood irrigation.

Surveys were carried out in late July 2016.

Figure 3. Recording of experimental data during summer pruning.

2.3. Laboratory Tests

Laboratory tests aimed to define an appropriate procedure to determine the centre of gravity and lateral stability of SPFHs.

Tests were carried out on two SPFHs (hereafter labelled A and B) the characteristics of which are summarized in Table 3.

Table 3. Main characteristics of the SPFHs adopted for the laboratory tests.

SPFH	Unit	Wheel-Base	Tire Index Radius	Width-Track (Front)	Width-Track (Rear)	Maximum Platform Height	Maximum Lateral Extension
A	mm	1700	360	1670	1690	2900	850
B	mm	1948	371	1653	1653	2650	750

2.3.1. Definition of the Machine Settings

In order to set the laboratory experimental conditions, a standard operator and standard fruit bin dimensions and positions had to be defined. It was therefore necessary to define the mass, the CoG and the positioning on the SPFH of the masses simulating the operators and the fruit bin. The size and position of the bins and of the operators in the laboratory tests were defined on the basis of standard practice observed during field tests and of specifications from appropriate international standards [19,20].

2.3.2. SPFH Centre of Gravity Assessment

Following the recommendation in ISO 16231-2, assessment of the centre of gravity was carried out on the two SFPHs in compliance with the ISO 789-6 "Agricultural tractors—Test procedures, Part 6: Centre of Gravity" standard [21,22] by measuring the variation of the mass at the ground after lifting one of the axles. The measurements were carried out using:

- a 16 t maximum capacity overhead crane (Demag Cranes Components Spa, Italy);
- a Fisco Solatronic EN 17 digital inclinometer (Solar Design Company, Machynlleth, UK) with 0.1° resolution,
- four Argeo DFWKR force plates (Dini Argeo S.r.l., Spezzano di Fiorano Modenese (MO), Italy).

ISO 16231-2, ISO 789-6, ISO 22915-1 and UNI-EN 1459 recommend that the machine be equipped and adjusted ready for work with tanks filled at their proper operating levels. The only issue that remains open is the level of fuel in the tank, which must be considered in light of the stability it induces (e.g., UNI-EN 1459 and ISO 22915-1 recommend that fuel tank is full in case stability is thereby decreased). A simulation highlighted the impact of 40 kg of fuel on the vertical position of the CoG, when the vehicle's platform is at the maximum height, in 11–14 mm depending on the setting. In this study all the experimental tests were carried out with a full tank. The machine had no auxiliary equipment for bin loading/unloading; ballasts simulated the presence of operators on the platform and the tire pressure was set at the manufacturer's recommended value. The CoG's position was assessed by means of the double weight method: weights were first recorded with the machine standing with all four wheels on flat pavement and then with one axle lifted until it reached a 20° inclination (Figure 4) On this occasion the wheelbase, the index radius of the rear wheel, and the front and rear track widths were also recorded.

The machines were tested with the platform completely lowered and with no mass on the platform (Setting "L") and in the setting with the platform at its maximum height and carrying different masses ("H" Settings). See Table 4 for the detailed scheme of the adopted H settings.

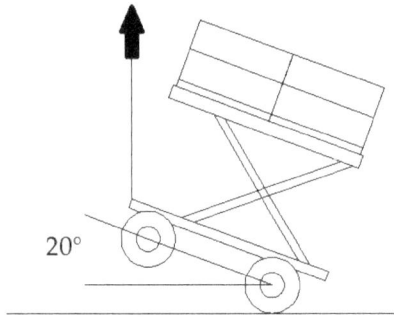

Figure 4. The double weight method (standing on the four wheels and with a 20° inclination) was adopted for the assessment of the centre of gravity.

Table 1. Settings of the SPFHs during the CoG determination.

Setting Code	Lateral Extensions	Weight on the Platform (kg)	Weight on the Upper Lateral Extension (kg)	Weight on the Sloped Down Lateral Extension (kg)
H0	both completely retracted	0	0	0
H1	both fully extended	400	200	200
H2	only the sloped down side fully extended	400	0	200
H3	only the sloped down side fully extended	0	0	200
L	both completely retracted	0	0	0

2.3.3. SPFH Stability Assessment

The stability overturning angle (SOA) was assessed following ISO-16231 standards [22,23] by placing two SPFHs (hereafter labelled A and B) on a tilting platform and determining the angle in the lengthwise direction of the machine only. The detailed description of the testing procedure hereafter reported was taken from UNI ISO 22915-1 [24] and UNI EN 1459 [20]:

- the tilting platform was in continuous motion;
- the platform was inclined slowly and continuously;
- the angle was measured by means of a Fisco Solatronic EN 17 digital inclinometer (Solar Design Company, Machynlleth, UK) having a 0.1° resolution;
- the platform did not undergo any significant deformation that might have affected results;
- tires were inflated to the manufacturer's recommended pressure;
- chains and ropes harnessed the machine, preventing it from overturning completely and from exiting the testing surface;
- the initial position of the machine on the testing surface was maintained using the parking brake and lateral constraints complying with ISO requirements (10% of the wheel diameter up to a maximum height of 50 mm).

The SPFHs were tested with the platform at the maximum height in the Hi settings presented in Table 4.

3. Results

This section reports the results of the performed analysis.

The results of the market survey are presented first. Field tests are then analysed to determine the most common operating conditions. Lastly, the paper reports the results of the rollover stability test using a tilting platform under the most common operating conditions and operators/fruit-bin settings.

3.1. Market Survey

The survey of the Italian market identified 20 different firms producing a good 110 different SPFH models. The results are in accordance with the findings in an MSA (Mutualité sociale agricole) study revealing that most of the SPFH models in France are from Italian manufacturers [4]. Italy is, in fact, a world leader in agricultural and forestry machinery production, with large global and small local companies active in the sector [25].

Three groups of machines were defined on the basis of the track width, one of the parameters affecting machine stability the most: the first (narrower machines) includes tracks 1100 mm to 1200 mm wide; the second, intermediate group, includes machines with track widths ranging between 1201 and 1600 mm, and the last group includes wider machines with track widths greater than 1600 mm. Track widths are in relation to the intended use of the machine and depend mostly on the layout of the orchard. The maximum lateral extension of the platform is 500 mm for the narrow machines, 700–800 mm for the intermediate ones, and 1200 mm for the wider models, whereas the maximum height of the platform is 2300 mm for the narrow models, 2300–2700 mm for the intermediate ones, and 2700–3000 mm for wider machines. Descriptive statistics of the collected dataset are summarized in Figure 5.

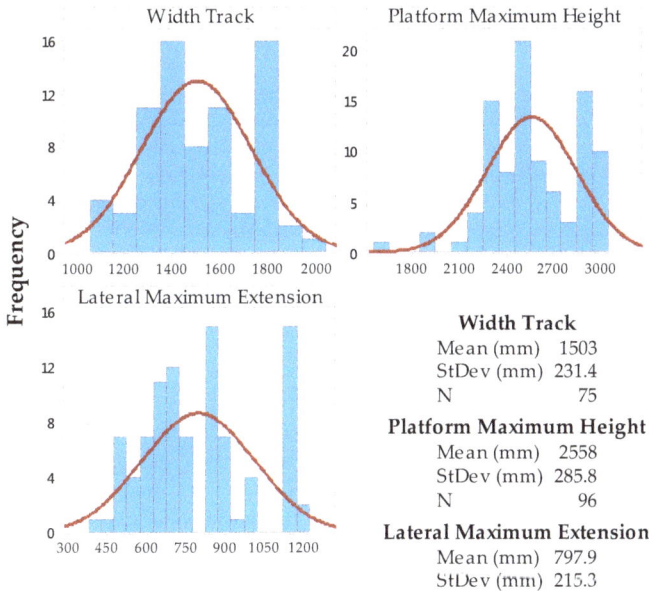

Figure 5. Descriptive statistics of the collected dataset.

Based on the characteristics of the 110 SPFHs it was decided that the models to be studied in laboratory conditions representing the worst case scenario should meet the following requirements:

- single platform without a self-levelling system
- extending structure that lifts the working platform up to 2900 mm;
- at least 800 mm lateral extension of the working platform;

In addition, among the SPFHs having such characteristics, those with a narrower track width (up to 1700 mm) were considered more suitable: the choice therefore fell on SPFH A, the characteristics of which are reported in Table 3. SPFH model B belongs to the same category and was selected for comparison in laboratory conditions: also its features are summarized in Table 3.

3.2. Experimental Activity in Open Field Conditions

SPFH operational conditions during fruit picking and pruning are considered static because of the very slow forward speed. The SPFHs involved in the study were operating on a levelled grassy surface with less than 1° longitudinal and lateral slopes. The mean value of the measured vehicle roll angles was 0.85°, with a maximum recorded value under all conditions of 2.9°. Data for the four SPFHs in operational conditions are summarized in Table 5.

Table 5. The measured roll angle of the tested SPFH during fruit picking and pruning.

SPFH	Maximum Roll Angle (°)	Maximum Forward Speed (km/h)	Measurement Time (s)
A	2.0	0.1	10
	1.8	1.1	52
	2.1	1.0	44
	1.2	0.9	42
C	2.7	2.1	101
	2.7	0.7	133
	2.9	0.4	120
	1.1	0.4	120
D	0.7	0.3	107
	0.7	0.4	196
	0.7	0.2	92
E	1.6	0.5	65
	0.6	0.5	29

Field tests revealed that during fruit picking operators normally work with different kinds of portable containers placed on the platform; these are filled directly and then placed at the centre of the platform for unloading. The estimated weight of these containers is 7–8 kg. Under normal working conditions, operators stand on the platform facing the trees and with the tip of their shoes touching the safety edge of the lateral extension.

3.3. Laboratory Tests

3.3.1. Definition of Machine Settings for the Laboratory Tests

For the laboratory tests 100 kg was selected as the standard weight of the operator. The value comprises the 90–95 kg weight of the operators considered in ISO standards [19,20] and the 7–8 kg basket that operators use during fruit picking. Note that SPFH user manuals consider a mass weighing 80–120 kg. Considering the weight of the standard operator, the CoG is 1000 mm above the platform [24].

The "standard operator" considered in the tests has the following characteristics:

- weight: 100 kg
- height of CoG: 1000 mm above the floor of the platform;
- lateral position of the CoG: 100 mm from the edge of the lateral extension; and
- longitudinal position of the CoG: in the hypothesis that there are two operators on the platform, they are supposed to stand 1300 mm from each other and in the middle of the platform.

When picking fruit, the bins are placed on specific rails at the centre of the platform. Bins have standard dimensions. Bins with the following characteristics were selected for use in laboratory tests:

- external dimension: 1200 × 1100 × 630 mm;
- internal dimensions: 1100 × 1010 × 475 mm;
- capacity: 0.56 m³; and
- weight when empty: 37 kg.

The capacity in kg depends on the fruit. Table 6 reports the weight capacity for different products.

Table 6. Mass capacity of the selected fruit bin.

Product	Capacity (kg)
Apples	260
Peaches	310
Oranges	310
Carrots	340
Olive	350
Tomatoes	360
Potatoes	360

The dimensions of the mass simulating the fruit bin placed on the platform are as follows:

- weight: 400 kg;
- height of the CoG: 400 mm above the floor of the platform; and
- position: at the centre of the platform.

Figure 6 illustrates the position of the standard operators and bin in laboratory tests for assessing the CoG.

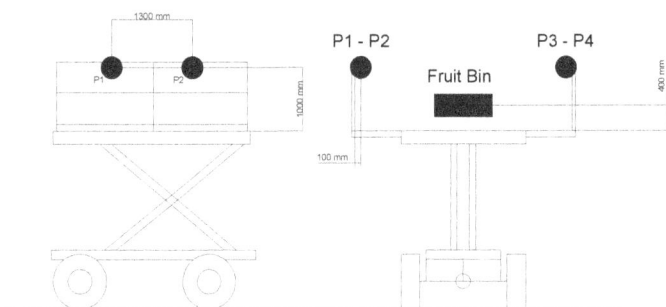

Figure 6. Scheme of the masses on the platform during CoG assessment (P1, P2, P3, and P4 are the operators).

Figure 7 illustrates the position of standard operators and the bin during the lateral stability tests, whereas Table 4 reports the test setting (H0–H3).

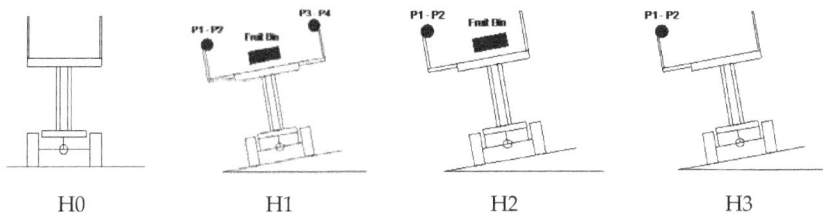

Figure 7. Settings during lateral stability assessment tests (front view).

3.3.2. Centre of Gravity Assessment

The coordinates of the CoG resulting from the ISO 16231 double weight method for two SPFHs (A and B) are reported in Table 7 for the different test settings (see Table 4).

Table 7. Centre of gravity coordinates (mm) in the two tested SPFHs.

Settings	A			B		
	X [a]	Y [b]	Z [c]	X [a]	Y [b]	Z [c]
H1	671	5	1994	805	46	1665
H2	677	134	1904	795	181	1621
H3	687	140	1692	772	206	1419
H0	703	10	1540	780	62	1256
L	611	10	701	823	55	492

[a] Along the transversal plane of the vehicle, from the rear axle positive values frontward; [b] along the median plane of the vehicle, from the centre line, positive values rightward; [c] along the vertical axis of the vehicle, from ground.

Tests on both SPFHs reveal that height of the CoG (Z axis) clearly depends on the masses (operators and bins) and their position on the raised platform. When the platform is fully extended, the CoG increases from H0 (no masses) to H3 (2 operators) to H2 (two operators on one side and a bin) and reaches the maximum value for H1 (four operators and a bin). The CoG shows a 30% increase from H0 to H1. The presence of additionally masses on the platform has little effect on the longitudinal position (X axis) of the CoG, whereas it greatly affects the transversal position (Y axis) of the CoG in settings H3 (2 operators on one side) and H2 (two operators on one side, plus a bin).

3.3.3. SPFH Lateral Stability Assessment

SPFHs A and B (see Table 3) were tested at the tilt table. Both SPFHs have one pivoting axle. SPFH "A" has a swivelling suspension without any limiting device. This means that the pivoting axle is free to rotate through a wide range of angles with respect to the vehicle chassis (i.e., until the axle/tyres touch the chassis). SPFH "B", instead, has the pivoting axle equipped with springs that limit swivelling of the suspension (self-levelling system).

For both vehicles, the first wheel to lose contact with the test bench during the tilting test is the one of the axle fixed to the frame on the upslope side of the vehicle. There are two phases of roll-over: initial detachment of the wheel from the tilted surface and subsequent complete detachment (vehicle rollover).

In the absence of a self-levelling mechanism for the pivoting axle, due to the inertia of the vehicle, detachment of the first wheel may lead to a complete roll-over of the machine, even if the angle of inclination remains constant. In contrast, in the presence of a limiting device for the pivoting axle, rollover stability of the vehicle is not compromised even if one wheel is detached from the ground. In the case of SPFH "B", although one tyre initially detached from the floor of the tilting platform, the pivoting axle touched the frame and stopped the tilting motion by acting as a damping system: complete loss of lateral stability only occurred when the inclination angle was increased further. In contrast, because SPFH "A" was not fitted with a damping system, initial detachment was followed by complete loss of lateral stability.

The results of the rollover tests on SPFH "A" and "B" at the tilting platform are reported in Table 8 and compared with SOAs calculated in accordance with ISO 16231-2. Although such standard is not applicable on SPFHs, this was done to check how much the output of the algorithm of the standard fits the overturning angles assessed experimentally.

Table 8. Roll-over angles (°) of the two tested SPFH.

Setting Code	A			B		
	I [1] and II [2]	SOA [3] I [1]	I [1]	SOA [3] I [1]	II [2]	SOA [3] II [2]
H0	20.0	20.7	20.3	23.3	26.3	33.5
H1	14.8	16.0	16.1	17.3	21.7	27.0
H2	11.9	12.9	12.2	13.0	18.7	23.7
H3	13.3	14.3	13.9	14.4	20.9	25.9

[1] Angle of first detachment; [2] angle of complete roll-over; [3] based on ISO 16231 calculation.

From such table it can be noticed that SOA I and SOA II values (that can be compared with the values the ISO 16231-2 defines α and σ) differ from measured ones of about 1° and 5°.

Lateral stability tests reveal that, although the two different SPFH models have platforms with different maximum vertical and lateral extensions, they have very similar angles of first detachment, but different angles of lateral stability. In vehicles equipped with pivoting axles, but no device acting as a dumping system, the inertia of the vehicle affects the roll-over dynamics. The conditions most likely leading to loss of lateral stability are those where the lateral extension on the downslope of the SPFH is fully extended and carries two operators with the fruit bin in the middle of the platform (setting H2). When there is no bin in the middle of the platform, loss of lateral stability occurs at slightly higher lateral inclinations (setting H3. According to manufacturer technical specifications, the fruit bin is considered a stabilizing factor when the SPFH is equipped with a lateral self-levelling platform. This means that the condition most likely to lead to a loss of lateral stability is that without a bin (setting H3).

Test results suggest that settings H2 and H3 are the ones mostly suitable for use in SPFH lateral stability testing.

4. Discussion

SPFHs are machines commonly adopted in orchards farms to carry out many operations efficiently. There are many different models manufactured by a large number of firms for use in different orchard settings.

Generally operated at reduced speeds on flat surfaces, SPHFs can lose stability, likely resulting in lateral rollover. This represents a serious risk for operators on the platform, as they do not have any means of protection.

Market survey results reveal that there is a lack of data both on the operating conditions of these machines and on SPHF rollover angles: although such information is reported in user manuals, there is no standard approach to measurement, and manufacturers adopt different methods and safety coefficients. It is, therefore, impossible to establish a reference rollover angle.

Experimental activity carried out in open field conditions (on perfectly levelled ground) provided information on the average (0.85°) and maximum (2.9°) roll angle experienced by such machines. It also highlighted the actual operating conditions that any standard test should take into consideration (e.g., placement and weight of operators and bins, if any). Testing was used to check SPFH compliance with international standards for CoG and rollover angles in agricultural machinery; it also allowed the definition of specific test settings for SPFHs.

The lowest rollover angle was recorded for harvesters working with the lateral platform fully extended, operators on one side only and a bin placed in the middle of the platform; as for self-levelling SPFHs, the stabilizing effect of the bin declared by some manufacturers should be carefully considered because it is not supported by any studies.

Laboratory investigation revealed the importance of the setting in which the vehicle is tested, especially the masses simulating the weight and position of operators and the fruit bin on the platform. The study revealed the influence of damping systems in vehicles equipped with pivoting axles:

the presence of a swivel limiting device on the swivelling axle restricts swivelling of the axle prior to the overturning of the machine.

The study identified a reliable procedure, consistent with the real operating condition of these vehicles, for assessing lateral stability under standard testing conditions enabling comparison.

Acknowledgments: The study was carried out within the "PROMOSIC" project, within the framework of the "BRIC 2015" call funded by INAIL (Italian National Institute for Insurance against Accidents at Work). The authors acknowledge the role of Ivan Carminati, Gianluigi Rozzoni, Alex Filisetti, and Elia Premoli for their valuable help in carrying out the measurements.

Author Contributions: Maurizio Cutini and Vincenzo Laurendi conceived the experiments; Maurizio Cutini, Carlo Bisaglia, and Edoardo Sabbioni designed the experiments; Maurizio Cutini, Massimo Brambilla, and Eugenio Cavallo performed the experiments; Maurizio Cutini, Massimo Brambilla, Stefano Melzi, Edoardo Sabbioni, and Michele Vignati analyzed the data; Maurizio Cutini, Carlo Bisaglia, and Eugenio Cavallo contributed materials tools; and Maurizio Cutini, Massimo Brambilla and Eugenio Cavallo drafted the manuscript.

Conflicts of Interest: The authors declare no conflict of interest. The founding sponsors had no role in the design of the study; in the collection, analyses, or interpretation of data; in the writing of the manuscript; and in the decision to publish the results.

References

1. EN 16952–2016 Agricultural Machinery—Rough-Terrain Work Platforms for Orchard's Operations (WPO)—Safety. Draft Edition: 1 February 2016. Available online: https://shop.austrian-standards.at (accessed on 7 June 2017).

2. Coppock, G.E.; Jutras, P.J. An investigation of the mobile picker's platform approach to partial mechanization of citrus fruit picking. *Proc. Fla. State Hort. Soc.* **1960**, *73*, 258–263.

3. Fathallah, A.F. Musculoskeletal disorders in labor-intensive agriculture. *Appl. Ergon.* **2010**, *41*, 738–743. [CrossRef] [PubMed]

4. MSA (2015) La sécurité des Plates-Formes de Récolte et de Taille en Arboriculture. Available online: http://www.msa.fr (accessed on 7 June 2017).

5. Cavallo, E.; Görücü, S.; Murphy, D.J. Perception of side rollover hazards in a pennsylvania rural population while operating an all-terrain vehicle (ATV). *Work* **2015**, *51*, 281–288. [CrossRef] [PubMed]

6. Görücü, S.; Cavallo, E.; Murphy, D.J. Perceptions of tilt angles of an agricultural tractor. *J. Agromed.* **2014**, *19*, 5–14. [CrossRef] [PubMed]

7. Franceschetti, B.; Lenain, R.; Rondelli, V. Comparison between a rollover tractor dynamic model and actual lateral tests. *Biosyst. Eng.* **2014**, *127*, 79–91. [CrossRef]

8. Vidoni, R.; Bietresato, M.; Gasparetto, A.; Mazzetto, F. Evaluation and stability comparison of different vehicle configurations for robotic agricultural operations on side-slopes. *Biosyst. Eng.* **2015**, *129*, 197–211. [CrossRef]

9. Wang, W.; Wu, T.; Hohimer, C.J.; Mo, C.; Zhang, Q. Stability analysis for orchard wearable robotic system. *IFAC PapersOnline* **2016**, *49*, 61–65. [CrossRef]

10. Jung, D.; Jeong, J.; Woo, S.M.; Jang, E.; Park, K.; Son, J. A study on the stability of a vehicle with lifting utility. *Adv. Mater. Res.* **2013**, *753–755*, 1169–1174. [CrossRef]

11. Myers, M.L. Ride-On Lawnmowers. The hazards of overturning. *Prof. Saf.* **2009**, *54*, 52–63. Available online: www.asse.org (accessed on 7 June 2017).

12. Liu, J.; Ayers, P.D. Off-road vehicle rollover and field testing of stability index. *J. Agric. Saf. Health* **1999**, *5*, 59–72. [CrossRef]

13. Molari, G.; Badodi, M.; Guarnieri, A.; Mattetti, M. Structural strength evaluation of driver's protective structures for self-propelled agricultural machines. *J. Agric. Saf. Health* **2014**, *20*, 165–174. [CrossRef] [PubMed]

14. Shulruf, B.; Balemi, A. Risk and preventive factors for fatalities in all-terrain vehicle accidents in New Zealand. *Accid. Anal. Prev.* **2010**, *42*, 612–618. [CrossRef] [PubMed]

15. Minitab 17 Statistical Software. Minitab, Inc.: State College, PA, USA, 2010. Available online: www.minitab.com (accessed on 7 June 2017).

16. Diemoz, M.; Vittone, G.; Pantezzi, T. Possibili evoluzioni nella potatura del melo (Potential improvements in apple tree pruning). *Inf. Agrar.* **2003**, *59*, 77–78.

17. Bargioni, G.; Loreti, F.; Pisani, P.L. Performance of peach and nectarine in a high density system in Italy. *HortScience* **1983**, *18*, 143–146.

18. Corelli-Grappadelli, L.; Marini, R.P. Orchard Planting Systems. In *The Peach: Botany, Production and Uses*; Layne, D.R., Bassi, D., Eds.; CAB International: Oxfordshire, UK, 2008; ISBN 978-1-84593-386-9.

19. *Industrial Trucks. Verification of Stability. Part 2: Counterbalanced Trucks with Mast*; UNI ISO 22915-1/2:2008; International Organization for Standardization: Geneva, Switzerland, 2008.

20. *Safety of Industrial Trucks—Self Propelled Variable Reach Trucks.9-6 (Agricultural Tractors—Test Procedures— Part 6: Centre of gravity)*; UNI EN 1459:2010; UNI EN: Milano, Italy, 2010.

21. *Agricultural Tractors—Test Procedures, Part 6: Centre of Gravity Standard*; ISO 789-6; International Organization for Standardization: Geneva, Switzerland, 1982.

22. *Self-Propelled Agricultural Machinery—Assessment of Stability—Part 1: Principles*; ISO 16231-1:2013; International Organization for Standardization: Geneva, Switzerland, 2013.

23. *Self-Propelled Agricultural Machinery—Assessment of Stability—Part 2: Determination of Static Stability and Test Procedures*; ISO 16231-2:2015; International Organization for Standardization: Geneva, Switzerland, 2015.

24. *Industrial Trucks. Verification of Stability. Part 1: General*; UNI ISO 22915–1/1:2008; International Organization for Standardization: Geneva, Switzerland, 2008.

25. Cavallo, E.; Ferrari, E.; Coccia, M. Likely technological trajectories in agricultural tractors by analysing innovative attitudes of farmers. *Int. J. Technol. Policy Manag.* **2015**, *15*, 158–177. [CrossRef]

agriculture

MDPI

Article

Phytotoxicity and Chemical Characterization of Compost Derived from Pig Slurry Solid Fraction for Organic Pellet Production

Niccolò Pampuro [1], Carlo Bisaglia [2], Elio Romano [2], Massimo Brambilla [2], Ester Foppa Pedretti [3] and Eugenio Cavallo [1,*]

[1]   Institute for Agricultural and Earth Moving Machines (IMAMOTER), Italian National Research Council (CNR), Strada delle Cacce, 73, 10135 Torino (TO), Italy; n.pampuro@ima.to.cnr.it

[2]   Consiglio per la ricerca in agricoltura e l'analisi dell'economia agraria (CREA), Centro di ricerca Ingegneria e Trasformazioni agroalimentari (CREA-IT), Sede di Treviglio, Via Milano, 43, 24047 Treviglio (BG), Italy; carlo.bisaglia@crea.gov.it (C.B.); elio.romano@crea.gov.it (E.R.); massimo.brambilla@crea.gov.it (M.B.)

[3]   Dipartimento di Scienze Agrarie, Alimentari ed Ambientali, Università Politecnica delle Marche, Via Brecce Bianche, 10, 60131 Ancona (AN), Italy; e.foppa@univpm.it

*   Correspondence: eugenio.cavallo@cnr.it; Tel.: +39-011-3977-724

Received: 26 September 2017; Accepted: 31 October 2017; Published: 4 November 2017

Abstract: The phytotoxicity of four different composts obtained from pig slurry solid fraction composted by itself (SSFC) and mixed with sawdust (SC), woodchips (WCC) and wheat straw (WSC) was tested with bioassay methods. For each compost type, the effect of water extracts of compost on seed germination and primary root growth of cress (*Lepidium Sativum* L.) was investigated. Composts were also chemically analysed for total nitrogen, ammonium, electrical conductivity and heavy metal (Cu and Zn). The chemicals were correlated to phytotoxicity indices. The mean values of the germination index (GI) obtained were 160.7, 187.9, 200.9 and 264.4 for WSC, WCC, SC and SSFC, respectively. Growth index (GrI) ranged from the 229.4%, the highest value, for SSFC, followed by 201.9% for SC, and 193.1% for WCC, to the lowest value, 121.4%, for WSC. Electrical conductivity showed a significant and negative correlation with relative seed germination at the 50% and 75% concentrations. A strong positive correlation was found for water-extractable Cu with relative root growth and germination index at the 10% concentration. Water-extractable Zn showed a significant positive correlation with relative root growth and GI at the 10% concentration. These results highlighted that the four composts could be used for organic pellet production and subsequently distributed as a soil amendment with positive effects on seed germination and plant growth (GI > 80%).

Keywords: compost quality; cress bioassay; organic pellet; phytotoxicity; pig solid fraction

1. Introduction

In several European countries, intensive pig production systems produce high quantities of liquid manure (slurry) in limited and specific geographic areas. With reference to Italy, the 6th Italian National Census of Agriculture indicates that the regions of Piedmont, Lombardy and Emilia-Romagna account for 90% of all pig breeding in the country [1]. In both Europe and Italy, slurry storage and subsequent land application is the predominant manure management practice, likely due to its simplicity, low cost, and potential to reduce the total cost of crop production as a chemical fertiliser replacement [2]. However, this technique carries several environmental pollution risks, including an excessive input of potentially harmful trace metals [3], an increase in nutrient—nitrogen and phosphorous—loss from soils through leaching, erosion and runoff [4], and the emission of ammonia and greenhouse gases

(GHG) [5]. In this context, the Nitrates Directive (91/676/EEC) introduced a limit of 170 kg ha^{-1} y^{-1} for application of animal manure nitrogen (N) in areas of the member countries particularly exposed to water pollution, the so-called Nitrate Vulnerable Zones (NVZ). As a result of this restriction, and considering that the agricultural surface available for land spreading is limited, the slurry has to be transported to fields over greater distances, increasing the costs of the logistics. Consequently, there is a growing need for technologies to competitively manage livestock slurries. The separation of the solid and liquid fractions simplifies handling, making possible to adopt different management technique for the two phases. The liquid fraction (LF), which is rich in soluble N [6], is generally applied in areas adjacent to the farm, while the solid fraction (SF), rich in nutrients (P and N) and organic matter (OM) [6], and containing less water, can be applied to lands at greater distances. According to recent investigations, (unpublished data), the SF can be economically transported to fields up to 25 km from the livestock farm.

A promising approach for increasing the benefits of pig slurry SF, as well as for creating a potential new market for pig slurry-derived fertiliser, is to pelletise it. Pelletising increases the bulk density of SF from an initial value of 400–450 kg m^{-3} to a final one of more than 1000 kg m^{-3} [7,8]. This allows better handling and transportation of SF at greater distances (even at hundreds of km as an order of magnitude) in order to move nitrogen (N) from Nitrate Vulnerable Zones to others less prone to pollution. Furthermore, Romano et al. [9] showed that pelletising homogenizes and further concentrates SF nutrients, thereby improving its fertilising and amending actions.

The moisture content of SF is the most important limiting factor for pelletising: a moisture content higher than 75–80% makes SF unsuitable for the process [10]. In previous studies [11,12], turning windrow composting has been proven as a simple and cheap technique to reduce the moisture content of SF. As a matter of fact, the heat generated by the composting process is able to reduce the moisture content of the substrate by 40%, hence suitable for pelletizing.

Composting is an aerobic process that involves the decomposition of organic matter (OM) under controlled temperature, moisture, oxygen and nutrient conditions [13]. Composting also implies OM sanitization regarding weeds and pathogens [14].

For optimising the composting, a bulking agent is generally added to SF. This makes it possible to adjust substrate properties such as air space, moisture content, C/N ratio, particle density, pH and mechanical structure, positively affecting the decomposition rate and, therefore, the development of the temperature [15]. Typical bulking agents used to compost N-rich wastes like animal manures are lignocellulosic agricultural and forestry by-products, such as cereal straw, cotton waste, and wood by-products [15]. Their low moisture and high C/N ratios can improve the benefits of animal manures [13].

Compost derived from pig slurry solid fraction can be re-used as a new resource material, such as soil fertiliser and conditioner, to replace the more expensive and less environmentally sustainable chemical fertilisers for crop production [16,17]. However, the presence of non-biodegradable and toxic heavy metals limits agricultural application of composted manure [18]. Pig slurry SF often contains high concentrations of copper (Cu) compared with other animal manures, because Cu supplements are normally added to pig rations to accelerate weight gain and increase the food conversion rates when fattening pigs [19]. In addition, zinc (Zn) is also added to pig diets to counteract any toxicity which might be caused by the high Cu content [20]. Only a small proportion (5–10%) of dietary Cu and Zn is absorbed by the pigs, while the rest is voided in the pigs faeces [20]. These elements, at high concentrations, can negatively affect seed germination, development of young seedlings, roots and plants growth.

In the present study, cress (*Lepidium sativum* L.) bioassays were used to evaluate the toxicity of four different composts derived from pig slurry solid fraction in order to examine if the organic pellet obtained by processing these composts can be recycled back to agricultural land without causing any negative effects on seed germination and plant growth.

2. Materials and Methods

2.1. Composting Trials

Four different windrows were realised for composting; pig slurry solid fraction by itself (SSFC) and with the addition of 3 types of vegetal materials as bulking agents. The 3 mixtures subjected to the composting process were obtained by mixing, on wet basis, pig slurry solid fraction with 18% sawdust (SC), 30% wood chips (WCC) and 14% wheat straw (WSC), respectively. The materials were mixed in these percentages to obtain a theoretical C/N ratio equal to 30 to optimise the composting process development [15]. In detail, the composting process took place by setting up four windrows as follows:

- SSFC: consisting of 6000 kg of pig slurry SF from screw press separator;
- SC: consisting of 5000 kg of pig slurry SF obtained from decanting centrifuge mixed with 900 kg of sawdust;
- WCC: consisting of 8000 kg of pig slurry SF from screw press separator mixed with 2400 kg of woodchips;
- W3C: consisting of 5000 kg of pig slurry SF from screw press separator mixed with 720 kg of wheat straw.

The windrows were placed on concrete floor under a covering, to avoid leaching and to protect from rain. The covering was not in contact with the surface of the windrow, allowing air to circulate and oxygen to be supplied. The ambient temperature and the temperatures inside the windrows at a depth of 0.4 m (T1), 0.8 m (T2) and 1.2 m (T3) from the surface of the windrows were continuously recorded (Figure 1) using thermocouple sensors (Type K) connected to a multichannel acquisition system (Grant, mod. SQ 1600, UK). To reduce the moisture content of the organic mixtures, making the materials suitable for pelletising, windrows were composted with a turning strategy: windrows were turned when the temperature of two of the three probes inside the composing material exceeded 60 °C [21]. The experimental composting process was observed for 130 days.

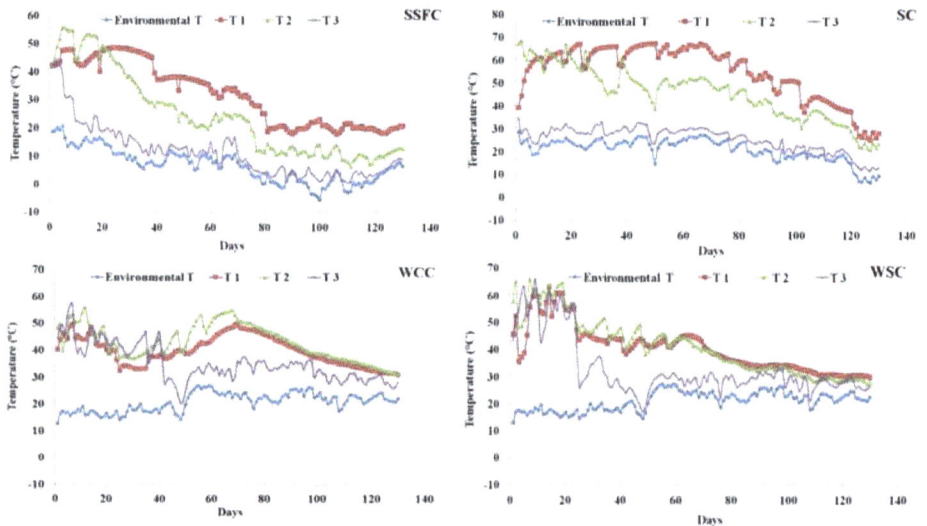

Figure 1. Average environmental temperature and temperatures development at a depth of 0.4 m (T1), 0.8 m (T2) and 1.2 m (T3) inside the SSFC, SC, WCC and WSC windrows.

The trial was carried out at the IMAMOTER (Institute for Agricultural and Earth Moving Machines) testing site in Turin, Italy (44°57′ N, 7°36′ E, 245 m above sea level).

2.2. Measuring Chemical Parameters

At the end of the composting process, for each investigated windrow, a sample of about 200 g was collected from 5 random locations and thoroughly mixed to generate a single composite sample [18]. The obtained samples were stored for 24 h in a cooling cell at 0–7 °C.

Dry matter (DM) was calculated after drying at 105 °C for 24 h (Table 1). Total nitrogen (TN) and ammonium (NH_4^+) were determined using the Kjeldahl standard method (BD40HT, Lachat Instruments). Water-extractable 1:10 (w/v) Cu and Zn were determined by atomic absorption spectrometry method (Elan 6000, Perkin-Elmer Corporation, Norwalk, CT, USA) [22]. (Table 1).

Table 1. Chemical characterisation of the four composts investigated (SSFC: slurry solid fraction compost, SC: sawdust compost, WCC: woodchip compost, WSC: wheat straw compost). Mean value of three replicates ± Standard Deviation.

Compost Samples	Compost Characteristics [a]									
	DM (%)		NH_4^+ (mg g^{-1})		Total N (mg g^{-1})		Ext. Zn [b] (µg g^{-1})		Ext. Cu [b] (µg g^{-1})	
SSFC	65.4	±0.15	2.9	±0.20	11.1	±0.19	24.0	±0.19	4.0	±0.12
SC	68.1	±0.12	5.2	±0.15	25.5	±0.17	22.0	±0.15	3.2	±0.23
WCC	67.9	±0.10	4.0	±0.28	17.3	±0.06	18.0	±0.06	1.9	±0.15
WSC	67.5	±0.06	2.9	±0.16	14.6	±0.17	16.0	±0.17	2.8	±0.12

[a] All characteristics are on dry weight basis; [b] Ext: water extractable.

2.3. Seed Germination Test

The effect of compost phytotoxicity on seed germination, root length and germination index was determined with cress (*Lepidium sativum* L.) bioassays.

After determining the dry matter content of the four composts, the moisture content of the samples was standardised at 85% by adding deionised water [23]. The water extracts were obtained by making a 75% concentration of the standardised sample and shaking this for 2 hours. After shaking, the flasks were centrifuged at 6000 rpm for 15 min and the supernatant was then again centrifuged for 15 min. [23]. Not much is known about the phytotoxic level of compost derived from pig slurry SF; for this reason, four different concentrations, 75%, 50%, 25% and 10%, of this supernatant were investigated. The pH and electrical conductivity (EC) of the extracts were determined (Table 2).

Table 2. Electrical Conductivity and pH of the four composts extracts (SSFC: slurry solid fraction compost, SC: sawdust compost, WCC: woodchips compost, WSC: wheat straw compost). Mean value of three replicates ± Standard Deviation.

Compost Samples	EC (dS m^{-1})				pH			
	75%	50%	25%	10%	75%	50%	25%	10%
SSFC	3.89 ± 0.02	2.83 ± 0.06	1.56 ± 0.03	0.75 ± 0.02	6.7 ± 0.01	6.5 ± 0.03	7.1 ± 0.01	6.4 ± 0.02
SC	7.96 ± 0.16	5.69 ± 0.08	1.97 ± 0.02	1.16 ± 0.02	7.4 ± 0.03	6.3 ± 0.01	6.3 ± 0.02	5.9 ± 0.01
WCC	1.69 ± 0.17	1.16 ± 0.01	0.61 ± 0.01	0.28 ± 0.01	5.5 ± 0.01	5.4 ± 0.01	5.7 ± 0.02	6.1 ± 0.01
WSC	1.90 ± 0.06	1.31 ± 0.05	0.69 ± 0.01	0.29 ± 0.01	6.6 ± 0.02	6.7 ± 0.02	7.2 ± 0.01	6.7 ± 0.02

Ten cress seeds were placed on layer of filter paper (Schleicher and Schuell no. 595, 85 mm round filters) in 90 mm Petri dishes and 5 mL of each concentration was added [23]. Distilled water was used as control. The experiment had a completely randomised block design with three blocks and two pseudo-replications (i.e., two Petri dishes with the same dilution). The Petri dishes were incubated in a growth chamber at 27 ± 2 °C and 70% relative humidity without photoperiod. At 24, 48 and 72 h after the beginning of the incubation, percentage of germination was recorded. A visible root was used as the operational definition of seed germination. After 72 h, also the length of the roots was measured.

The percentages of relative seed germination (RSG) after 24, 48 and 72 h, relative root growth (RRG) and germination index (GI) after 72 h of exposure to compost extracts were calculated as follows [24]:

$$RSG\ (\%) = (n\ of\ seeds\ germinated\ in\ compost\ extract/n\ of\ seeds\ germinated\ in\ control) \times 100 \quad (1)$$

$$RRG\ (\%) = (mean\ root\ length\ in\ compost\ extract/mean\ root\ length\ in\ control) \times 100 \quad (2)$$

$$GI\ (\%) = (RSG \times RRG)/100 \quad (3)$$

2.4. Plant Growth Bioassy

The plant growth bioassay was carried out on *Lepidium sativum* L. using the 4 composts investigated (SSFC, SC, WCC and WSC) mixed with sand and peat.

The substrate was prepared by mixing sand and peat with volume ratio 1 to 1 [25]. The composts were added to the substrate in two doses equal to 75 and 150 g of dry matter (DM) for L of substrate [25].

The different mixtures obtained were placed in plastic pots of volume equal to 0.5 L. On the bottom of the pots, a layer of expanded clay was placed to permit drainage. Initially, all pots were moistened with deionised water to attain a 60% water filled pore space (WFPS). The water added to each pot was calculated to supply 70% of the water holding capacity. Thereafter, soil water content was adjusted via a drop irrigation system every two to five days as required for the crop. All pots were kept in a greenhouse for 21 days at about 22 °C [25].

The experiment had a completely randomised block design with six replicates for each of the substrates. A replicate of pots without compost was included into the study as control. .

The Growth Index (GrI) was calculated according to the following equation:

$$GrI\ (75\ or\ 150\ g\ L^{-1})\% = (Gt/Gc) \times 100 \quad (4)$$

$$GrI\% = ((GrI75 + GrI150)/2) \times 100 \quad (5)$$

where:

Gt = mean production of plants in treatment;
Gc = mean production of plants in control.

2.5. Statistical Analysis

Analysis of variance (ANOVA) was performed to compare the effect of compost type and its concentration on RSG, RRG, GI and GrI; post-hoc Tukey's test was used. The normality of data distribution and assumption of equal variance were checked using the Shapiro-Wilk and Levene test, respectively. The effect of the chemical properties of the compost extracts within the concentrations was evaluated by correlation analyses. Statistical analysis was performed using SPSS software (IBM SPSS Statistics for Windows, Version 21.0, IBM Corp, Armonk, NY, USA).

3. Results and Discussion

3.1. Relative Seed Germination

Composts and concentrations analysed in this study did not affect seed germination and the germination percentages were higher ($p < 0.05$) than those found in the control (deionised water).

The ANOVA highlighted that neither compost type nor concentration affected ($p > 0.05$) RSG after 24 h (RSG-24), 48 h (RSG-48) and 72 h (RSG-72).

Furthermore, no differences ($p > 0.05$) were found between RSG obtained at 24, 48 and 72 h. The mean values of RSG obtained were 95.6, 95.0 and 96.4% after 24 h, 48 h and 72 h, respectively (Figure 2).

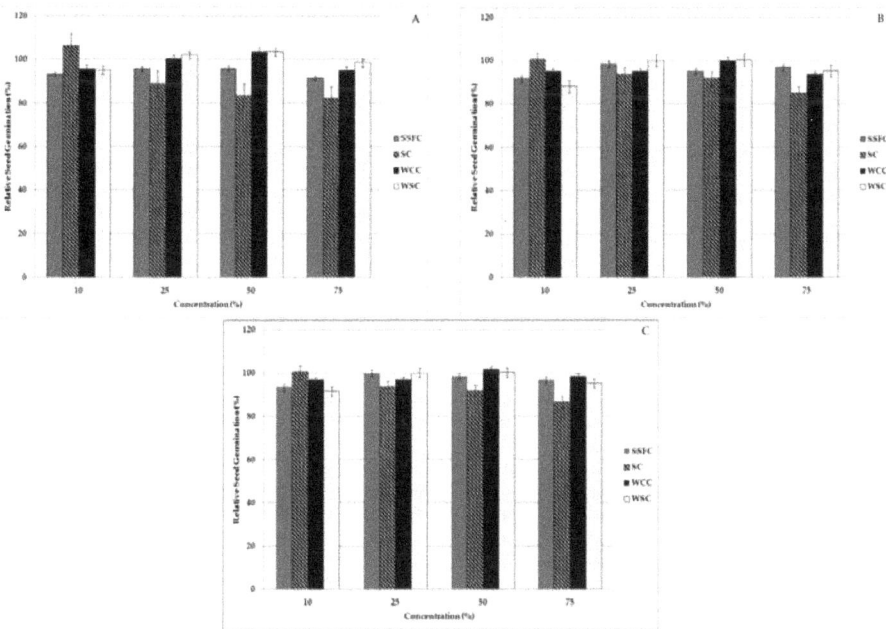

Figure 2. RSG of cress seeds in water extract of four compost (WSC: wheat straw compost; WCC: woodchips compost; SC: sawdust compost; SSFC: slurry solid fraction compost) in four concentrations after 24 h (**A**); 48 h (**B**) and 72 h (**C**). Error bars indicate standard error ($n = 6$).

3.2. Relative Root Growth and Germination Index

Table 3 shows the results of relative root growth (RSG). The rank of mean RRG for the compost extracts was SSFC > SC > WCC > WSC. At all concentrations, RRG of all composts exceeded 100%, suggesting a stimulating effect on root growth (Table 3). At the 10% concentration, the RRG of SSFC was higher ($p < 0.05$) than WSC and WCC. At the 25%, 50% and 75% concentrations, the RRG values were not different ($p > 0.05$).

Table 3. RRG of cress seeds as affected by water extracts of four compost (WSC: wheat straw compost, WCC: woodchip compost, SC: sawdust compost, SSFC: slurry solid fraction compost) in four concentrations after 72 h. Data are the mean of six replicates.

Compost	Concentration				Mean
	75%	50%	25%	10%	
WSC	185.5 [a]	199.0 [a]	157.0 [a]	119.7 [a]	165.3
WCC	231.7 [a]	212.1 [a]	197.1 [a]	120.0 [a]	190.2
SC	226.5 [a]	235.4 [a]	226.0 [a]	179.7 [ab]	216.9
SSFC	275.8 [a]	264.0 [a]	270.1 [a]	278.3 [b]	272.1
Mean	229.9	227.6	212.6	174.4	

RRG mean values followed by the same letter (a or b) within columns are not significantly different ($p > 0.05$).

Table 4 presents the relationship between the germination index and compost extracts. Growth stimulation was observed at all concentrations of composts extracts. The germination indices were always greater than the control (water only with GI = 100%). The increase in GI was due to longer root length when compared with the control. The presence of adequate amounts of NH_4^+ and other nutrients in composts extracts could be the cause of the high GI obtained [26].

Table 4. GI of cress seeds as affected by water extracts of four compost (WSC: wheat straw compost, WCC: woodchip compost, SC: sawdust compost, SSFC: slurry solid fraction compost) in four concentrations after 72 h. Data are the mean of six replicates.

Compost	Concentration				Mean
	75%	50%	25%	10%	
WSC	176.6 [a]	199.4 [a]	157.0 [a]	109.7 [a]	160.7
WCC	228.3 [a]	216.0 [a]	190.9 [a]	116.2 [a]	187.9
SC	195.9 [a]	215.8 [a]	211.4 [a]	180.4 [ab]	200.9
SSFC	267.1 [a]	260.1 [a]	270.1 [a]	260.3 [b]	264.4
Mean	217.0	222.8	207.4	166.7	

GI mean values followed by the same letter (a or b) within columns are not significantly different ($p > 0.05$).

As reported by Zucconi et al. [27], the compost is phytotoxin-free when GI values are higher than 80%. The WSC, WCC, SC and SSFC showed GI values higher than this limit and, therefore, they can be considered phytotoxin-free.

3.3. Plant Growth Bioassay

Table 5 shows the results of the plant growth bioassay (GrI).

The ANOVA highlighted that compost type affects ($p < 0.05$) GrI. The order in mean GrI for the four composts investigated was SSFC > SC > WCC > WSC (Table 5). For all composts Growth Index was higher ($p < 0.05$) than that found in the control (without compost) suggesting a stimulating effect on plant growth.

Table 5. Growth Index (GrI) values. Data are the mean of six replicates.

Compost	GrI75	GrI150	GrI
	(g L^{-1})	(g L^{-1})	(%)
WSC	85.7	157.1	121.4 [a]
WCC	166.7	219.6	193.1 [b]
SC	170.5	233.3	201.9 [b]
SSFC	189.4	269.5	229.4 [b]

GrI mean values followed by the same letter (a or b) within columns are not significantly different ($p > 0.05$).

According to some authors [25], compost with GrI values greater than 100% is considered not phytotoxic. All the composts investigated showed GrI values higher than this limit and, therefore, they can be considered phytotoxin-free.

3.4. Linear Correlations

As reported in Table 6, ammonium appeared not to affect ($p > 0.05$) seed germination and root growth; these results are in line with those reported by Hoekstra et al. [28].

Table 6. Linear correlations (shown by letters) between RSG after 24 h (RSG-24), RRG and GI at four compost concentrations with five chemical parameters of the compost extracts.

Concentration		NH_4^+	Total N	Ext. Zn	Ext. Cu	EC
	RSG-24	0.21 NS	0.22 NS	−0.08 NS	−0.14 NS	0.19 NS
10%	RRG	−0.14 NS	−0.24 NS	0.52 [A]	0.63 [A]	0.34 NS
	GI	−0.06 NS	−0.16 NS	0.49 [a]	0.56 [A]	0.37 NS
	RSG-24	−0.21 NS	−0.19 NS	−0.05 NS	−0.01 NS	−0.29 NS
25%	RRG	0.03 NS	−0.04 NS	0.34 NS	0.31 NS	0.20 NS
	GI	−0.01 NS	−0.08 NS	0.34 NS	0.32 NS	0.16 NS
	RSG-24	−0.34 NS	−0.32 NS	−0.02 NS	0.04 NS	−0.46 [a]
50%	RRG	0.01 NS	−0.04 NS	0.20 NS	0.20 NS	0.13 NS
	GI	−0.05 NS	0.09 NS	0.20 NS	0.21 NS	0.05 NS
	RSG-24	−0.27 NS	−0.25 NS	−0.06 NS	0.02 NS	−0.43 [a]
75%	RRG	−0.01 NS	−0.11 NS	0.37 NS	0.32 NS	0.08 NS
	GI	−0.08 NS	−0.17 NS	0.38 NS	0.34 NS	0.01 NS

[a] $p < 0.05$; [A] $p < 0.01$.

Unlike results of other phytotoxicity experiments [29,30], ammonium appeared not to affect seed germination and root growth. However, ammonium in solution can be toxic to plant growth. The toxicity results mainly from ammonia (NH_3), which affects plant growth and metabolism at low concentration levels at which NH_4^+ is not harmful [31]. The concentration of ammonia depends on the concentration of NH_4^+ via the equilibrium NH_4^+ (aq) = NH_3 (aq) + H^+ and on the volatilisation of NH_3. A concentration of NH_3 of 13 mM has been proved to be toxic [32]. However, concentrations of NH_3 (as calculated from the pH and concentration NH_4^+ by means of the equilibrium equation) in the composts extracts of the experiment were below this value.

EC showed a statistically significant negative correlation with RSG-24 at the 50% and 75% concentrations (Table 6). Salinity can have a detrimental effect on seed germination and plant growth, especially in the seedling stage, though the response of various plant species to salinity differs considerably. In general, salinity effects are mostly negligible in extracts, with EC readings of 2.50 dS m^{-1} or less [33]. This critical level was exceeded in the SC and SSFC extracts in the 50% and 75% concentrations.

Water-extractable Cu, which was highest in SSFC, appeared to be positively correlated with RRG and GI at the 10% concentration. However it is known that heavy metals can cause a marked delay in germination, and that they can severely inhibit plant growth. Concentration of water-extractable Cu in the compost extracts was maximally 0.21 µg mL^{-1}, though according to results from a previous study [28], 0.04 µg mL^{-1} of Cu inhibit root growth of plants. However, it should be mentioned that critical concentrations of heavy metals for toxicity in compost extracts are likely to be higher than critical values mentioned in literature, because of the relatively high amount of organic compounds, which can bind heavy metals [28].

Water-extractable Zn showed a high and significant positive correlation with RRG and significant but less high correlation with GI at the 10% concentration (Table 6). Concentration of water-extractable Zn was below phytotoxic levels as mentioned in the literature. The maximum concentration of water-extractable Zn in the compost extracts was 1.2 mg L^{-1} compared to critical values ranging from 75 to 600 mg L^{-1} as reported by Hoekstra et al. [28]. This might explain the fact that no significant negative correlations of water-extractable Zn with RSG-24, RRG and GI were found.

4. Conclusions

Four different composts, resulting from pig slurry SF composting with three vegetal bulking agents, underwent bioassays to evaluate their potential toxicity following cress (*Lepidium sativum* L.) germination index and root length assessments.

The mean values of germination index obtained were 160.7%, 187.9%, 200.9% and 264.4% for WSC, WCC, SC and SSFC, respectively. The growth index values of all composts investigated were >100%—121.4%, 193.1%, 201.9% and 229.4% for WSC, WCC, SC and SSFC, respectively—suggesting a stimulating effect on plant growth.

The outcomes of the investigation suggest that compost from pig slurry solid fraction (SSFC) and mixtures of pig slurry solid fraction with different vegetal materials as bulking agents (WSC, WCC, SC) after 130 days of composting, are phytotoxic-free. For this reason, it can be concluded that the four composts could be used for organic pellet production and subsequently distributed as a soil amendment without risk on seed germination and plantlet growth.

Acknowledgments: This work was carried out within the framework of the "FITRAREF" project, funded by the Italian Ministry of Agriculture and Forestry (GRANT NUMBER, DM29638/7818/10).

Author Contributions: Niccolò Pampuro, Carlo Bisaglia, Ester Foppa Pedretti and Eugenio Cavallo conceived and designed the experiments; Niccolò Pampuro performed the experiments; Elio Romano and Massimo Brambilla analyzed the data; Niccolò Pampuro and Eugenio Cavallo wrote the paper.

Conflicts of Interest: The authors declare no conflict of interest. The funding sponsors had no role in the design of the study; in the collection, analyses, or interpretation of data; in the writing of the manuscript, and in the decision to publish the results.

References

1. ISTAT—Italian National Institute of Statistics (2012). Preliminary Results of the 6th General Census of Agriculture. Available online: http://censimentoagricoltura.istat.it (accessed on 28 March 2016).
2. Kunz, A.; Miele, M.; Steinmetz, R.L.R. Advanced swine manure treatment and utilization in Brazil. *Bioresour. Technol.* **2009**, *100*, 5485–5489. [CrossRef] [PubMed]
3. Lu, L.-L.; Wang, X.-D.; Xu, M.-H. Effect of zinc and composting time on dynamics of different soluble copper in chicken manures. *Agric. Sci. China* **2010**, *9*, 861–870. [CrossRef]
4. Gomez-Brandon, M.; Lazcano, C.; Dominguez, J. The evaluation of stability and maturity during the composting of cattle manure. *Chemosphere* **2008**, *70*, 436–444. [CrossRef] [PubMed]
5. Salazar, F.J.; Chadwick, D.; Pain, B.F.; Hatch, D.; Owen, E. Nitrogen budgets for three cropping systems fertilized with cattle manure. *Bioresour. Technol.* **2005**, *96*, 235–245. [CrossRef] [PubMed]
6. Fangueiro, D.; Lopes, C.; Surgy, S.; Vasconcelos, E. Effect of the pig slurry separation techniques on the characteristics and potential availability of N to plants in the resulting liquid and solid fractions. *Biosyst. Eng.* **2012**, *113*, 187–194. [CrossRef]
7. Pampuro, N.; Facello, A.; Cavallo, E. Pressure and specific energy requirements for densification of compost derived from swine solid fraction. *Span. J. Agric. Res.* **2013**, *11*, 678–684. [CrossRef]
8. Pampuro, N.; Bagagiolo, G.; Priarone, P.C.; Cavallo, E. Effects of pelletizing pressure and the addition of woody bulking agents on the physical and mechanical properties of pellets made from composted pig solid fraction. *Powder Technol.* **2017**, *311*, 112–119. [CrossRef]
9. Romano, E.; Brambilla, M.; Bisaglia, C.; Pampuro, N.; Foppa Pedretti, E.; Cavallo, E. Pelletization of composted swine manure solid fraction with different organic co-formulates: Effect of pellet physical properties on rotating spreader distribution patterns. *Int. J. Recycl. Org. Waste Agric.* **2014**, *3*, 101–111. [CrossRef]
10. Alemi, H.; Kianmehr, M.H.; Borghaee, A.M. Effect of pellet processing of fertilization on slow-release nitrogen in soil. *Asian J. Plant Sci.* **2010**, *9*, 74–80.
11. Pampuro, N.; Dinuccio, E.; Balsari, P.; Cavallo, E. Gaseous emissions and nutrient dynamics during composting of swine solid fraction for pellet production. *Appl. Math. Sci.* **2014**, *8*, 6459–6468. [CrossRef]
12. Pampuro, N.; Dinuccio, E.; Balsari, P.; Cavallo, E. Evaluation of two composting strategies for making pig slurry solid fraction suitable for pelletizing. *Atmos. Pollut. Res.* **2016**, *7*, 288–293. [CrossRef]
13. Nolan, T.; Troy, S.M.; Healy, M.G.; Kwapinski, W.; Leahy, J.J.; Lawlor, P.G. Characterization of compost produced from separated pig manure and a variety of bulking agents at low initial C/N ratios. *Bioresour. Technol.* **2011**, *102*, 7131–7138. [CrossRef] [PubMed]

14. Parkinson, R.; Gibbs, P.; Burchett, S.; Misselbrook, T. Effect of turning regime and seasonal weather conditions on nitrogen and phosphorus losses during aerobic composting of cattle manure. *Bioresour. Technol.* **2004**, *91*, 171–178. [CrossRef]

15. Bernal, M.P.; Alburquerque, J.A.; Moral, R. Composting of animal manures and chemical criteria for compost maturity assessment. A review. *Bioresour. Technol.* **2009**, *100*, 5444–5453. [CrossRef] [PubMed]

16. Chrysargyris, A.; Saridakis, C.; Tzortzakis, N. Use of municipal solid waste compost as growing medium component for melon seedlings production. *J. Plant Biol. Soil Health* **2013**, *2*, 1–5.

17. Papamichalaki, M.; Papadaki, A.; Tzortzakis, N. Substitution of peat with municipal solid waste compost in watermelon seedling production combined with fertigation. *Chil. J. Agric. Res.* **2014**, *74*, 452–459. [CrossRef]

18. He, M.-M.; Tian, G.-M.; Liang, X.-Q. Phytotoxicity and speciation of copper, zinc and lead during the aerobic composting of sewage sludge. *J. Hazad. Mater.* **2009**, *163*, 671–677. [CrossRef] [PubMed]

19. Liu, S.; Wang, X.-D.; Lu, L.-L.; Diao, S.-R.; Zhang, J.-F. Competitive complexation of copper and zinc by sequentially extracted humic substances from manure compost. *Agric. Sci. China* **2008**, *7*, 1253–1259. [CrossRef]

20. Tam, N.F.Y.; Tiquia, S. Assessing toxicity of spent pig litter using a seed germination technique. *Resour. Conserv. Recy* **1994**, *11*, 261–274. [CrossRef]

21. Caceres, F.; Flotats, X.; Marfa, O. Changes in the chemical and physiochemical properties of the solid fraction of cattle slurry during composting using different aeration strategies. *Waste Manag.* **2006**, *26*, 1081–1091. [CrossRef] [PubMed]

22. Page, A.L.; Miller, R.H.; Keeney, D.R. *Methods of Soil Analysis*; Part 2; American Society of Agronomy, Inc. Soil Science of America: Madison, WI, USA, 1982.

23. Piemonte, R. Metodi di analisi dei compost. *Collana Ambient.* **1998**, *6*, 84–87.

24. Fuentes, A.; Llorens, M.; Saez, J.; Aguilar, M.I.; Ortuno, J.F.; Meseguer, V.F. Phytotoxicity and heavy metals speciation of stabilised sewage sludges. *J. Hazad. Mater.* **2004**, *108*, 161–169. [CrossRef] [PubMed]

25. Piemonte, R. Il compostaggio: Processo, tecniche ed applicazione. *Collana Ambient.* **2001**, *25*, 83–88.

26. Romero, C.; Ramos, P.; Costa, C.; Marquez, M.C. Raw and digested municipal waste compost leachate as potential fertilizer: Comparison with a commercial fertilizer. *J. Clean. Prod.* **2013**, *59*, 73–78. [CrossRef]

27. Zucconi, F.; Pera, A.; Forte, M.; De Bertoldi, M. Evaluating toxicity of immature compost. *BioCycle* **1981**, *22*, 54–57.

28. Hoekstra, N.J.; Bosker, T.; Lantinga, E.A. Effects of cattle dung from farms with different feeding strategies on germination and initial root growth of cress (*Lepidium sativum* L.). *Agric. Ecosyst. Environ.* **2002**, *93*, 189–196. [CrossRef]

29. Tiquia, S.M.; Tam, N.F.Y. Elimination of phytotoxicity during co-composting of spent pig-manure sawdust litter and pig sludge. *Bioresour. Technol.* **1998**, *65*, 43–49. [CrossRef]

30. Wong, M.H.; Cheung, Y.H.; Cheung, C.L. The effects of ammonia and ethylene oxide in animal manure and sewage sludge on the seed germination and root elongation of *Brassica parachinensis*. *Environ. Pollut.* **1983**, *30*, 109–123. [CrossRef]

31. Mengel, K.; Kirkby, E.A. *Principles of Plant Nutrition*, 4th ed.; International Potash Insitute: Horgen, Switzerland, 1987; p 745.

32. Bennet, A.C.; Adams, F. Concentration of NH_3 (aq) required for incipient NH_3 toxicity to seedlings. *Soil Sci. Soc. Am. J.* **1970**, *34*, 259–263. [CrossRef]

33. Ofosu-Budu, G.K.; JHogarh, J.N.; Fobil, J.N.; Quaye, A.; Danso, S.K.A.; Carboo, D. Harmonizing procedures for the evaluation of compost maturity in two compost types in Ghana. *Resour. Conserv. Recycl.* **2010**, *54*, 205–209. [CrossRef]

agriculture

MDPI

Technical Note

Mechatronic Solutions for the Safety of Workers Involved in the Use of Manure Spreader

Massimo Cecchini [1,*], Danilo Monarca [1], Vincenzo Laurendi [2], Daniele Puri [2] and Filippo Cossio [1]

[1]   Department of Agricultural and Forestry Sciences (DAFNE), University of Tuscia, Via S. Camillo De Lellis, 01100 Viterb, Italy; monarca@unitus.it (D.M.); f.cossio@unitus.it (F.C.)
[2]   National Institute for Insurance against Accidents at Work (INAIL), Via di Fontana Candida, 1, 00078 Monte Porzio Catone, Italy; V.laurendi@inail.it (V.L.); d.puri@inail.it (D.P.)
*   Correspondence: cecchini@unitus.it; Tel.: +39-0761-357353

Received: 25 September 2017; Accepted: 30 October 2017; Published: 6 November 2017

Abstract: An internationally acknowledged requirement is to analyze and provide technical solutions for prevention and safety during the use and maintenance of manure spreader wagons. Injuries statistics data and specific studies show that particular constructive criticalities have been identified on these machines, which are the cause of serious and often fatal accidents. These accidents particularly occur during the washing and maintenance phases—especially when such practices are carried out inside the hopper when the rotating parts of the machine are in action. The current technical standards and the Various safety requirements under consideration have not always been effective for protecting workers. To this end, the use of SWOT analysis (Strengths, Weaknesses, Opportunities, and Threats) allowed us to highlight critical and positive aspects of the different solutions studied for reducing the risk due to contact with the rotating parts. The selected and tested solution consists of a decoupling system automatically activated when the wheels of the wagon are not moving. Such a solution prevents the contact with the moving rotating parts of the machine when the worker is inside the hopper. This mechatronic solution allowed us to obtain a prototype that has led to the resolution of the issues related to the use of the wagon itself: in fact, the system guarantees the stopping of manure spreading organs in about 12 s from the moment of the wheels stopping.

Keywords: manure spreader; safety; decoupler; mechatronic; SWOT analysis

1. Introduction

The risk of injuries related to the use of agricultural machinery has always been of primary importance, as evidenced by the high incidence of work-related accidents resulting from the improper use of agricultural machinery and equipment [1–3].

The aim of this work focuses on the needs—recognized at national and European levels—to provide technical solutions against the risk of crushing, catching, and cutting during the use of self-propelled or towed manure spreaders [4–6]. These are agricultural machines used to distribute manure or other materials over a field [7]. Their use is fairly widespread in livestock farms, but it could be even more widespread in the future because climate change may require more organic matter inputs to the soil over Vast areas of the globe [8,9].

Sector-specific studies and surveys [6,10,11] have identified particular constructive critical issues on some machines currently in the market and/or already in use that involve the above-mentioned risks and determine the occurrence of a significant number of serious or fatal accidents. In particular, the access of an individual into the loading hopper when the rotating parts of the machine are in motion is not prevented.

Although specific safety procedures such as lockout or energy dissipation are required by occupational health and safety (OHS) regulation in many countries [12], the workers often work in hazardous conditions because in this way they are facilitated in cleaning operations. With the rotating parts in motion, the debris are removed more easily compared with the operation carried out only with the aid of water jets. In this operating mode, workers show greater attention to reducing working times and fatigue rather than increasing safety levels, thus demonstrating a risk underestimation [13]. Another reason for safety procedures not always being observed is that agricultural machinery are often owned and operated by families and not companies; thus, the OHS regulations are rarely observed.

The technical standards in force concerning this type of machine or the risks arising from its use [14] have not always been effective for the protection of workers. A critical point which is often observed regards the washing and maintenance operations carried out by operators located inside the hopper when the rotating parts of the machine are in motion. Similar problems also occur in other machines, such as forest chippers [15].

In European and international literature, statistical data regarding accidents during the use and maintenance of this type of machine are available [16–22]. In France, between years 2002 and 2012, eight injuries during the use of manure spreaders were recorded by IRSTEA (Institut National de Recherche en Sciences et Technologies pour l'Environnement et l'Agriculture) [20], three of which were fatal; the common cause of these accidents is the trapping of the operator between the spreading organs. Moreover, these accidents occurred during three different stages of work: cleaning, maintenance, and unlocking the rotor. In Germany, between 1998 and 2008, 12 fatal accidents were recorded by LSV-SpV (Spitzenverband der landwirtschaftlichen Sozialversicherung) [16] during the use of this machine. The common cause for eight of these accidents was the same: catching of operator between the spreading organs. These accidents occurred during Various machining steps: three during cleaning, two during maintenance, two during unlock, and one during a non-defined working phase. There were a total of 17 accidents occurring in Italy during the use or maintenance of such equipment which were recorded by INAIL (Istituto Nazionale per l'Assicurazione contro gli Infortuni sul Lavoro) [17,19], all occurring between 2002 and 2015: nine fatal accidents were caused by the overturning of the tractor to which the manure spreader was attached, due to the excessive slopes of the ground (so cannot be counted among those the manure spreaders are responsible for); two cases with the same dynamics and tragic outcome, but involving self-propelled spreaders; two cases (of which one was fatal) occurred during the replacement and maintenance of the trailer wheels; two fatal cases were due to the crushing caused by the not-inserted handbrake; two cases (of which one was fatal) occurred during rotor maintenance. Other data are available outside Europe: in Ontario (Canada), six fatal accidents due to manure spreaders were recorded by CAIR (Canadian Agricultural Injury Reporting) between 1990 and 2008 [18]; in California (USA), the OSHA (Occupational Safety and Health Administration) recorded one fatal accident in 2015, during a cleaning operation [22].

Since 2009, the Health and Safety Office of the French Ministry of Agriculture and Food started Various feasibility studies with regard to the improvement of the safety of manure spreaders during the washing operations, with the aim of a revision of the harmonized Standard EN 690 + A1 (Safety of Manure Spreader) [14]. The results of these studies confirm both the possibility of cleaning the moving parts of the machine (e.g., rollers and conveyor belt) without the need for the operator to be inside the load compartment while carrying out this operation, and the possibility of providing the machine of a system that prevents the movement of rotating parts when the machine itself is not in motion (steady wheels), thereby eliminating the risk of trapping the operator inside the rotating parts [6,10,11]. Moreover, the development of a safety indicator during machine design, and an associated algorithm for the assessment and optimization of productivity, could improve the safety of the machine itself [23–25].

2. Materials and Methods

With the aim of selecting the best solution in terms of risk reduction, SWOT analysis (Strengths, Weaknesses, Opportunities and Threats analysis) was applied to different solutions proposed by different authors [6,10,11]. The choice of this method is due to the fact that the authors of the different analyzed solutions have already carried out a risk assessment in accordance with ISO 12100 [26]: SWOT analysis also considers external parameters such as economics and feasibility of the proposed solutions.

The SWOT analysis is a support analysis that responds to a need for rationalization of decision-making processes [27]. In practice, this type of study is a logical process, originally used in business economics and then applied to other areas [28–30], which makes it possible to make systematic and useful information collected about a specific theme. The amount of data collected with this system is crucial to outline the policies and lines of action that result from enhancing strengths and reducing weaknesses in light of the opportunities and risks that normally arise from the external situation.

The advantages of this analysis are: depth analysis of the context in the definition of strategies; Verification of matching between strategy and needs effective improvements, it allows for consensus on the strategies (if all parties involved in the intervention participate in the analysis); and flexibility.

The disadvantages of this analysis are: the risk of subjective procedures by the evaluation team in the selection of the actions; it can describe reality in a way which is too simplistic; if there is no implementation in the context of partnership, there is a risk of discrepancy between a pragmatic scientific and political plan [31–33].

3. Results

The mechatronic solution resulted in a system that minimizes the risks for the operator's safety. From a mechatronic point of View, the decoupler consists of a magneto-mechanical mechanism [34] that prevents motion to all moving parts of the wagon if the machine is not in motion. The reset of the movement is possible Via a hold-to-run control applied in a secure area of the wagon itself.

The basic elements of the system are:

- wagon wheel movement detectors (wheels);
- a motor disengagement device (clutch);
- a torque limiter to limit the torque during overloads;
- a programmable logic controller (PLC);
- a man-made command for manual resetting of conveyor and distributor systems, located in a safe area;
- a hydraulic distributor or a solenoid Valve for conveyor control.

Considering the existing electromagnetic clutches on the market, there is little availability of clutches suitable for electric Voltages that correspond to those of the tractor (12 V), and above all suitable to withstand the dissipation of rotations with the torque Values of the machines rating. The minimum data for the correct sizing of the clutches are Very Variable. The only information currently available are shown in Table 1.

Table 1. Data for clutch dimensioning.

Rotation Speed [1] (rpm)	Transmitting Torque (Nm)	Supply Voltage
540	2200	12
1000	1600	12

[1] depending on the model.

The Variability of the characteristics of the wagons on the market is Very wide; other Variables are to be considered that would not allow a fair uniformity of adoption. Possible Variables are due to:

- transmission shaft type below the loading platform;
- geometric shape;
- length;
- diameter;
- mass of the entire axis;
- any Vibrations and/or movements.

Since transmission shafts are Very similar to the cardan shafts commonly used on agricultural Vehicles, we have come to the following conclusions.

The SWOT analysis conducted (Table 2) shows how the application of the decoupler is a mechatronic solution applicable on a great scale which ensures an optimal result to remedy the safety problems related to this machine. A similar solution has recently been applied to other machines [15].

Table 2. SWOT (Strengths, Weaknesses, Opportunities, and Threats) analysis regarding the decoupling device.

Strengths	Weaknesses
Working organs stopped during cleaning and maintenance operations.	High component costs. High installation, assembly, and setup costs.
Possibility of a manual reset in a safe area.	Necessity of regulatory transposition and any objections by manufacturers.
Possibility to break the movements of the rotating parts with the aid of the tractor hydraulics.	Difficulties in adapting machines already on the market and in use.
Flow solenoid Valve that facilitates the adjustment of the speed of the conveyor belts.	Not easily adaptable by small/middle builders.
The rotating sensor detects the wagon's motion.	Need to adjust the speed of the conveyor belt.
Minor space displacement in the case of downstream positioning of the clutch.	Possible malfunctions and/or breaks of the Various components.
Possibility to break the movements of the two transmission organs.	With clutch located downstream of the hydraulic unit, provide a stop mechanism for conveyor belts (risk of injury of the lower limbs if the chains remain in motion).
Electromagnetic clutches that can be powered by the electric Voltage (12 V) of the tractor.	Requirements of a separate hydraulic circuit if hydraulic clutches are used.
Less expensive, less bulky, and easier to install and integrate electromagnetic clutch.	Low availability of 12 V clutches.
Separate tractor/wagon hydraulic circuits.	The need for a torque limiter.
Difficult system inactivation.	The need for a programmable logic controller (PLC).
	Possible conflicts with electronic regulation systems.
	Not easily inserted in the ISOBUS technology.
Opportunities	**Threats**
Robust, durable, and reliable system.	Procedural distortion in the production line.
Polyfunctional system for other types of machines (e.g., round baler that, together with manual reset, must only engage the machine when it is in motion).	Needing of specialized technical personnel.
Improved safety.	Possible rearmament of the system with the help of a second person or thing that keeps the hold-to-run control inserted.
Possible rearming of the motion of the rotating parts by means of hold-to-run control.	High risks if the movements are not disrupted.
Probable reduction of sensor costs.	Sensors relatively fragile.
	Hydraulic pump driven by the PTO (power take off).
	The second stop mechanism makes the total costs rise.
	In the absence of a hydraulic unit, the application of a circuit causes considerable additional costs.
	PLCs and other microcomponents make the set relatively expensive and complex.
	Total costs not within the reach of all manufacturing companies.
	Field operational problems and possibility of frequent blockages due to clogging.

A decoupling system (Figures 1 and 2) has been designed and developed thanks to the cooperation of the company Ren Mark Snc (San Polo d'Enza, Italy). A prototype of the system was applied to a towed model of manure spreader (Ren Mark RP140).

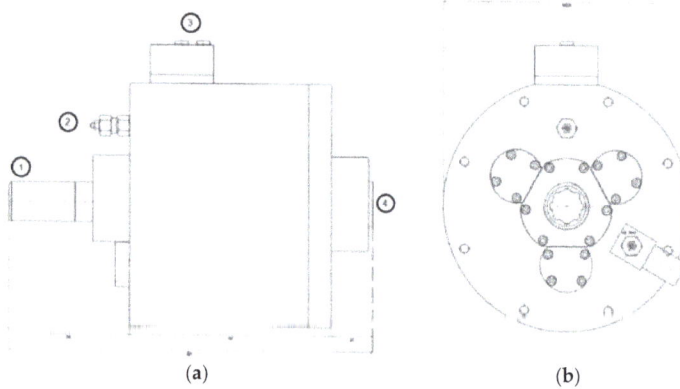

(a)   (b)

Figure 1. Device scheme: (**a**) Lateral View; (**b**) Frontal View. 1. PTO cardan shaft attachment; 2. main Valve; 3. electric control; 4. connection to the wagon.

Figure 2. The prototype device.

A system diagram is shown in Figure 3.

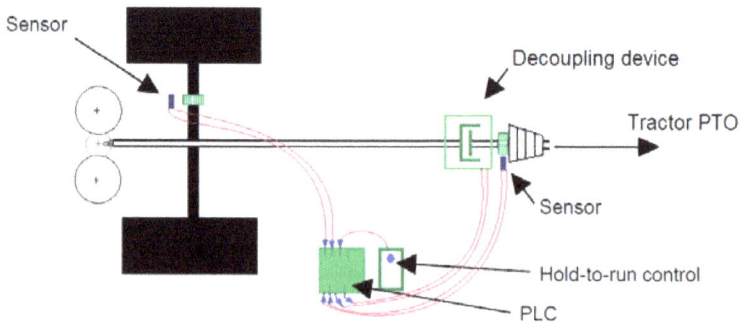

Figure 3. Scheme of the system (modified from [10]).

Wheel motion detection is achieved by means of a magnetic proximity sensor that detects the passage of the metal surface of the nuts mounted on the wheel drum (Figure 4), which, passing through a distance of 1 to 2 mm, make that sensor generate an electrical pulse that is detected by the microprocessor (Figure 5). Five dices are mounted on the wheel to detect even low rotation speeds.

Figure 4. Sensor mounted on the wheel.

Figure 5. Microprocessor and beeper.

When the sensor no longer detects the metal surface on the drum for a time less than or equal to 6 seconds, the system activates the blinking and deactivates the output to release the movement of the mechanical organs by stopping its movement as a function of the motion detection by the sensor mounted on the tractor PTO (Figure 6). Consequently, the microprocessor determines the disengagement of the multidisc clutch as a result of the internal pressure loss of the decoupler generated by the electric pump and thus allows decoupling transmission to the manure spreader that stops while the tractor PTO continues to be active.

Figure 6. Sensor on PTO.

In order to prevent the motion transmission in case of failure in connecting to the 12 V power supply, the used clutch is of the "normally open" type.

In this way, the manure spreading organs, located behind the chassis on which the decoupling device is mounted, are no longer connected to the power take-off and rotate to neutral until they are stopped in a short time.

A series of field trials showed an average time of stop of the rotors equal to 12 s (Table 3). The 12-s time can be set when programming the PLC: in particular, the decoupler will operate 6 s after the wheels are stopped and the other 6 seconds are due to the inertia of the rotors.

Table 3. Stop time recorded during test.

Test No.	Stop Time (s)	Average Time (s)	Standard Deviation (s)
1	13		
2	12		
3	11	12	0.707
4	12		
5	12		

A Video showing the operation of the device is available at the link [35].

As soon as the manure spreader connected to the tractor runs again, the sensors—specifically the ones on the wheel's drum—resume signaling, thus resulting in the rearm of the multidisc clutch, and the resumption of rotating parts' motion.

As said before, the system is also equipped with a hold-to-run control which allows to engage or disengage the clutch when the operator has the need to intervene at a standstill. The command must be positioned at a safe distance from the working organs and in a position which allows good Visibility of the danger zones.

The technical characteristics of the prototype device installed on the wagon used for the tests are given in Table 4.

Table 4. Technical characteristics.

Microprocessor
Software configuration: - Alarm sound in case of moving organs in the absence of wagon advancement - Block of the moving parts in the event of wagon stoppage
Hardware configuration: - One standard output for the light or sound signal - One standard output for unlocking the movement of the mechanical parts - One digital input with 1 to 100 Hz bandwidth and 0/12 V amplitude for the running sensor - One digital input with 1 to 100 Hz bandwidth and 0/12 V amplitude for active PTO sensor
Operating conditions: - Power supply: 10/18 V DC with reverse polarity protection and overvoltage impulse - Absorption: 5 mA (excluding signalling devices) - Temperature: from $-20\ °C$ to $+60\ °C$ - Maximum humidity: 90% non-condensing - Protection: IP-65
Proximity sensors
XS612 Sensor: - Section: 53 mm - Rated detection distance: 0.16 (4 mm) - Discrete output function: 1 NO - Output circuit type: AC/DC - Rated Voltage: 24 to 240 V AC/DC (50/60 Hz) - Switching capacity current: 5 to 200 mA AC/DC - Power supply limits: 20 to 264 V AC/DC - Residual current $\leq$0.8 mA, open condition - Switching frequency: $\leq$1000 Hz DC; $\leq$25 Hz AC - Voltage drop: $\leq$5.5 V, closed condition
XS618 Sensor: - Section: 62 mm - Rated detection distance: 0.31 (8 mm) - Discrete output function: 1 NO - Output circuit type: AC/DC - Rated Voltage: 24 to 240 V AC/DC (50/60 Hz) - Switching capacity current: 5 to 200 mA DC–5 to 300 mA AC - Power supply limits: 20 to 264 V AC/DC - Residual current: $\leq$0.8 mA, open condition - Switching frequency: $\leq$1000 Hz DC; $\leq$25 Hz AC - Voltage drop: $\leq$5.5 V, closed condition

Applicability

The prototype decoupler can be applied to Various models of manure spreader: the only technical trick that needs to be adopted is to change the internal solenoid Valve pressure. The tested prototype—built according to the power absorbed by the machine (60 kW)—had an operating pressure of 15 bar.

Depending on the absorbed power, the pressure must be adjusted according to the following Values:

- absorbed power 60 kW → operating pressure 15 bar;
- absorbed power 74 kW → operating pressure 20 bar;
- absorbed power 88 kW → operating pressure 30 bar.

4. Discussion

The tested mechatronic system would be the most effective and safe in terms of the safety of the operator working inside the hopper of the wagon. In fact, the basic concept of "firm wheels–static working organs" would prevent—or would definitely decrease—any type of risk of contact with rotating parts.

Nevertheless, as mentioned above, the implementation issues are many. The considerable additional cost that this device would entail for the manufacturers of such machines should also be considered, which would in some cases raise the sales prices (and hence the purchasing cost for consumers) with a possible drop in sales.

In addition, given the complexity of designing and implementing the system, it is likely that there will be a wide dissent from manufacturers, which would probably be opposed to the proposal for adapting the technical standard relating to the safety of manure spreader wagons.

However, this technical solution gives the opportunity—depending on the dimensional types of wagons and therefore of the decoupling system—to adapt and improve the safety of the machinery fleet present throughout the European territory.

To date, the field-tested decoupler is the best solution to overcome the major problems that arise when using the manure spreader wagon. The 12 s from the wheel stopping are sufficient to ensure that it is impossible to enter the hopper when the rotors are still in motion.

In this way, specific activities of the workers that were previously made in the absence of safety conditions (and in a way that does not comply with current health and safety regulations) could be carried out in complete safety.

Finally, the reliability of the solution will need to be addressed. Standards such as ISO 13849 [36,37] and IEC 62061 [38] describe the design of safety control systems. The solution which combines sensors, a programmable logic controller, and hydraulics must meet a specific performance level in terms of integrity. It must be considered that in this case the risk is high: in fact, the severity of harm is high, the frequency of exposure is high, and the possibility of avoiding the harm is low (rotating speed, lack of space); thus, the required performance level is high.

The last step should be the Validation of the solution by standardization bodies. This activity is already planned thanks to the support of INAIL.

Acknowledgments: Project realized with the financial support of INAIL. Authors wish to thank Ren Mark di Fontana e Genitoni Snc for contributing to the prototype and experimental trials.

Author Contributions: All authors contributed equally to the realization of the work.

Conflicts of Interest: The authors declare no conflict of interest.

References

1. Pawlak, H.; Nowakowicz-Dębek, B. Agriculture: Accident-prone working environment. *Agric. Agric. Sci. Procedia* **2015**, *7*, 209–214. [CrossRef]
2. Svendsen, K.; Aas, O.; Hilt, B. Nonfatal occupational injuries in Norwegian farmers. *Saf. Health Work* **2014**, *5*, 147–151. [CrossRef]
3. Suutarinen, J. Tractor accidents and their prevention. *Int. J. Ind. Ergon.* **1992**, *10*, 321–329. [CrossRef]
4. Damas, S. *Amélioration de la Sécurité des Épandeurs de Fumier. Rapport Stage Irstea*; IRSTEA: Antony, France, 2010; pp. 1–11.
5. Le Formal, F.; Tricot, N. *Feasibility Study: Improvement of Manure Spreader Safety*; Cemagref Report; Cemagref: Antony, France, 2009; pp. 1–15.
6. Al Bassit, L.; Le Formal, F.; Tricot, N. *Improvement of Manure Spreaders Safety: Feasibility Study*; Cemagref Report; Cemagref: Antony, France, 2010; pp. 1–16.
7. Bacenetti, J.; Lovarelli, D.; Fiala, M. Mechanisation of organic fertiliser spreading, choice of fertiliser and crop residue management as solutions for maize environmental impact mitigation. *Eur. J. Agron.* **2016**, *79*, 107–118. [CrossRef]

8. Colantoni, A.; Ferrara, C.; Perini, L.; Salvati, L. Assessing trends in climate aridity and Vulnerability to soil degradation in Italy. *Ecol. Indic.* **2015**, *48*, 599–604. [CrossRef]

9. Stoate, C.; Boatman, N.D.; Borralho, R.J.; Rio Carvalho, C.; De Snoo, G.R.; Eden, P. Ecological impacts of arable intensification in Europe. *J. Environ. Manag.* **2001**, *63*, 337–365. [CrossRef]

10. Al Bassit, L.; Tricot, N. *Improvement of Manure Spreaders Safety—Feasibility Study*; Irstea Report; IRSTEA: Antony, France, 2013; pp. 1–20.

11. Al Bassit, L.; Tricot, N. *Improvement of Manure Spreader Safety in the Cleaning Phase—Feasibility Study. Action No. 2—Addendum*; Irstea Report; IRSTEA: Antony, France, 2014; pp. 1–12.

12. Poisson, P.; Chinniah, Y.; Jocelyn, S. Design of a safety control system to improve the Verification step in machinery lockout procedures: A case study. *Reliab. Eng. Syst. Saf.* **2016**, *156*, 266–276. [CrossRef]

13. Westaby, J.D.; Lee, B.C. Antecedents of injury among youth in agricultural settings: A longitudinal examination of safety consciousness, dangerous risk taking, and safety knowledge. *J. Saf. Res.* **2003**, *34*, 227–240. [CrossRef]

14. European Committee for Standardization (CEN). *Agricultural machinery—Manure spreaders—Safety*; EN 690:2013; European Committee for Standardization: Brussels, Belgium, 2013.

15. Colantoni, A.; Mazzocchi, F.; Laurendi, V.; Grigolato, S.; Monarca, F.; Monarca, D.; Cecchini, M. Innovative solution for reducing the run-down time of the chipper disc using a brake clamp device. *Agriculture* **2017**, *7*, 71. [CrossRef]

16. Accident Statistics LSV-SpV. Available online: http://www.svlfg.de/suche/index.html?wm=sub&m=all& ps=10&q=Unf%C3%A4lle+Landwirtschaft&Submit=finden (accessed on 21 June 2017).

17. Banca Dati Statistica INAIL. Available online: http://bancadaticsa.inail.it/bancadaticsa/login.asp (accessed on 21 June 2017).

18. Canadian Agricultural Injury Reporting. Agricultural Fatalities in Canada 1990–2008. Available online: http://www.cair-sbac.ca/wp-content/uploads/2012/03/National-Report-1990-2008-FULL-REPORT-FINAL-EN.pdf (accessed on 21 June 2017).

19. INAIL. Gli Infortuni Sul Lavoro E Il Sistema Informo. Available online: https://www.inail.it/cs/internet/ comunicazione/pubblicazioni/catalogo-generale/gli-infortuni-sul-lavoro-e-il-sistema-informo.html (accessed on 21 June 2017).

20. Institut National De Recherche En Sciences Et Technologies Pour L'environnement Et L'agriculture. Available online: http://www.irstea.fr/search/node/base%20de (accessed on 21 June 2017).

21. Kogler, R.; Quendler, E.; Boxberger, J. Analysis of occupational accidents with agricultural machinery in the period 2008–2010 in Austria. *Saf. Sci.* **2015**, *72*, 319–328. [CrossRef]

22. Occupational Safety and Health Administration. Reports of Fatalities and Catastrophes—Archive. Available online: https://www.osha.gov/dep/fatcat/dep_fatcat_archive.html (accessed on 21 June 2017).

23. Azadeh, A.; Shams Mianaei, H.; Asadzadeh, S.M.; Saberi, M.; Sheikhalishahi, M. A flexible ANN-GA-multivariate algorithm for assessment and optimization of machinery productivity in complex. *J. Manuf. Syst.* **2015**, *35*, 46–75. [CrossRef]

24. Booth, R.T. Machinery safety: Progress in the prevention of technological accidents. *Saf. Sci.* **1993**, *16*, 247–248. [CrossRef]

25. Sadeghi, L.; Mathieu, L.; Tricot, N.; Al Bassit, L. Developing a safety indicator to measure the safety level during design for safety. *Saf. Sci.* **2015**, *80*, 252–263. [CrossRef]

26. International Organization for Standardization (ISO). *Safety of Machinery—General Principles for Design—Risk Assessment and Risk Reduction*; ISO 12100:2010; International Organization for Standardization: Geneva, Switzerland, 2010.

27. Atkinson, K. SWOT analysis: A tool for continuing professional development. *Int. J. Ther. Rehabil.* **1998**, *5*, 433–435. [CrossRef]

28. Bull, J.W.; Jobstvogt, N.; Bohnke-Henrichs, A.; Mascarenhas, A.; Sitas, N.; Baulcomb, C.; Lambini, C.K.; Rawlins, M.; Baral, H.; Zahringer, J.; et al. Strengths, weaknesses, opportunities and threats: A SWOT analysis of the ecosystem services framework. *Ecosyst. Serv.* **2015**, *17*, 99–111. [CrossRef]

29. Lin, F.; Chen, X.; Yao, H. Evaluating the use of Nash-Sutcliffe efficiency coefficient in goodness-of-fit measures for daily runoff simulation with SWAT. *J. Hydrol. Eng.* **2017**, *22*. [CrossRef]

30. Marek, G.W.; Gowda, P.H.; Evett, S.R.; Baumhardt, R.L.; Brauer, D.K.; Howell, T.A.; Marek, T.H.; Srinivasan, R. Calibration and Validation of the SWAT model for predicting daily ET over irrigated crops in the Texas High Plains using lysimetric data. *Trans. ASABE* **2016**, *59*, 611–622. [CrossRef]
31. Haile, M.; Krupka, J. Fuzzy evaluation of SWOT analysis. *Int. J. Supply Chain Manag.* **2016**, *5*, 172–179.
32. Is SWOT analysis still fit for purpose?: The management tool has been exploring strengths, weaknesses, opportunities and threats for decades. *Strateg. Dir.* **2015**, *31*, 13–15. [CrossRef]
33. Pandya, S. Improving the learning and developmental potential of SWOT analysis: Introducing the LISA framework. *Strateg. Dir.* **2017**, *33*, 12–14. [CrossRef]
34. Rimediotti, M.; Sarri, D.; Cavallo, E.; Lombardo, S.; Lisci, R.; Vieri, M. Innovative mechatronic solutions for decoupling in agricultural machinery. *Chem. Eng. Trans.* **2017**, *58*, 91–96. [CrossRef]
35. Promosic: Improving the Safety of Manure Spreaders. Available online: https://youtu.be/w5vDZhzcvZY (accessed on 21 June 2017).
36. International Organization for Standardization (ISO). *Safety of Machinery—Safety-Related Parts of Control Systems—Part 1: General Principles for Design*; ISO 13849-1:2015; International Organization for Standardization: Geneva, Switzerland, 2015.
37. International Organization for Standardization (ISO). *Safety of Machinery—Safety-Related Parts of Control Systems—Part 2: Validation*; ISO 13849-2:2012; International Organization for Standardization: Geneva, Switzerland, 2012.
38. International Electrotechnical Commission (IEC). *Safety of Machinery–Functional Safety of Safety-Related Electrical, Electronic and Programmable Electronic Control Systems*; IEC 62061:2005; International Electrotechnical Commission: Geneva, Switzerland, 2005.

agriculture

MDPI

Review

The Analysis of the Cause-Effect Relation between Tractor Overturns and Traumatic Lesions Suffered by Drivers and Passengers: A Crucial Step in the Reconstruction of Accident Dynamics and the Improvement of Prevention

Carlo Moreschi [1], Ugo Da Broi [1,*], Sirio Rossano Secondo Cividino [2], Rino Gubiani [2], Gianfranco Pergher [2], Michela Vello [3] and Fabiano Rinaldi [3]

[1] Department of Medical Area, Forensic Medicine Section, University of Udine, Piazzale S. Maria della Misericordia 15, 33100 Udine, Italy; carlo.moreschi@uniud.it
[2] Department of Agricultural, Food, Environmental and Animal Sciences, Agricultural Engineering Section, University of Udine, Via delle Scienze 208, 33100 Udine, Italy; agricolturasicura@gmail.com (S.R.S.C.); rino.gubiani@uniud.it (R.G.); gianfranco.pergher@uniud.it (G.P.)
[3] Sofia & Silaq Corporate Spin-Off, University of Udine, Via Zanon 16, 33100 Udine, Italy; michela.vello@uniud.it (M.V.); sofia.innovazione@gmail.com (F.R.)
*   Correspondence: ugo.dabroi@uniud.it; Tel.: +39-0432-554-363; Fax: +39-0432-554-364

Received: 5 October 2017; Accepted: 27 November 2017; Published: 2 December 2017

Abstract: The evaluation of the dynamics of accidents involving the overturning of farm tractors is difficult for both engineers and coroners. A clear reconstruction of the causes, vectorial forces, speed, acceleration, timing and direction of rear, front and side rollovers may be complicated by the complexity of the lesions, the absence of witnesses and the death of the operator, and sometimes also by multiple overturns. Careful analysis of the death scene, vehicle, traumatic lesions and their comparison with the mechanical structures of the vehicle and the morphology of the terrain, should help experts to reconstruct the dynamics of accidents and may help in the design of new preventive equipment and procedures.

Keywords: farm tractor; occupational accidents; prevention

1. Introduction

Farm tractors are heavy, large, powerful vehicles. If they are used (a) without the right safety equipment as Roll Over Protective Structures (ROPS), seat belts, helmets, crush proof walls of cabs, (b) without a correct evaluation of operational risks (mechanical peculiarities of the vehicle, unstable terrain, towing an excessive load, driving on sloping and irregular or slippery ground where there is a low coefficient of traction) or (c) when the overturn angle is exceeded, they may overturn and throw the occupant(s) onto the ground and crush them [1–8].

The influence of engineering research and techniques on the construction of farm tractors has, in recent years, been seen mainly in the development of and improvements to ROPS, on the morphology and protection of the Deflection Limiting Volume (DLV), and on safety equipment such as seat belts, helmets, crush-proof cab walls and safety shields for the power take-off (PTO) but nearly 50% of fatal farm accidents still involve tractors, while a significant number of rollovers (50–60%) result in the death of the drivers or passengers [1–5].

Etherthon et al. reported that 59% of tractor-related fatalities occur in agriculture, forestry and fishing, with the remaining cases occurring in the manufacturing, services and construction sectors [1].

A variety of agricultural activities have been identified by the US Government Centers for Disease Control and Prevention as being frequently associated with tractor rollovers. These include using rotary mowers (32%), transporting equipment or farm products (21%), checking livestock or property (14%), hauling logs (11%) and planting, ploughing or cutting hay (11%) [9].

Pickett et al. also looked at the incidence of fatal injuries in work-related farming accidents and found that 9.6% of deaths occurred because of sideways overturns and 6.4% in rear or front overturns [3].

The Directives of the European Community and the Organization for Economic Co-operation and Development (OECD) are constantly striving to improve the technical manufacturing parameters and research procedures to analyze the ability of vehicles to withstand impacts and crushing and to devise new regulations which aim to prevent overturning, to guarantee the solidity of cabins and to preserve the DLV: types approvals refer to each category of tractor and involve specific variants as the number of powered axles, steered axles, and braked axles, ROPS and preservation of the DLV [7,10,11].

However, despite ongoing engineering research, there are still no exhaustive analytical procedures or new investigative methodologies which might enable us to evaluate and thus prevent injuries or fatalities to operators and passengers when tractors overturn.

The analytical procedures normally used by coroners and forensic pathologists to describe the morphology and the anatomical location of traumatic lesions and to correlate them with the causative vectorial forces acting on the human body during accidents, appear to be useful for agricultural engineers in order to clarify the dynamics of tractor overturns and to plan preventive devices and operative procedures [12].

2. Tractor Overturn Risk Factors

The main reasons why tractors overturn, are:

1. Human behavioral factors in which tractor drivers:

    (a) ignore or fail to observe correct standards of conduct when behind the wheel
    (b) corner abruptly and at speed
    (c) are working alone for long periods, in adverse environmental and weather conditions, in isolated, rural areas (82% on farms and only 18% on public roads, as reported by the US Centers for Disease Control and Prevention) where it may be extremely difficult to get rapid access to emergency services and medical aid; such work may also be performed at night without an efficient lighting system [9,12].
    (d) may have been drinking or taken drugs, thus affecting reaction times
    (e) may be elderly and have cardiovascular or neurological issues which dangerously affect reaction times and the ability to recover from trauma

2. Factors involving ground and weather conditions:

    When maneuvering the tractor on a slope at more than the α overturn angle, side, rear or front rollover will result (the α overturn angle and the % gradient of the slope are represented mathematically by the following equation: $i = a_t/2\ h\alpha$, where i is the percentage gradient, a_t is the tractor's wheel track, h is the height of the center of gravity, and α the angle between the incline and the horizontal ground line) [12–18].
    In wet or icy conditions:

    (a) the ground can become slippery with the tractor operating in conditions where there is a dangerous coefficient of traction which leads to side, rear or front overturns (The coefficient of traction between two surfaces, e.g., rubber tyre and ground surface, is expressed by the following equation: $A_f = C_a \times C_f$, where A_f is the frictional force which resists the relative motion between two surfaces (tyre and ground), C_a is the coefficient of traction between those two surfaces, and C_f is the compression force involving two opposing surfaces (i.e., the weight bearing upon the wheel)) [18].

(b) verges, escarpments and the banks of waterways may become waterlogged and give way.

3. Factors involving the technical or functional characteristics of tractors: farm tractors have a high center of gravity and/or a narrow axle track; they may be rather old and not equipped with adequate or upgraded safety systems; they may also be poorly maintained and have the wrong tyre pressures.

4. Factors due to the behavior of machinery and equipment towed by a tractor and coupled to the PTO:

(a) excessive loads towed by a tractor.

(b) excessive loads may be towed by a tractor and coupled to a functioning PTO; in both cases the operator may fail to consider the fact that the PTO coupling and the heavy load will cause the vehicle to behave differently when, for example, cornering or traversing a slope.

Besides human, technical, environmental and weather risk factors there are other critical risk factors which may increase the number of injuries and fatalities: adults and minors transported as passengers (Purschwitz et al. reported that victims of tractor rollovers range in age from less than 1 year to over 90) may be seated in inappropriate places on the vehicle and not be wearing seat belts or helmets; in these cases, tractor rollovers can result in very serious trauma and extensive crushing; obviously children, because they are physically smaller and have less resistant tissue, can suffer devastating trauma, with crushing, bone fractures, and severed limbs in various areas of the anatomy, with consequent polytraumatic shock [9,14].

Dogan et al. reported that tractor rollovers cause more fatalities among passengers, both adults and children, than among drivers: this is probably since adult drivers are afforded greater protection by the ROPS, and are physically more robust than children [12,15–17].

Even when the vehicle is equipped with ROPS, drivers can suffer fatal injuries because of tractor rollover, especially when they are not wearing a seat belts. Such injuries result from violent impact of the head, chest, spinal column or limbs against the internal surfaces of the cabin or the steering wheel, or may occur when external objects, rocks, branches or tree trunks penetrate the cabin safety zone (DLV) during or at the end of the overturn [2,7,10,12].

Researchers and engineers are currently trying to fully understand the dynamics of tractor overturns, which may involve different impact points on the body, and to explain the presence of lesions in different anatomical areas, to be able to identify critical safety issues. Unfortunately, experimental tests involving the use of dummies or prototypes still fail to properly explain the dynamics of an overturn, and the impact upon and deformations of the cabin and ROPS through the absorption of kinetic energy. As a result, the tests do not help us to understand how we can counteract the action of those vectorial forces responsible for driver injuries or death [18].

Over the last few decades manufacturers have tried to produce specific safety systems incorporating inclinometers or position-sensors to alert the driver to an increasing risk of rollover, and have connected such devices to recording devices, which function rather a like aircraft flight recorder. However, technical improvements still should be made to these instruments if they are to supply exhaustive data to researchers, engineers and manufacturers about the dynamics of tractor rollovers [18].

Coroners, forensic pathologists and agricultural engineers should work together closely not only to complete investigations required by the courts but also to support research on the dynamics of tractor overturns and the design of new vehicles, preventive equipment and operational guidelines: the knowledge of physical, mechanical and pathophysiological risk factors should be synergistically taken into account by technical and medico legal experts when investigating the consequences of tractor rollovers or researching preventive equipments or procedures.

3. Genesis of Traumatic Lesions Caused by Tractor Overturn

Morphological analysis of the lesions suffered by the victims and their compatibility with the mechanical structures of the tractor or with features of the terrain play a key role in the reconstruction of accidents. This procedure is normal practice for coroners but is rarely carried out by agricultural engineers [19–21].

Traumas or injuries caused by farm vehicles and equipment have specific and recognizable characteristics and can provide useful information not only for coroners but also for technicians, engineers, builders and researchers [19–21].

When part of a tractor strikes the human body, it produces macroscopic and microscopic modifications to both the superficial and deep tissue, depending on both the amount of energy absorbed and on the shape of the part or parts of the vehicle, ground, branches, rocks, etc., which are in contact with the driver's body. Obviously, the damage inflicted upon those tissue structures which absorb the energy is also determined by the resilience, elasticity and deformability of the tissue itself [19–21].

The same results are seen when the moving human body strikes a stationery or moving object.

The forces which act upon the human body are the same as those which acts on any physical structure (e.g., breaking or deforming parts of the vehicle, the ground or trees in the immediate area of the accident), and mainly involve mechanisms of compression, traction, bending and torsion.

The human body is equipped with various tissue components, each of which has specific characteristics of strength, elasticity and deformability and so the effect of damaging, external forces, whether single or multiple, opposed or synergetic, can generate widely varying traumatic results.

In the case of the compression and traction of human tissue, as with all other inanimate materials, the resulting deformation is expressed by the following equation: $E = \sigma/\varepsilon$ where:

E is Young's Modulus expressed in Newtons/surface area in m^2 of the body involved
σ = force/surface, is the ratio between the applied force and the surface area of the body involved, orthogonally to the force applied
$\varepsilon = \Delta l/l$ is the ratio between the length of the body after and before the load is applied.
Deformation caused by bending processes is expressed by the following equation:
$H = \sigma/\varepsilon$ where:
H is the Flexural Modulus expressed in Newtons/ surface area in m^2 of the body involved
σ = force/surface is the ratio between the applied force and the surface area of the body involved
$\varepsilon = \Delta l$ is the amount of flexion.
The deformation produced by torsion is expressed by the following equation:
$K = \sigma/\theta$ where:
K is the Torsion Modulus expressed in Newtons/ surface area in m^2 of the body involved
σ = force momentum/surface is the ratio between the applied force and the surface area of the body involved
$\theta = \Delta°$ is the torsion angle.

Another important factor to consider is the length of time that the force (be it compression, traction, bending or torsion) is acting upon the body. These forces can be constant, increasing or decreasing, and release different amounts of energy [18–21].

4. Traumatic Injury Patterns Due to Tractor Overturn

The injuries resulting from tractor rollover which are of interest from a medico-legal and an engineering point of view, similarly to those caused by other vehicles, are well-known in forensic pathology and are usually revealed during the external examination of the body, and fall into the following categories:

- excoriation
- ecchymosis
- brush burn abrasions

- blistering
- tearing
- tearing and bruising
- cuts
- sharp injuries
- cuts and sharp injuries
- tissue loss.

These lesions are normally caused by impact with surfaces which have:

- flat surfaces [i.e., side walls of the cabin, mudguards, engine covers, as well as the ground (farmland, tracks or roads)].
- uneven surface which may be rounded, pointed, sharp or irregular (i.e., the ROPS, the steering wheel, uncovered parts of the engine, type tread, rocks, branches or tree-trunk slying on the ground) [12].

In addition to superficial lesions, serious deep bone fractures and organ ruptures may occur, owing to the significant amounts of energy they have absorbed.

According to where they are located, characteristic morphological features may be observed: wounds near broken-off bone stumps which look like cuts because the bone slices through the skin (as well as the muscles, blood vessels and nerves) from the inside and tend to produce wounds with neat edges [12,16,17].

Bone tissue offers poor resistance to torsion, traction and bending, but copes better with compression; fractures can be caused by means of a mechanism of direct absorption of an external force or by transmission of an external force absorbed in a specific area of the skeletal structure distant from the point of fracture (for example, the fracture high on the femur, at pelvis level, due to impact with the foot, lower leg or knee). The skull is of interest and importance here: cranial fractures can be caused by impact with large flat surfaces (the most common scenario is that of a fall) or impact with edges which may be rounded, sharp, pointed, or irregular (parts of the tractor, rocks, stones, branches or tree trunks).

While different areas of the cranium vary in thickness and strength, it has been estimated that the cranial vault can withstand deflections of several millimeters without fracturing. Impact with flat surfaces can cause linear fractures of the skullcap which radiate out from the point of impact (caused by the bending first of the inner and then the outer cranial tables) and circular fractures (caused by the bending first of the outer and then the inner cranial layers). In the case of uneven surfaces, the fracture may be depressed, with the size and shape corresponding to the impacting object or structure [12,16,19–21].

The rupture of internal organs is more frequent in cases of massive trauma, when the forces acting upon the body are single or multiple and synergetic but of high intensity, as is the case with violent impact, crushing, traction or fragmentation [12,16,19–21].

The internal organs also have specific characteristics of resistance, elasticity and shock absorption, while the solid and hollow organs behave differently.

The mechanisms involved may be direct or indirect: the transmission of the force, the acceleration and deceleration produced by the impact can cause the detachment of muscles, tendons and vascular peduncles of organs [12,16,19–21].

The typologies of fatal injury which can occur because of farm tractor rollovers, may involve the driver or passengers being:

- thrown to the ground and crushed by the vehicle with lethal injuries to the chest, head or limbs
- thrown to the ground and suffering serious or fatal injuries due to the fall and collision with rocks, tree trunks, branches or the ground/road

- thrown to the ground and crushed more than once by the machine in cases of multiple rollovers and then found fatally crushed at some distance from the machine
- thrown to the ground and crushed more than once by the tractor in the event of multiple rollovers, with the victim found crushed under the vehicle
- thrown to the ground and run over by the still moving tractor
- thrown into water (streams, irrigation channels, ponds, lakes, etc.), crushed and drowned
- entangled in and/or strangled by the moving tractor parts resulting in lethal mutilation (i.e., from the PTO or Power Take-Off)
- injured by foreign objects, such as rocks, branches or tree trunks penetrating the cabin safety zone during single or multiple rollovers
- burned after the contact with hot parts of the engine or the exhaust, or burned/carbonized after the vehicle caught fire [6,7,10,11,15,16,19,20,22–26]

5. Pathophysiology of Traumas Caused by Tractor Overturn

From the pathophysiological point of view Goodman et al. reported that fatalities due to tractor accidents may result in the chest being crushed (82.6%), exsanguination due to thoracic or extrathoracic lesions (4.4%), strangulation or asphyxia (4%) and drowning (3%) [20].

In rare cases of fire, the victims may display evidence of burning or carbonization, in addition to trauma injuries.

As reported by Bernhardt el al. the intensity of the vectorial forces acting upon the bodies of the driver or passengers in tractor rollover accidents is demonstrated by the fact that nearly three-fourts of the victims die in the first hour after the accidents while nearly 87% die within the first 24 h [12,20,27].

Moreover, as reported by Myers et al., the percentage of deaths resulting from side overturns in ROPS-equipped tractors is less than half those occurring when non ROPS-equipped tractors overturn (1.6% versus 3.7%) [7,20].

Similarly, Cole et al. reported that 1.12% of deaths involved ROPS-equipped tractor overturns (to the side, rear and front) as against 5.42% in tractors not equipped with ROPS [28].

Cole et al. also underlined that surviving victims of rollover accidents may suffer temporary disability in 13.5% and permanent disability in 3.16% of cases after accidents involving non-ROPS-equipped tractors [28].

In cases where the victim is crushed, either by the vehicle or other objects, if the weight is concentrated on the chest area, death is caused progressively by crush asphyxia, following compression of the rib cage and the arrest of respiratory movements and alveolar ventilation. In such cases, external post-mortem examination may reveal the "ecchymotic mask" phenomenon, characterized by conjunctival and facial petechiae, and intense purple congestion and swelling of the head, face, neck, upper chest and sometimes the upper limbs [16,29–32].

Petechiae may also be found in the oral mucosa while bulging of the eyeballs and epistaxis may also occur. The presence of cutaneous and mucous petechiae, purple congestion and swelling of the upper body is caused by: (a) an increase in venous capillary pressure owing to the reduced return flow to the right chambers of the heart (this increase is facilitated by the absence of valves in the main veins of the neck and head); (b) persistent arterial flow towards neck and head, c) the crush victim performing an involuntary Valsalva manoeuvre, which produces a further increase in intrathoracic pressure and a further reduction in the return flow to the left heart chambers [16,29–32].

Obviously, this compression of the chest does not only produce haemodynamic effects (increase in the peripheral venous pressure in the upper body, associated with the continued arterial blood flow towards the periphery) but also respiratory consequences, with hypoxaemia caused by the arresting of breathing movements and alveolar ventilation [18,21,31,33 37].

If the chest is immobilized but not crushed, the discoloration of the skin associated with crush asphyxia will be absent since there will be no increase in venous capillary pressure; but because the breathing movements and alveolar ventilation will be obstructed, arterial hypoxaemia will occur and lead rapidly to death by cardiac arrest [18,29–31,33,34].

Another type of injury is that caused by the victim being trapped under the wheels of the still-moving tractor: significant lesions in the affected parts of the body, here we see serious damage to internal organs, bone fractures and severe surface wounds (lacerated and contused wounds, ecchymosis, patterned excoriations, tissue loss which may mirror the shape of tractor tyres or treads); such wounds are normally caused by the crushing action of a rubber tyre or continuous track [16,29–31].

It is also important to distinguish, during the post-mortem examination, between injuries caused when the body of the victim is thrown to the ground and those which follow in rapid succession when the victim, already on the ground, is crushed by part of the tractor: there will be superficial wounds caused by the impact with the ground (which may be flat or irregular with stones, rocks, branches or tree-trunks) and both superficial and deep wounds produced by crushing under the tractor and by contact with specific structural parts of the vehicle [12,16,29–31].

6. Morphology, Anatomical Location and Cause-Effect Relation of Traumatic Lesions Due to Tractor Overturn

The observation at post-mortem of the morphological characteristics, the topographic anatomy of the lesions caused by the rollover and the mechanisms by which they are produced, is an essential part of the analysis of the dynamics of accidents caused by tractors and agricultural machinery [12,18].

Depending on the vectorial forces involved and the shape of the vehicle and its component parts, the lesions may superficially have varying morphologies and be attributable to single or multiple points of impact (POI) with specific structural or mechanical parts of the vehicle [16,19–21].

Macroscopic impacts, both superficial and deep, against the structural and mechanical parts of the vehicle, and with the ground surface and foreign objects outside the vehicle, must always be identified, studied and interpreted during the post-mortem examination to understand accident dynamics [16,19–21].

In cases of impact (with or without dragging) or compression the following may be observed: patterned excoriations or ecchymosis, and wounds (lacerated, lacerated/contused, cuts and cuts/sharp injuries), or tissue loss corresponding to parts or surfaces of the vehicle in question, and these can be useful in reconstructing the dynamics of the accident. The shapes and patterns of the excoriation and ecchymosis can in any case point to contact with structural or mechanical parts of the tractor, be they large or small, and which are analogous to those produced trucks in road traffic accidents.

The coroner during the necroscopic examination, and the engineers during the technical examination of the vehicle, should always consider the fact that accidents caused by farm tractors overturning involve phenomenally high forces, rapid acceleration and significant mass so that the impact of singular or multiple vectorial actions on the body of victims is often devastating and with lethal consequences [18].

As reported by Goodman et al., such vectorial actions may cause, especially in cases where the victim's chest is crushed, a variety of traumatic lesions associated with thoracic immobilization and compression, such as fractured ribs (non-displaced fractures, compound fractures, flail chest with serious ventilatory and haemodinamic impairment), sternal and clavicular fractures, spinal and scapular fractures, bruising and detachment of large areas of tissue from the external surface of the chest [12,18,20].

Such injuries may also cause endothoracic, parenchymal and vascular lesions. Organs located inside the mediastinum may also be damaged. There may be various intrathoracic consequences such as lung collapse, haemopneumothorax, congestion, contusion or lacerations of lungs sometimes associated with subcutaneous emphysema [18,20].

Extrathoracic anatomical areas may also suffer crushing, such as the head, maxillo-facial structures, cervical or lumbar spinal cord, abdominal and pelvic structures and limbs, which lead to fatal traumatic and haemorrhagic shock; the loss of limbs may cause massive exanguination [3,20].

Dogan et al. reported a frequency of lethal lesions in different anatomical areas as follows: head (33%), chest (10.5%), abdomen (2.3%), the extremities (1.2%) [12].

Gassend et al. reported that, in cases of tractor rollovers, 43% of fatalities involved extrathoracic injuries and 21% involved a combination of head and pelvic traumas. Dogan et al. presented the following results for a variety of combined lesions:

- head and chest (16.3%)
- chest and abdomen (12.8%)
- head, chest and abdomen (9.3%)
- head, chest, abdomen and extremities (5.8%)
- chest, abdomen and extremities (2.3%)
- head, chest and extremities (1.2%)
- head and abdomen (1.2%)
- head, abdomen and extremities (1.2%)
- head and extremities (1.2%)
- abdomen and extremities (1.2%) [2,12].

The variability of the locations of single and combined lesions reported by Dogan et al. confirms that the dynamics of tractor rollovers involve a multiplicity of vectorial forces [12].

Rees also confirmed that the trunk (chest, spinal column and pelvis) is more likely to be injured than the head or extremities and reported that injuries due to tractor overturns may cause the death of the driver in one out of four cases [38].

These data were confirmed by Gassend et al. who reported that 81% of victims of a tractor overturn normally die at the scene of the accident, 8% on the way to hospital and 11% after reaching the hospital [2].

Ince et al. similarly reported a significant frequency (48.8%) of deaths at the scene of the accident or during transportation to hospital [39].

Furthermore, Cogbill et al. reported that in the event of multiple injuries involving different anatomical areas of the body, the sum of the frequencies of all cases involving the chest amounts to 40%. This is due to the large number of different lesions, both superficial and deep, which are caused when the structural parts of the machine hit the surface of the chest. The contour of the lesions, especially those seen on the chest, may match parts of the engine block, the edge of the rear tyre or the mudguard; the shape of the superficial lesions may also match an even ground surface or any stones, rocks, tree-trunks, branches lying upon the area where the victim falls, before being hit and crushed by the vehicle [12,40].

Therefore, a key part of the post-mortem examination performed by coroners is the identification, description and comparison of the impact points on the body with the ground, with the tractor's structure and mechanical parts. The presence of side, front or rear impact points on the body compatible with contact with the structure or parts of the vehicle is very important in the reconstruction of the tractor's direction of roll, i.e., to the side, forwards or backwards [18].

7. Medicolegal and Technical Implications

Coroners may be required to answer the courts' questions about the whole dynamic profile of the accident and its consequences to determine the circumstances of the accident, the traumatic lesions inflicted and the cause of death. Therefore, post-mortem investigations should aim to prove that a specific vehicle was involved in the accident, to explain why and how it overturned, and to evaluate the cause-effect relation between the dynamics of the overturn and the lethal injuries caused by the impact of structural or mechanical parts of the vehicle upon the victim's body.

The main task of coroners and medico-legal investigators in cases of farm tractor overturn is to ascertain the real occurrence of the accident and to demonstrate a mechanistic cause-effect relationship between the action of specific parts of the vehicle and the lesions observed.

To be sure that a death due to fatal traumatic injuries occurred after a tractor overturn, the following investigative steps are of crucial importance to integrate the necroscopic findings with what was observed at the scene of the accident:

- to interact with experts in the field of agricultural engineering to evaluate and discuss any technical and mechanical issues which may help to understand the dynamics of the event (single sideways rollover within 90°, single/multiple sideways rollover more than 90°, single or multiple rear or front rollovers).
- to evaluate the death scene, the structural, mechanical and technical features of the vehicle, its direction of travel before and during the accident, the gradient of the slope and the morphology of the ground, weather and light conditions at the time of the accident and the type of work being performed at the time and its setting, i.e., whether it is (a) agricultural or zootechnical work, in fields or wooded areas, involving pruning, or sawing tree-trunks and branches; (b) normal field work, cultivation of a vegetable garden or arable land; (c) maintenance work such as hedge and grass cutting on farms or in parks and gardens; or (d) processes such as harvesting, haymaking, pruning or irrigation;
- to verify whether certified ROPS were fitted, whether a helmet and seat belt was fitted and in use;
- to analyze whether the morphology and characteristics of the various lesions, both superficial and deep, and in any anatomical area of the corpse, match any specific parts of the machine, ground or objects external to the cabin [12];
- to search for all specific signs of crush asphyxia (distinguishing the distribution of post-mortem lividity from the position of the ecchymotic mask or any ecchymosis in other areas of the body) and all thoracic and extrathoracic traumatic lesions caused by the accident;
- to reconstruct the medical history of the deceased and his/her psycho-physical condition when of the accident;
- to analyze toxicological data for signs of alcohol or drug use;
- to exclude any causes of death other than the lesions produced during the rollover and evaluate the vitality of wounds present at the moment of death, in order to be sure that it was not a homicide made to look like an accident, nor was it a death from other causes not covered by insurance, that was made to look like a fatality caused by tractor rollover;
- to exclude any natural cause of death (i.e., stroke, cardiovascular acute pathologies) responsible for the loss of control of the vehicle and its rollover;
- ascertain how isolated the scene of the accident was, and investigate the involvement of the rescue services (when they were alerted, the distance covered, and the time of arrival at the scene);
- ascertain if the victim died when of the accident, on the way to hospital or after admission to hospital [18].

To summarize, coroners and agricultural engineers need to co-operate to provide, by means of the analysis of the traumatic lesions and the dynamics of farm tractor rollovers, not only technical responses to the questions put by the courts, but also information which may prove to be useful in drawing up preventive criteria and finding solutions which may help us to avoid or mitigate the consequences of the overturning of a farm tractor or other self-propelled farm machinery [18].

8. Concluding Remarks

The co-operation between coroners, forensic pathologists and agricultural engineers can generate innovative methodologies through specific observation and research projects:

- retrospective analysis and statistical description phases with analysis of (a) the level of preparedness and perception among farm tractor drivers of the risk of accidents, (b) the causal dynamics of serious and fatal accidents, including the analysis of fatal or disabling injuries and further work on the demographic characteristics and age of the drivers, on the type of work performed and the topography of the scene of the accident [24]
- experimental simulation phases, with the definition of test scenarios and their relative models (prototype-vehicles, dummies, etc.)
- final phases of proposals and solutions with the design of innovative tractors, equipment/devices and new models of driver behaviour [41].

There remain, however, the following problems regarding tractor rollovers, which demand innovative solutions:

- tractor drivers are often unaware of the risks of an accident while driving, and of the need to adhere to a code of conduct when driving and to use modern safety equipment [11,25];
- even when inside a ROPS-equipped cabin and when wearing seat belt and helmet, the driver can still suffer serious, and sometimes fatal, injury because of a single or multiple rollover beyond 90°, when foreign objects (rocks, branches, tree-trunks, etc.) penetrate the driver's safety zone or, in the event of a multiple rollover, if the ROPS collapses [11,25];
- during rapid acceleration and deceleration when the vehicle is rolling, even when the tractor is equipped with ROPS and the driver is wearing helmet and seat belt, serious trauma can occur, resulting in injuries to the head, chest, abdomen and limbs due to the body hitting the front, rear or side of the cabin interior or foreign objects which intrude into the Deflection Limiting Volume [11,12,25].

In view of the inability of current safety features to properly protect the driver within the DLV, innovative new systems and devices need to be designed and we would accordingly like to make the following recommendations:

- improve experimental observations regarding the dynamics of tractor rollovers and the genesis of the different injuries caused by such accidents (the mechanical characteristics of the vehicle, the kind of accident and its location, the typology and location of lesions, the relative final positions of the victim and the vehicle) [12];
- foster close cooperation between coroners and engineers;
- strive to develop new preventive devices, equipment and procedures (wrap-around seats which reduce lateral movement, compulsory fitting of audio alarms in the cabin, cushioning systems to offer greater protection to front and rear for the head, chest and pelvis, inclinometers which electronically control engine shutdown and braking systems);
- ensure that only properly trained people can drive tractors and other agricultural equipment;
- ensure that people with psychophysical impairments are not allowed to drive tractors and other agricultural equipment; this may include elderly, infirm or retired farmers or members of farming families; this issue is of particular importance nowadays when many countries in Europe are encouraging or forcing people to remain active and keep working longer, with the result that there tend now to be more people driving these vehicles in advanced age, and this may, in turn, increase the number of fatalities due to tractor rollovers [1,8,12,42–44].

We need to improve training and raise awareness among drivers regarding safe driving and the use of existing safety equipment, but at the same time work with industry to promote research into new devices/equipment and operational guidelines.

Only in this way will we be able to reduce the number of deaths, life-threatening injuries and permanent disabilities caused by tractor overturns [1,42–47].

Acknowledgments: No funds or grants were received in support of our research work.

Author Contributions: Carlo Moreschi and Ugo Da Broi researchers of the Forensic Medicine Department evaluated and discussed the dynamics and morphology of traumatic lesions due to tractor overturns while Sirio Rossano Secondo Cividino, Rino Gubiani, Gianfranco Pergher researchers of the Agricultural Engineering Department and Michela Vello, Fabiano Rinaldi researchers of the Sofia & Silaq Corporate Spin-Off examined the technical aspects of such accidents.

Conflicts of Interest: The authors declare no conflict of interest.

References

1. Etherton, J.R.; Myers, J.R.; Jensen, R.C.; Russel, J.C.; Braddee, R.W. Agricultural Machine-Related Deaths. *Am. J. Public Health* **1991**, *81*, 766–768. [CrossRef] [PubMed]
2. Gassend, J.L.; Bakovic, M.; Mayer, D.; Strinovic, D.; Skavic, J.; Petrovecki, V. Tractor driving and alcohol—A highly hazardous combination. *Forensic Sci. Int. Suppl. Ser.* **2009**, *1*, 76–79. [CrossRef]
3. Pickett, W.; Hartling, L.; Brison, R.J.; Guernsey, J.R. Fatal work-related farm injuries in Canada 1991–1995. *Can. Med. Assoc. J.* **1999**, *160*, 1843–1848.
4. Day, L.M. Farm work related fatalities among adults in Victoria, Australia: The human cost of agriculture. *Accid. Anal. Prev.* **1999**, *31*, 153–159. [CrossRef]
5. Bunn, T.L.; Slavova, S.; Hall, L. Narrative text analysis of Kentucky tractor fatality reports. *Accid. Anal. Prev.* **2008**, *40*, 419–425. [CrossRef] [PubMed]
6. Myers, M.L.; Cole, H.P.; Westneat, H.P. Seat belt use during tractor overturn. *J. Agric. Saf. Health* **2006**, *12*, 43–49. [CrossRef] [PubMed]
7. Myers, M.L.; Cole, H.P.; Westneat, S.C. Injury severity related to overturn characteristics of tractors. *J. Saf. Res.* **2009**, *40*, 165–170. [CrossRef] [PubMed]
8. Mariger, S.C.; Grisso, R.D.; Perumpral, J.V.; Sorenson, A.W.; Christensen, N.K.; Miller, R.L. Virginia agricultural health and safety survey. *J. Agric. Saf. Health* **2009**, *15*, 37–47. [CrossRef] [PubMed]
9. Department of Health and Human Services—US Government. Farm Tractor Related Fatalities, Kentucky 1994. Centers for Disease Control and Prevention. *MMWR Wkly.* **1995**, *44*, 481–484.
10. Myers, J.R.; Hendricks, K.J. Agricultural Tractor Overturn Deaths: Assessment of Trends and Risk Factors. *Am. J. Ind. Med.* **2010**, *53*, 662–672. [CrossRef] [PubMed]
11. Rondelli, V.; Guzzoni, A.L. Selecting ROPS safety margins for wheeled agricultural tractors based on tractor mass. *Biosyst. Eng.* **2010**, *105*, 402–410. [CrossRef]
12. Dogan, K.H.; Demirci, S.; Sunam, G.S.; Deniz, I.; Gunaydin, G. Evaluation of Farm Tractor-Related Fatalities. *Am. J. Forensic Med. Pathol.* **2010**, *31*, 64–68. [CrossRef] [PubMed]
13. Fulcher, J.; Noller, A.; Kay, D. Framing tractor fatalities in Virginia: An 11-year retrospective review. *Am. J. Forensic Med. Pathol.* **2002**, *33*, 377–381. [CrossRef] [PubMed]
14. Purschwitz, M.A.; Field, W.E. Scope and magnitude of injuries in the agricultural workplace. *Am. J. Ind. Med.* **1990**, *18*, 179–192. [CrossRef] [PubMed]
15. Beer, S.R.; Deboy, G.R.; Field, W.E. Analysis of 151 agricultural driveline-related incidents resulting in fatal and non-fatal injuries to US children and adolescents under age of 18 from 1970 through 2004. *J. Agric. Saf. Health* **2007**, *13*, 147–164. [CrossRef] [PubMed]
16. Byard, R.W.; Gilbert, J.; Lipsett, J.; James, R. Farm and tractor-related fatalities in children in South Australia. *J. Paediatr. Child Health* **1998**, *34*, 139–141. [CrossRef] [PubMed]
17. Darcin, E.S.; Darcin, M. Fatal tractor injuries between 2005 and 2015 in Bilecik, Turkey. *Biomed. Res. (India)* **2017**, *28*, 549–555.
18. Moreschi, C.; Da Broi, U.; Fanzutto, A.; Cividino, S.; Gubiani, R.; Pergher, G. Medicolegal Investigations into Deaths Due to Crush Asphyxia After Tractor Side Rollovers. *Am. J. Forensic Med. Pathol.* **2017**, *38*, 312–317. [CrossRef] [PubMed]
19. Jones, C.B.; Day, L.; Staines, C. Trends in tractor related fatalities among adults working on farms in Victoria, Australia, 1985–2010. *Accid. Anal. Prev.* **2013**, *50*, 110–114. [CrossRef] [PubMed]

20. Goodmann, R.A.; Smith, J.D.; Sikes, R.K.; Rogers, D.L.; Mickey, J.L. Fatalities Associated with Farm Fractors: An Epidemiologic Study. *Public Health Rep.* **1985**, *100*, 329–333.

21. Kumar, A.; Mohan, D.; Mahajan, P. Studies on Tractor Related Injuries in Northern India. *Accid. Anal. Prev.* **1998**, *30*, 53–60. [CrossRef]

22. Ertel, P.W. *The American Tractor: A Century of Legendary Machines*; MBI Publishing Company: Osceola, WI, USA, 2001.

23. Arndt, J.F. *Roll-Over Protective Structures for Farm and Construction Tractors: A 50 Years Review*; Society of Automotive Engineers: Peoria, AZ, USA, 1971.

24. Caffaro, F.; Lundqvist, P.; Micheletti Cremasco, M.; Nilsson, K.; Prinzke, S.; Cavallo, E. Machinery-related perceived risks and safety attitudes in senior Swedish farmers. *J. Agromed.* **2017**. [CrossRef] [PubMed]

25. Caffaro, F.; Roccato, M.; Micheletti Cremasco, M.; Cavallo, E. Part-Time Farmers and Accidents with Agricultural Machinery: A Moderated Mediated Model on the Role Played by Frequency of Use of and Unsafe Beliefs. *J. Occup. Health* **2017**. [CrossRef] [PubMed]

26. Zardawi, I.M. Coronial autopsy in a rural setting. *J. Forensic Leg. Med.* **2013**, *20*, 848–851. [CrossRef] [PubMed]

27. Bernhardt, J.H.; Langley, R.L. Analysis of tractor-related deaths in North Carolina from 1979 to 1988. *J. Rural Health* **1999**, *15*, 285–295. [CrossRef] [PubMed]

28. Cole, H.P.; Myers, M.L.; Westneat, S.C. Frequency and severity of injuries to operators during overturns of farm tractors. *J. Agric. Saf. Health* **2006**, *12*, 127–138. [CrossRef] [PubMed]

29. Sekizawa, A.; Yanagawa, Y.; Nishi, K.; Takasu, A.; Sakamoto, T. A case of thoracic degloving injury with flail chest. *Am. J. Emerg. Med.* **2011**, *29*, 841.e1–841.e2. [CrossRef] [PubMed]

30. Brinkmann, B. Zur Pathophysiologie and Pathomorphologie bei Tod durch Druckstauung. *Z. Rechts Med.* **1978**, *81*, 79–96. [CrossRef]

31. Sklar, D.P.; Baack, B.; McFeeley, P.; Osler, T.; Marder, E.; Demarest, G. Traumatic asphyxia in New Mexico: A five-year experience. *Am. J. Emerg. Med.* **1988**, *6*, 219–223. [CrossRef]

32. Friberg, T.R.; Weinreb, R.N. Ocular manifestations of gravity inversion. *JAMA* **1985**, *253*, 1755–1757. [CrossRef] [PubMed]

33. Byard, R.W.; Wick, R.; Simpson, E.; Gilbert, J.D. The pathological features and circumstances of death of lethal crush/ traumatic asphyxia in adults a 25-year study. *Forensic Sci. Int.* **2006**, *159*, 200–205. [CrossRef] [PubMed]

34. Williams, J.S.; Minken, S.L.; Adams, J.T. Traumatic asphyxia-reappraised. *Ann. Surg.* **1968**, *167*, 384–392. [CrossRef] [PubMed]

35. Ely, S.F.; Hirsch, C.S. Asphyxial deaths and petechiae: A review. *J. Forensic Sci.* **2000**, *45*, 1274–1277. [CrossRef]

36. Ollivier, D.A. Relation médicale des événements survenus au Champs de Mars le 14 Juin 1837. *Ann. d'Hyg.* **1837**, *18*, 485–489.

37. Perthes, G. Ueber aus gedehnte Blutextravasate am Kopf infolge von Compression des Thorax. *Deutsche Zeitschrift für Chirurgie* **1899**, *50*, 436–443. [CrossRef]

38. Rees, W.D. Agricultural tractor accidents: A description of 14 tractor accidents and a comparison with road traffic accidents. *Br. Med. J.* **1965**, *2*, 63–66. [CrossRef] [PubMed]

39. Ince, H.; Erzengin, O.U. Analysis of Tractor Related Deaths. *J. Agromed.* **2013**, *2*, 87–97.

40. Cogbill, T.H.; Steenlage, E.S.; Landercasper, J.; Strutt, P.J. Death and disability from agricultural injuries in Wisconsin: A 12-year experience with 739 patients. *J. Trauma* **1999**, *31*, 1632–1637. [CrossRef]

41. Chisholm, C.J.A. Mathematical model of tractor overturning and impact behavior. *J. Agric. Eng. Res.* **1979**, *24*, 375–394. [CrossRef]

42. Cummings, P.H. Farm accidents and injuries among farm families and workers: A pilot-study. *Am. Assoc. Occup. Health Nurses J.* **1991**, *39*, 409–415.

43. Rautiainen, R.H.; Ledolter, J.; Donham, K.J.; Ohsfeldt, R.L.; Zerling, C. Risk factors for serious injury in Finnish agriculture. *Am. J. Ind. Med.* **2009**, *52*, 419–428. [CrossRef] [PubMed]

44. Li, Z.; Mitsuoka, M.; Inoue, E.; Okayasu, T.; Hirai, Y.; Zhu, Z. Parameter sensitivity for tractor lateral stability against Phase I overturn on random road surfaces. *Biosyst. Eng.* **2016**, *150*, 10–23. [CrossRef]

45. Tiwari, P.S.; Gite, L.P.; Dubey, A.K.; Kot, L.S. Agricultural injuries in Central India: nature, magnitude, and economic impact. *J. Agric. Saf. Health* **2002**, *8*, 95–111. [CrossRef] [PubMed]

46. Ballesteros, T.; Arana, I.; Ezcurdia, A.P.; Alfaro, J.R. E2D-ROPS: Development and tests of an automatically deployable, in height and width, front-mounted ROPS for narrow-track tractors. *Biosyst. Eng.* **2013**, *116*, 1–14. [CrossRef]

47. Sanderson, W.T.; Madsen, M.D.; Rautianen, R.; Kelly, K.M.; Zwerling, C.; Taylor, C.D.; Merchant, J.A. Tractor overturn concerns in Iowa: Perspectives from the Keokuk county rural health study. *J. Agric. Saf. Health* **2006**, *12*, 71–81. [CrossRef] [PubMed]

agriculture

MDPI

Article

Safety Improvements on Wood Chippers Currently in Use: A Study on Feasibility in the Italian Context

Giorgia Bagagiolo [1] , Vincenzo Laurendi [2] and Eugenio Cavallo [1,*]

[1] Institute for Agricultural and Earth Moving Machines (IMAMOTER), National Research Council of Italy (CNR), Strada delle Cacce 73, 10135 Torino, Italy; g.bagagiolo@ima.to.cnr.it
[2] National Institute for Insurance Against Accidents at Work (INAIL), Via di Fontana Candida 1, Monte Porzio Catone, 00078 Roma, Italy; v.laurendi@inail.it
* Correspondence: eugenio.cavallo@cnr.it; Tel.: +39-01-1397-7724

Received: 30 September 2017; Accepted: 28 November 2017; Published: 3 December 2017

Abstract: Following formal opposition by France, the harmonized safety standards regarding manually-loaded wood chippers (EN 13525:2005+A2:2009) which presumed compliance with the Essential Health and Safety Requirements (EHSR) required by the Machine Directive (Directive 2006/42/EC), have recently been withdrawn, and a new draft of the standard is currently under revision. In order to assess the potential impact of the expected future harmonized standards within the Italian context, this study has examined the main issues in implementing EHSRs on wood chippers already being used. Safety issues regarding wood chippers already in use were identified in an analysis of the draft standard, through the observation of a number of case studies, and qualitative analysis of the essential technical interventions. A number of agricultural and forestry operators and companies participated in the study, pointing out the technical and economic obstacle facing the safety features requested by the pending new standard. It emerged that the main safety issues concerned the implementation of the reverse function, the stop bar, and the protective devices, the infeed chute dimension, the emergency stop function, and the designated feeding area. The possibility of adopting such solutions mainly depends on technical feasibility and costs, but an important role is also played by the attitude towards safety and a lack of adequate information regarding safety obligations and procedures among users.

Keywords: safety; wood chippers; standards; machinery

1. Introduction

Farm machinery is an important contributor to the high rates of occupational injury in agriculture [1–4]. The tractor is the leading cause of accident in agriculture and the major proportion of injuries and deaths are associated with the rollovers [5,6]. Regarding non-tractor agricultural machinery, the most common causes of injury or death in farming are entanglement, crushing and shearing in machines [7]. Moreover, reports on accidents dynamics confirm that a very high number of occupational accidents in agriculture are caused by contact with moving parts such as rollers, conveyors and rotators [8–10]. In these circumstances, a lack of safety features such as manufacturer-made shields, guards, lids, and covers, generally defined as safety devices, contributes to machinery operators injuring upper and lower limbs in moving parts [11–13].

An additional factor regarding occupational injuries is the incorrect behavior of machine operators during field adjustment [9,14,15]; for example, when removing obstructions from the machinery without turning off the machine or after removing protective devices [16]. This brings operators into close contact with components that may present a risk of entanglement [15,17,18]. Risky behavior is also the result of operators not reading operation manuals, particularly the safety warnings [19,20] and often not noticing or understanding the safety instructions and the pictograms affixed on the machine [21,22].

Another significant determining factor that is statistically associated with injuries caused by farm machinery is the age of the machine in question [14]. There are a number of explanations for this: older units usually require more frequent maintenance interventions bringing users into intimate contact with hazardous parts (e.g., moving parts, cutting blades) which indirectly leads to a greater risk for injury [14,23,24]. Furthermore, a study by Baker et al. [25] demonstrated that for each year increase in the age of the machine the odds of injury rise by 4%, particularly in machinery purchased second-hand. The same study concluded that older machinery is more likely to lack certain safety devices, or that those present are somehow deficient. Another study [11] found a correlation between the absence of safety devices and the age of the machine. Likewise, a study conducted in Italy [26], pointed out that among the main causes of accidents in the agricultural sector, the machinery itself is often intrinsically deficient and does not meet the safety requirements due to its age. Indeed, according to literature reports, a large proportion of in-use farm machinery is not equipped with the most up-to-date features required by safety standards [27].

In the Italian context, this scarce compliance with safety standards is mostly due to the economic size of farms. Although the average size of the Italian business has increased, the agricultural sector continues to be characterized by a large number of very small holdings [28]. In 2013, the average size of Italian farms ranged from 7.9 to 8.4 hectares [29], and farms smaller than 5 hectares accounted for 68% of the total [30] while in 2014, holdings with a standard value production of less than € 15,000 represented 54.3% of the total number of farms in Italy [31].

1.1. Manually Loaded Wood Chipper and Specific Hazards

A number of studies indicate that the agro-forestry biomass production sector is characterized by a high incidence of injuries [32–35]. Focusing on this sector, wood chipping machines are the most common cause of crushing, entanglement and shearing hazards [36–38]. Wood chippers are employed in forestry, agriculture, horticulture, and landscaping, and turn wood into "wood chips" in order to reduce volume for subsequent disposal, or for use in bio-energy production [39,40]. There are a number of sizes available: the bigger units have a greater capacity, are usually equipped with their own engine and they are mechanically loaded by a telescopic arm. The smaller units are generally mobile types, trailed or carried by the tractor's rear three-point linkage and coupled with the rear power take-off (PTO) and are manually loaded [41].

Wood chippers basically consist of: (i) a horizontal or near horizontal infeed chute; (ii) infeed components such as rollers or conveyors; (iii) rotating chipping components (made of knives fitted on a drum or a disk), and (iv) a discharge chute. The wood logs or branches are loaded into the infeed chute, and feed rollers at the end of the infeed chute grasp the material and force it into the chipper cutting unit where the knives chip the wood and force the chips through a discharge chute. The chipper knives generally rotate between 1000 and 2000 revolutions per minute [37]. The chipping components container is usually equipped with a removable hood to allow access to the components for maintenance and repair. These smaller machines are generally employed by small businesses and contractors to reduce the volume of logs and branches of limited diameter.

Many of the manually-fed wood chippers are equipped with a mechanical feed control bar that activates the feed rollers when it is pulled [37]. The bar is a pressure-sensitive device usually mounted across the bottom and/or along the sides of the infeed chute for quick and easy activation. The bar should be designed and placed to avoid unintentional activation by a part of the operator's body in the event of entanglement, whereby the infeed action can be stopped. Agro-forestry wood chipping machines and operations present specific occupational hazards [42]; indeed they may be extremely dangerous and potentially life-threatening for operators should they become entangled in the chipping mechanism [36]. This risk is particularly high in manual infeed wood chippers since the operator works close to the infeed chute increasing the chance of contacts with the feeding or chipping components. Indeed the main hazards are related to careless contact with an unguarded infeed, chipping, and power transmission components, but operators can also become caught or snagged by material entering a wood chipper [36]. In many cases, if entangled, I is almost impossible to free oneself as, in an emergency situation, a number of factors—such as the speed of infeed system—may limit the operator's ability to access or activate a feed-stop device [36].

Little information is available regarding accidents involving manually-fed wood chippers in Italy [43]. On the European level, France released some recent data in which the "Bureau Santé Securité au Travail" (Occupational Health and Safety Office) of the French Ministry of Agriculture, Agrifood, and Forestry reported that a severe accident related to the use of wood chipper occurs at least once a year [44]. In most of the reported cases operators involved in the accidents were very young apprentices who did not follow the correct safety conduct. Additional figures refer to the United States where, in the last five years, 20 cases of accidents have been recorded, 15 of which were fatal [45].

1.2. Motivation of the Study

Following a formal opposition by France due to the number of severe and fatal accidents caused by the use of manually-fed wood chippers, on the 17 December 2014 the European Commission withdrew the harmonized standard EN 13525:2005+A2:2009 "Forestry machinery—Woodchipper safety" provided by CEN (the European Committee for Standardization). As a consequence of this European decision, application of this standard by a manufacturer no longer confers presumption of conformity [1] with the essential safety requirements according to the Machine Directive (Directive 2006/42/EC).

Conformity problems, however, also arise for wood chippers already on the market or in-use at the time the standard was withdrawn. This question is particularly pertinent in Italy where farm machinery is on average more than 20 years old [46] (ranging from 15 to 30 years depending on the farm size) and often older units, especially those manufactured before September 1996, without the CE mark, do not comply with some of the Essential Health and Safety Requirements (EHSR) of Machine Directive.

As is the case in other countries in the European Union, in Italy the implementation of fundamental safety requirements is mainly regulated by the Machine Directive and the Italian decree on Occupational Health and Safety (Legislative Decree 81/2008), in application of the European Framework Directive on Health and Safety at Work (Directive 89/391 EEC). In these regards, after 15 May 2008, all machinery, including agricultural and forestry machines, that no longer complies with safety standards and safety regulations should not be employed further. Therefore, regarding units currently in use, the Machine Directive urges farm employers, manufacturers, distributors, rental firms and dealers to assess whether their machinery complies with safety requirements and to adapt them according to the specific harmonized standards. The harmonized standards establish technical specifications considered suitable or sufficient in order to comply with the technical requirements provided by EU legislation [47]. Though European harmonized standards are not mandatory, application thereof is recommended since they provide compliance solutions and confer a presumption of conformity with the relevant essential health and safety requirements of the Machine Directive [48]. Within this framework, the compliance with the directive often leads to confusion amongst operators due to the lack of precise constructive directions, and this represents a serious issue for them, both in terms of technical and economic feasibility.

Based on these considerations, and in the light of the ongoing revision of the EN 13525 safety standard and considering the Italian situation of in-use farm machinery, it is important to verify the practical implementation of minimal safety measures on wood chippers currently in use, especially concerning the risk of getting caught or being pulled into the machinery and to draft measure to correct nonconformity issues.

For this reasons, a study was undertaken for the following points:

1. to check the actual condition and hazards associated with in-use wood chippers and to assess the level of conformity with safety standards;
2. to analyze appropriate solutions to eliminate and/or reduce risks due to contact, entanglement, dragging, cutting and crushing with in-use wood chippers to ensure minimal safety requirements;
3. to point out the main technical and economic issues observed at the local level in implementing the possible technical solutions to achieve minimal safety requirements on machines currently in use.

The outcomes of the study will give a clearer picture of the safety level of machines currently in use in Italy, to check the potential impact of the envisaged new harmonized standards on them, and the possible adoption of technical solutions to achieve the minimal safety requirements outlined in Annex I of Machine Directive and in the harmonized safety standard.

2. Materials and Methods

The study lasted one year, from February 2016 to January 2017 in the Piedmont Region (North-West of Italy). This region is Italy's second for energy production from renewable resources (11.6% of total national production in 2014) and the fourth for biomass production (10.5%). The investigation was carried out in the provinces of Torino and Cuneo where 68.2% of the region's biomass is produced [49].

The study was divided into three main stages, specifically focusing commercial manually-loaded horizontal wood chippers: (i) analysis of regulations and standards for wood chippers, (ii) onsite inspection of in-use wood chippers, and (iii) individual in-depth interviews with operators in the sector.

2.1. Analysis of Regulations and Standards for Wood Chippers

The most relevant regulations and standards related to wood chippers and the identification of major hazards and typical accidents have been examined. A deeper analysis focused on EN 13525, the European harmonized standard specifically regarding manually-loaded horizontal wood chippers, it represents the main reference to assess potential conformity of units currently in use. Special attention was paid to the technical note on 28–29 May 2015 by the French Minister of Agriculture in which the proposals for the implementation of the standard are explained in 10 points, and the new features introduced through the first available drafts of the revised standard. The version of the standard used for this study was the sixth draft revision of the EN 13525, updated on 10 August 2016. The aforementioned documents are unpublished works that were made available to the researchers in the framework of the "Protection of machinery operators against crush, entanglement and shearing" (PROMOSIC) project funded by the Italian National Institute for Insurance against Accidents at Work (INAIL).

2.2. Onsite Inspection of In-Use Wood Chippers

A sample of mobile manually-fed wood chippers used in farms and in forestry cooperatives in Northwest Italy was examined. In-use machines were recruited through direct contacts and referrals by representatives of the sector. During the recruitment phase, a series of characteristics were considered in order to meet the targets of the study. Machines had to be mobile, manually-fed, with a horizontal infeed chute, powered by the tractor's power take off, equipped with integrated infeed components (rollers), and with disk-type chipping components. Thence, a notable number of potential participant were not included in the survey since, especially the forestry cooperatives, as they used larger machines with mechanically-fed systems, which therefore did not meet the targets of the study. Finally, six machines were selected for onsite inspection; those considered the most representative models of in-use manually-fed chippers generally available on farms and used by contractors. A range of brands and manufacturing years was chosen in order to take into consideration a wider array of cases. For this study, the machines were evaluated in terms of compliance with the most recent version of the revised draft of the EN 13525, in expectation of the future publication of a new harmonized standard. Therefore, following the standard requirements and updated sections, onsite measurements and pictures were collected to analyze older units and to carry out a qualitative analysis.

The most significant examples of nonconformity were pointed out and technical interventions to adapt the machinery to the safety requirements were identified. The machines were measured and inspected, with particular focus on the dimensions and the positions of safety devices, while functional tests to check the controls were filmed with a camera.

At this stage, one of the case studies was used to check the speed of the wood during infeed and reversing as an operator loaded a machine with tree logs. The analysis was set by recording how the different controls of the infeed components were activated. The measurements were taken while chipping three logs of pine (*Pinus strobus* L.), three, four, and five meters in length respectively at the machine's top infeed speed. Infeed rate was constant and the time period was measured from the moment the wood hit the blades, until all the wood had been completely chipped.

2.3. Individual In-Depth Interviews with Operators of the Sector

Finally, a number of individuals (*n* = 8) operating in the forestry sector—including manufacturers, suppliers, users and mechanical workshops operators—were interviewed to highlight major technical and economic obstacles against adopting solutions, in compliance with safety standards.

The participants to individual interviews were recruited from farms, forestry cooperatives and manufacturing companies; they were identified through direct contacts and referrals from farmers and agricultural services. Potential respondents were contacted by telephone and given a short description of the study. Prior to each interview, subjects were briefed on the purpose of the survey and their rights as research participants [7]. All respondents provided informed consent.

Key informant interviews were conducted by two researchers [50] through guided semi-structured interview techniques [7], designed to raise the following key points:

- perspectives on risks related to in-use wood chippers and potential accidents;
- opportunity to put align wood chippers currently in use with the requirements requested by the Machine Directive and related standards the already in-use wood chippers;
- observations on the technical and economic implementation of revised safety standards.

During the interviews, one of the researchers played the role of moderator asking some open questions, while the second researcher took notes and asked further questions.

For the manufacturers—including representatives of two of Italy's biggest wood chippers manufacturers—a different approach was taken. They were asked about the main obstacles preventing them from aligning their products with safety standards and the draft of the standard was further discussed. A joint analysis was carried out on the various sections of the safety standard draft and the technical and economic feasibility of potential solutions to increase machinery safety was evaluated. The interviews lasted between 1 and 1.5 h.

Finally, some more recurrent topics were selected to organize and summarize the results of individual interviews.

3. Results

3.1. Regulations and Standards Analysis

The most relevant detected innovations introduced in the draft (sixth revised draft of EN 13525) regarding the withdrawn harmonized standard (EN 13525:2005+A2:2009) are in relation to:

- the functional and positional requirements of operator controls, such as infeed controls, lower and side protective device, top protective devices and emergency stops;
- the positional requirements of infeed components and chipping components, such as infeed chute, infeed conveyors, infeed rollers, and chipping mechanisms.

The safety aspects that the standard draft intends to improve are meant to fulfil essential health and safety requirements of Annex I of the Machine Directive, particular regarding points "1.3.7 Risk related to moving parts", and "1.3.8.2 Moving parts involved in the process" [51].

The revised draft provides a more extended version of Section 4.2 "Operator controls" and an additional Annex which summarizes the main features of the different Stop controls configurations. In particular, this section introduces a new safety device, the "Emergency stop", which the draft defines as a "manually actuated control device used to stop the hazardous functions of the machine as quickly as possible".

3.2. Onsite Wood Chipper Inspections

The recruited manually-fed wood chippers were all equipped with integrated infeed components (rollers) and disk-type chipping components with a variable number of knives (Figure 1). The mean age of the studied machinery was about 14.5 years. The oldest one was manufactured in 1989, and the most recent was manufactured in 2014.

All the six inspected machines had at least one element of nonconformity with regards to the risk protection standards required by the draft of the revised harmonized standard (Table 1).

Table 1. Summary of results of the inspections on in-use wood chippers with regard to how they conform to the revised version (6th revision draft) of EN 13525 standard.

Section of the Standard	Chipper 1 (28 Years)	Chipper 2 (19 Years)	Chipper 3 (12 Years)	Chipper 4 (6 Years)	Chipper 5 (6 Years)	Chipper 6 (3 Years)
§ 4.2.4 Infeed controls	NC	NC	PC	PC	PC	PC
§ 4.2.5 Location of lower and side protective device(s)	NC	NC	NC	PC	PC	PC
§ 4.2.6 Top protective device	NC	NC	NC	NC	PC	PC
§ 4.2.7 Emergency stop	NC	NC	NC	NC	NC	NC
§ 4.3.3.1 Hazard related to infeed and chipping components	NC	NC	PC	C	C	C
§4.3.3.5 Designated feeding area	NC	NC	NC	NC	NC	NC
§ 4.3.4.2 Risks due to infeed speed and reversing of infeed components	C	C	C	C	C	C
§ 4.3.4 Risk due to ejected objects	NC	C	NC	C	NC	C
§ 4.3.5 Protection against access to moving power transmission parts	NC	NC	PC	C	C	C
§ 4.4.2 Hydraulic components	PC	PC	C	C	C	C
§ 4.5 Preparation for transport and maintenance	NC	NC	C	NC	NC	NC

C = conform, PC = partially conform, NC = non-conform.

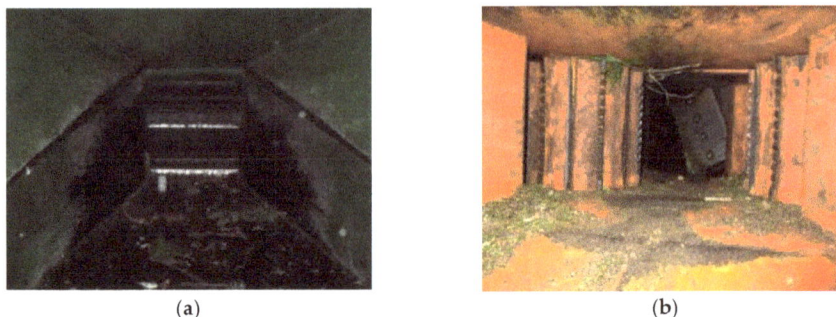

Figure 1. Detail of infeed components (rollers) (a) and knife on the disk shaft (b) of the surveyed in-use wood chippers.

3.2.1. Nonconformities Relating to Entanglement, Crushing and Shearing Hazards

- Infeed controls (§ 4.2.4—6th revision draft of EN 13525)

Although the control mechanism of all machines included the three compulsory functions ("Feed", "Stop", and "Reverse") some position and functional requirements were disregarded. The infeed area was not always visible due to the design of the chute and the control bar. Indeed, in one instance (Figure 2) the upper part of the chute was prolonged by a plate and was encased by a bulging safety bar that did not allow a direct view of the infeed area. Moreover, the positions and functions of operator controls were not always clearly indicated, while these should have been affixed near the control device on both sides of the chute. Especially in older units, respective safety pictograms were deteriorating or missing altogether (Figure 2).

With regard to functional requirements, even in the most recent examined units, the "Reverse" function was not provided with the "hold to run" function.

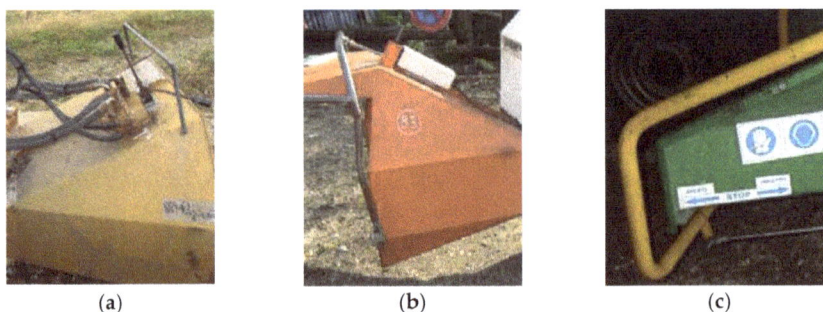

Figure 2. Examples of non-compliant infeed controls for in-use wood chipper. Details shown in figure (a) and (b) highlight the lack of any indication of machine operator controls; while figure (c) shows that the pictograms are placed in the correct position, but the functional requirements are not. Each of the examples included the reverse function but there was no "hold to run control" as requested by the paragraph 4.2.4.3 of the 6th revision draft of EN 13525.

- Location of lower and side protective device(s) relative to the infeed chute edges (§ 4.2.5—6th revision draft of EN 13525)

Half of the machines inspected completely lacked or presented incomplete lower horizontal and side vertical protective devices around the infeed chute edges (Figure 3). These protective devices should be designed and positioned in such a way that minimizes inadvertent activation by wood and branches as they are fed into the machine; at the same time, if entangled in the feed chute, the operator should be able to engage the lower and side protective devices involuntarily with a body part, not just their hands.

The safety devices in the oldest of the units were not red and yellow as recommended by the standard nor were they in bright colors contrasting with the background color of the infeed chute. In those cases, when lower and side protective devices were present, these did not always comply with the location regulations. In particular, regarding side protective devices, in some cases the distance from the outermost edge of the infeed chute was shorter than the standard 150 mm and did not cover at least the 75% of the maximum vertical opening of the infeed chute.

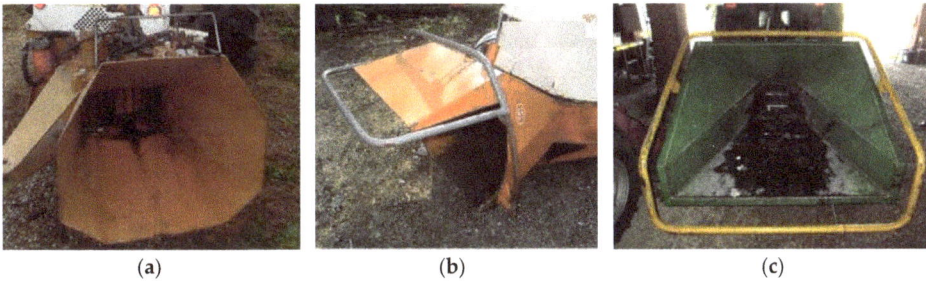

(a) (b) (c)

Figure 3. Some examples of protective devices that do not comply with standard requirements. (**a**) This machine presented no safety devices around the edges of infeed chute. (**b**) Side and upper protective devices were present, but there was no lower protective device (mandatory for the standard). (**c**) The infeed chute was equipped with side, lower and top protective devices, but in order to comply with the safety standard draft, the top protective bar should have a separate control or be replaced by an emergency stop.

- Top protective device (§ 4.2.6—6th revision draft of EN 13525)

The standard revised draft obliges all machinery to be equipped with a Top protective device consisting of one or more different devices that have the exclusive function of halting the infeed components. Unlike lower and side protective devices, the top horizontal bar is for intentional activation from both sides outside of the chute and from the ground. Among the examined case studies, a third of the verified machines did not have any top protective device, while in other cases where a top bar was provided, it was not completely independent from the other controls, meaning that it defaulted on one of the main functional requirements (Figure 4). The combination of a top protective device with other controls prevents a further condition: the operator cannot restart the infeed process merely by returning the top stop protective device to any position. In those machines surveyed that were equipped with the top protective device, the top bar is connected to the functional requirements and follows the same control patterns of the lower and side protective devices, meaning that the infeed action cannot be activated by a separate control. Likewise, for the lower and the side protective bars, the color of the protective device was not always respected.

(a)   (b)   (c)

Figure 4. Examples of the different configurations of protective devices. On the left (**a**) the chute was found non-compliant due to the lack of a top protective device while in a more recent unit (**b**) the top protective bar is combined with side and lower protective devices and these respond to the same controls (**c**). This solution does not conform to the standard either as the top protective device should be independent and there are no emergency stops.

- Emergency stop (§ 4.2.7—6th revision draft of EN 13525)

The emergency stop is a manually-activated control device that differs from other protective devices. Of all the machines surveyed, including the most recent, none of them was equipped with an emergency infeed stop, which is expected to be introduced in the revised standard.

- Hazard relating to infeed components and chipping components (§ 4.3.3.1—6th revision draft of EN 13525)

According to the standard, the height of the chute floor should be 600 mm from the ground and the chute itself should be 1500 mm deep, machines with a lower chute height and/or depth are hazardous as the operator can easily get their hands or feet stuck in the moving parts. In half of the infeed chutes inspected, the lower edge of the infeed chute was insufficiently distant from the ground, and in one instance the horizontal distance from the outer edge of the chute to the reference plane—corresponding to the feeding rollers—was not deep enough (Figure 5).

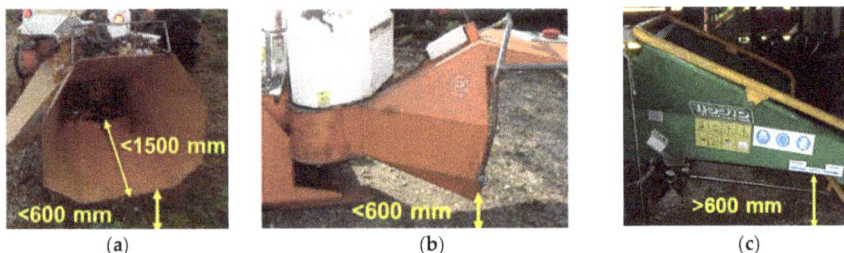

(a)   (b)   (c)

Figure 5. Case studies of infeed chute of in-use wood chippers. Following onsite measurements, in the first case (**a**) the lower edge of the infeed chute was lower than 600 mm from the ground and the chute was less than 1500 mm deep, in the second example (**b**) only the height of chute did not conform to standard requirements, while, in the last example (**c**) all positional measurements of infeed components were satisfactory.

- Designated feeding area (§4.3.3.5—6th revision draft of EN 13525)

The designated feeding area is defined in the standard as "one or more safe areas around the machine indicated by the manufacturer when the operator is manually feeding the machine" and it is particularly important since as it is used as a reference point in defining the positional requirements of operator controls. This area should be determined by the manufacturer, but in the samples examined, the machines lacked clear indications as requested by the standard (Figure 6).

(a)　　　　　　　　　　　　　　　　(b)

Figure 6. Specimens of the same model of wood chipper manufactured in different years on which lacked feeding area identification and safety pictorials were worn or incomplete. Moreover on (**a**) the hydraulic components are freely accessible while in (**b**) they are housed in a metal protection case.

3.2.2. Other Nonconformities Relating to Mechanical and Non-Mechanical Hazards

During the on-site verifications, a number of nonconformities, regarding hazards other than those related to infeed and chipping devices were detected, namely: risks due to ejected objects, risks due to moving power transmission parts, and risks due to non-mechanical hazards (e.g., hydraulic components, see figure 6). Though these kind of risks were not the core target of the study, they are worth reporting as they contribute to a clear overall view of the complete range of cases in which harmonized standard requirements are not observed by machines currently in use.

- Risks due to infeed speed and reversing of infeed components (§ 4.3.4.2—6th revision draft of EN 13525)

On the machines under inspection, the speed of wood tested below the limit of the maximum nominal speed of 1.0 m s^{-1} defined in the standard. The average speed of the tested logs was 0.2 m s^{-1} (Figure 7). The time wood takes to travel from the outermost edge of the chute to the rollers is particularly significant as it is closely tied with the time required by the standard for protective devices to stop infeed components. Increase in wood speed increases the speed of entanglement and reduces the reaction times for dealing with a dangerous situation.

Figure 7. Images of the procedures followed during the wood speed test, operated with a 28-years-old wood chipper.

- Risk due to ejected objects (§ 4.3.4 6th revision draft of EN 13525)

When chipping, discharge chute rotation should be limited to 20 degrees from a line drawn through the center of rotation and the outer edge of either side of the infeed chute. In some specimens of wood chippers inspected, it was possible to direct the discharge chute over the infeed chute, and in one case in particular, no obstacles were present to limit the rotation (Figure 8).

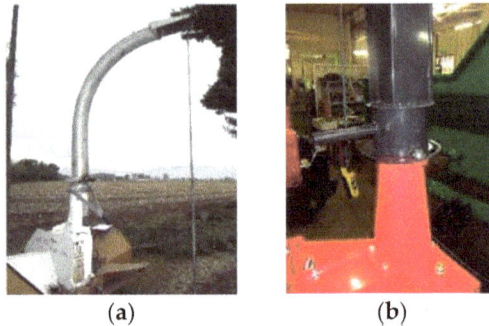

(a)   (b)

Figure 8. In some of the study cases (**a**) and (**b**), the discharge chute could be fully rotated, therefore well beyond the limit defined by the standard draft.

- Protection against access to moving power transmission parts (§ 4.3.5 6th revision draft of EN 13525)

All the machines inspected were powered by a tractor PTO. In two cases, moving power transmission parts were not adequately shielded by fixed repair such as a PTO safety shield (Figure 9).

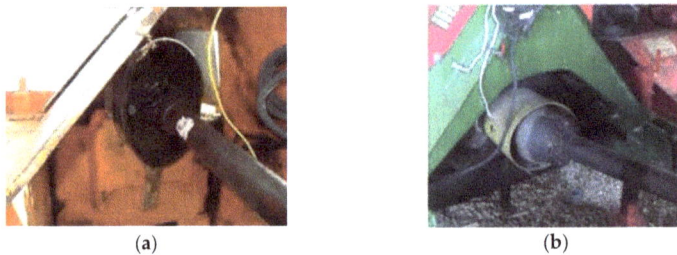

(a)   (b)

Figure 9. Example of observed unshielded power take-off (PTO) shafts (**a**) and partially shielded PTO (**b**) Hydraulic components (§ 4.4.2 6th revision draft of EN 13525).

In all the units inspected, hydraulic components were protected by a hood, but in the case of the two oldest units, the protective hood was merely hinged rather than bolted with interlocks, meaning that it was not firmly fixed to the machine (Figure 10).

(a)   (b)   (c)

Figure 10. Observed case studies of wood chippers: pictures (**a**) and (**b**) show chippers equipped with an easily removable hood that, when open, leaves the hydraulic lines without protection, on the right (**c**) a case study with a fixed repair (bolted hood).

- Preparation for transport and maintenance (§ 4.5 6th revision draft of EN 13525)

The infeed and discharge chutes/conveyors lacked handles near the articulation point to be easily folded for transport or maintenance (Figure 11). Even though this solution is not a hazard prevention device in any way, in terms of safety standard compliance, the implementation of required features has economic implications.

(a) (b) (c)

Figure 11. The discharge chute of the examined machine (**a**) and (**b**) was lacking in the two handles requested by the standard draft. These should be located at a distance of at least 300 mm from the nearest articulation point as shown in the picture on the right (**c**).

3.3. In-Deph Interviews

In-depth interviews highlighted a number of critical aspects that hindered the implementation of safety standards on in-use wood chippers. Based on the analysis of the participants' interviews, five prominent themes appeared to be highly relevant for users and manufacturers of wood chippers and could provide a helpful recommendation for future interventions. These themes were: "the reverse function", "the stop bar and the protective devices", "the infeed chute dimension", "the emergency stop", and "the designated feeding area". In addition to these themes, a feasibility and costs analysis of some solutions was provided (Table 2).

Table 2. List of the most probable costs to adapt in-use wood chippers to the most recent available revision draft of harmonized standards. Costs refers to average costs proposed for the Northwest Italy market.

Section of the Standard EN 13525	Solution for Adaptation to Standard	Costs	% on Average Purchasing Price [1]
§ 4.2.7 of 6th revision draft	Installation of electrovalve "no stress" device	€ 250	3.6%
	Installation of emergency stop push-button control	€ 200	2.9%
	Installation of complete emergency stop system (including hydraulic lines and labor)	€ 1.000–1.500	14.5–21.7%
§ 4.3.3.1 of 6th revision draft	Extension of the plate machine's infeed chute	€ 600	8.7%
§ 4.3.5 and § 4.4.2 of 6th revision draft	Application of a bolted hood to protect from hot components and moving power transmission parts	€ 150–200	2.2–2.9%
§ 4.2.4 of 6th revision draft	Substitution of hydraulic distributors	€ 300	4.3%
§ 4.5 of 6th revision draft	Application of a hinge in order to ease folding for transport or maintenance of the discharge chutes/conveyors	€ 300	4.3%

[1] The average price refers to chippers similar to those surveyed in this study.

3.3.1. The "Reverse" Function

According to the last revision draft of the standard, the "reverse" function should always be hold-to-run, but for the moment, this condition is not available on current machines. In fact, as observed during the onsite inspection, in most cases the "reverse" function is activated by a maintained position and it is mechanically controlled. With regard to this issue, both users and manufacturers expressed some perplexities about the standard's required implementation but also about its functionality. Generally speaking, respondents found difficulty in understanding the section of the revised draft that shows the combination of infeed controls allowed for lower protective devices. In particular, a couple of final users interviewed contested the worth of an optional reverse function beyond the chute edge, but this was probably because they were accustomed to other control configurations.

Finally, one of the two manufacturers consulted proposed a feasible solution involving working on the control of the existing hydraulic distribution system. If the hydraulic distributors are to adapt to the safety requirements imposed by the draft, they must be replaced in a sustained position by others, thereby providing the "hold to run" control.

3.3.2. Stop Bar and Protective Devices

The manufacturers reported that, currently, the safety bars (the lower, side protective devices and the top stop bar) are mechanical devices that insist on hydraulic lever distributors. In a number of current wood chippers, the protective bars respects positional requirements: they cover the full width of the infeed chute and up to a minimum of 75% of the vertical opening of infeed chute; but regarding functional requirements, many systems would need to be revised. The combined controls associated with the protective devices may vary according to the design adopted by the different manufacturers. Among some of the companies surveyed, infeed chute models work as a "swinging bar": the lower bar, if pushed forward, allows the conveyors to stop, while the upper one works in reverse as it needs to be pulled to halt the machine. Manufacturers explain this choice as the top bar, in accordance with the previous version of the harmonized safety standard, currently acts as a "connection and reinforcement" of the protective device rather than as an emergency stop. This mechanism does not meet the safety requirements of the latest revised standard as it may prove confusing during an emergency.

3.3.3. Infeed Chute Dimension

The standard demands set dimensions for the infeed chute and precise distances from the ground. The manufacturers interviewed confirmed that they were aware of the standard conditions and dimension requirements regarding wood chippers, as referred to in the previous version of the standard. Most of the recently designed models of manually-fed wood chippers do fulfil such conditions, while the older ones require intervention for compliance of different level of complexity. Both users and manufactures pointed out potential issues related to the stability of the machine in the event of interventions, such as increased infeed chute height and extended chute depth.

3.3.4. Emergency Stop

At present, almost no existing wood chippers are equipped with a separate emergency stop device as set out by the standard. However, some users interviewed reported having seen in agricultural machinery exhibitions that a number of manufacturers have already equipped their machines with emergency stop devices.

The manufacturers and technicians interviewed reported that the majority of operator controls are mechanical or hydraulic. The installation of an emergency stop push-button control, which is maintained by a separate control until reset, requires an electronic component. The addition of this safety system requires the installation of an electro-hydraulic valve and an electric power supply. Currently, some machines are already equipped with the so-called "no-stress" device. This device acts on the power supply of the machine by inverting the direction of rotation of the feed rollers when the rotation speed of the cutting device, the drum or the disk, drops below a set point. The device's electronic system act on the electrovalve that controls the activation of chipping components or of infeed conveyors. On those machines already equipped with this device, the installation of a safety system does not entail particularly costly interventions.

3.3.5. Designated Feeding Area

According to the majority of respondents, manufacturers, workshops and end-users, the position from which the operator should feed the chipper, as outlined in the draft, is in practice difficult to achieve, especially if the chipping material is particularly heavy. This suggests that even though the manufacturer may clearly mark the area with specific pictorials on the machine, the position itself is often disregarded. In order to prevent loading from the front of the chute, which increases the risk of the operator being caught in the machine, the only effective solution would be to impose side-loading by modifying the design of the infeed chute. According to the respondents this would be hardly feasible on in-use wood chippers.

3.3.6. Identified Solutions and Cost Analysis

As a result of the information provided by the interviews with professionals, an array of possible interventions to ensure essential safety requirements was identified and for some of these solutions possible adjustment costs were obtained (Table 2). The proposed costs relate to the average costs proposed by the workshops visited and hypothesized by the technical staff of some manufacturers. Even though the proposed costs appeared relatively affordable for most businesses, these may vary considerably according to the age and the state of the machine.

4. Discussion

Agricultural machine design is in continuous evolution, and attention has increasingly been paid to safety in recent years [52]. Huge progress has been made in safety and ergonomics since the 1980s [53]. Regarding machinery this evolution culminated in the introduction of the first Machine Directive (89/392/EEC), the first set of regulations meant to ensure a common safety level in manufactured machinery. In following years, a new edition of the Machine Directives came into force, safety regulations have become more demanding [54] and machinery safety has become one of the targets of the technological evolution [53]. The development of safety standards and regulations in recent years contributed to a higher level of safety in new machines compared to their older counterparts [27]. In this context, in compliance with the Machine Directive the same level of safety should be guaranteed for agricultural machinery currently in use, even that which is technologically inferior.

The focus of this study was: (i) to develop a more in-depth understanding of the current conditions and hazards associated with in-use manually fed wood chippers, assessing the conformity thereof with the revised draft of safety standard EN 13525; (ii) to evaluate within this context the possible consequences of new pending standard implementation, pointing out the main issues, actual feasibility, and the costs that this implementation involves. Moreover, the study highlighted solutions to achieve the required level of safety for operators, while verifying the effective technical feasibility and the economic impact of some.

Results confirmed that, in the area under survey (considered representative of the Italian context), the majority of wood chippers currently in use would not comply with the most recent available draft of standard, especially regarding protection against the risk of entanglement. This condition is not limited to the Italian context; indeed, international literature related to forestry industry reveals other examples of partial compliance to safety standards. Some studies carried out in New England, in USA, [55,56], assessing the adherence to the American National Standards for Arboricultural Operations (ANSI Z133.1—2006), found low levels of compliance to chipper safety standards across all surveyed arborists company types.

Interviews confirmed that the potential impact of the pending new standard is of notable significance for companies. Individual interviews with operators identified the following items as the most problematic: the reverse function", "the stop bar and the protective devices", "the infeed chute dimension", "the emergency stop", and "the designated feeding area". These items proved difficult to put into practice mainly due to the technical feasibility and economic issues involved, but the attitude towards safety and lack of information also play an important role.

4.1. Technical Feasibility

With regards to the technical feasibility of pending safety standard requirements, the manufacturers interviewed were able to identify some possible and operative solutions. Nevertheless, while admitting the potential feasibility thereof, they also suggested that the intervention could become more difficult as the machine aged. In particular, some solutions proposed in the harmonized standard regarding emergency stops were only possible on electronic machines able to operate on hydraulic lines controlling the infeed components. In this case, technical implementation proved very difficult on the oldest units, since any modifications may be incompatible with mechanical components or structural parts.

4.2. Economic Issues

Though the costs proposed for the constructive modification hypothesized in this study could be considered affordable for most businesses, these may vary considerably depending on the condition of the specific machine. Generally speaking, the economic issues and the size of the farm/company often represents a barrier to the adoption of safety measures. In practice, debate with users and producers, confirmed that, as in Italy most companies and farms are small or medium size, are unable to bear additional costs for machinery interventions [14,57]. Hagel et al. [58] identified associations between higher levels of "economic worries" and the absence of safety shields on grain augers. Moreover two studies by Cavallo et al. [28,54] demonstrated that in fact the larger the farm, the more interest shown in technological innovations aimed towards improving safety for machinery operators. A study by Fargnoli et al. [27] confirmed that small Italian agricultural and forestry companies are less willing to invest in initiatives aimed at improving safety at work.

4.3. Safety Attitude

Both the machinery inspections and interviews established that operators are aware of the hazards but at the same time, they perceive standards and regulations as a bureaucratic encumbrance rather than a means to improving working conditions in terms of safety. Operators appeared skeptical about the actual efficacy of newly-introduced standard requirements and proved particularly frustrated by safety solutions interfering in the management of their working activities on the operative level. This complies with a study by Weil et al. [7] on PTO driveline shielding, in which farmers were interviewed and reported that limited time and resources make work safety unfeasible and that "anything that interferes with getting the job done, or that costs more time and money, has a definite impacts on the livelihood of the farmers". These factors encourage farmers to believe that it is the better to rely on common sense, best practices and experiences rather than technical wood chipper protective devices.

Similar outcomes were found by Caffaro et al. [59] during the survey on perceived machinery-related risks and safety attitudes in senior Swedish farmers: respondents mainly referred to the common sense and previous experience as the best safety practices. These dynamics were very close to and consistent with many other studies related to the use of protective devices such as PTO shielding and use of Roll Over Protective Structures (ROPS) on tractors [60–63].

4.4. Lack of Information

As it emerged from the study, a lack of information and of precise constructive instructions, especially among farmers and users, makes it difficult for operators to consistently conform to up-to-date regulations. Additionally, it may be the case that the harmonized standard is not entirely clear for all professionals and whether or not a safety feature should be considered compliant with the Machine Directive becomes a matter of discretion [57]. With this regards, Fargnoli et al. [27] also pointed out the lack of knowledge and expertise in both risk assessment and safety management among operators and ascribed this to the large number of elderly farmers or foreigners, who have rarely received professional training. In fact, as some authors suggest [3], educational programs are the main approach to undertake in order to improve the safety practices of farmers.

Generally speaking, the operators interviewed turned out to be completely extraneous to the legislative background and the revision process of safety standards. Only manufacturers proved aware of the standard withdrawal as, in their role as machine producers, they had participated in European discussion boards. With regard to this last consideration, thanks to this study it was possible to obtain an important perspective (especially from manufacturers) and highlight the gap between European regulation and the execution of safety requirements on the machines currently in use.

4.5. Limitations of the Study

This study intended to reflect the current scenario of the potential application of revised safety standard in Italy, but it does presents some limitations. The sample group was represented by a limited number of case studies from the North West of Italy. Nevertheless, with regard to the category of manually-fed wood chippers, it is representative of in-use models of machinery generally available on farms and used by contractors in Italy. Inferences to larger scales or other context should consider this limitation. Currently, individual interviews were only carried out with key informants from the Piedmont region and although it is the second highest producer in Italy and the manufacturers involved are two of the most important Italian companies in the sector, the qualitative information collected during the interviews and the frequency of the issues raised cannot be generalized.

Additional limitations are given to the fact that occupational safety in agricultural in-use machines is a very sensitive topic. In fact, information collected about the respondents' safety behaviors and perspectives on safety issues and regulations may have been subject to bias towards more "socially desirable" answers.

As far as onsite verifications are concerned, a further functional test could have been carried out. The force required in activating protective devices should subject to testing. According to the standard draft this should not exceed 150 N on the horizontal parts of infeed controls and 200 N on other parts along the length. The functional tests carried out during this study on infeed controls and protective devices, let suppose that in some of inspected in-use machines the force required was higher than the maximum values stated by the standard.

Moreover, the study just concentrated on the revised draft of the harmonized safety standard for wood chippers (EN 13525) and improved safety for operators regarding the risk of getting caught and drawing into the machinery without taking into consideration other significant risks related to the use of manually-fed wood chippers, such as physical exposure to ergonomic hazards. In fact, operators loading the machines are prone to musculoskeletal disorders since they often undergo awkward postures, repetitive movements and frequent lifting of loads thus could; for this reason, further studies in this matter in particular could prove useful.

5. Conclusions

This study examined various technical features for reducing or eliminating risks related to entanglement, crush, and shearing on in-use wood chippers. Solutions were obtained based on the observation of a number of case studies and on the qualitative analysis of the essential technical interventions needed to increase the intrinsic safety level of machinery currently in use.

The study confirms that the majority of interventions required for wood chippers currently in use are technically feasible and affordable for most companies. Nevertheless, the complexity and costs of interventions increase with the age of the machinery. In particular, some solutions proposed by the reference standard regarding emergency stops would only be possible on machines equipped with an electronic system that can operate on hydraulic lines controlling the infeed components. In this case, technical implementation would be very difficult on older units, since any modification may prove incompatible with mechanical components or structural parts. Moreover, adaptation to current safety standard could prove economically unfeasible on smaller units with little market value, despite being most likely of wide-spread use in small-medium sized farms and cooperatives throughout Italy.

Acknowledgments: This study was carried out within the framework of "Protection of agricultural machinery operators from crush, entanglement, shearing" (PROMOSIC) project, funded by INAIL.

Author Contributions: Vincenzo Laurendi and Eugenio Cavallo conceived the study, Giorgia Bagagiolo and Eugenio Cavallo designed the study, Giorgia Bagagiolo and Eugenio Cavallo performed the study, analyzed the data collected, and drafted the manuscript. Eugenio Cavallo coordinated the study and critically revised the manuscript for its theoretical and intellectual content.

Conflicts of Interest: The authors declare no conflict of interest.

References and Note

1. Alt, N. International agricultural machinery standards for the benefit of agriculture and industry. In Proceedings of the 24th Annual Meeting of Club of Bologna, Hannover, Germany, 10–11 November 2013; pp. 1–4.
2. McCurdy, S.A.; Carroll, D.J. Agricultural injury. *Am. J. Ind. Med.* **2000**, *38*, 463–480. [CrossRef]
3. Layde, P.M.; Nordstrom, D.L.; Stueland, D.; Brand, L.; Olson, K.A. Machine-related occupational injuries in farm residents. *Ann. Epidemiol.* **1995**, *5*, 419–426. [CrossRef]
4. Day, L.; Voaklander, D.; Sim, M.; Wolfe, R.; Langley, J.; Dosman, J.; Hagel, L.; Ozanne-Smith, J. Risk factors for work related injury among male farmers. *Occup. Environ. Med.* **2009**, *66*, 312–318. [CrossRef] [PubMed]
5. Görücü, S.; Cavallo, E.; Murphy, J.D. Perceptions of tilt angles of an agricultural tractor. *J. Agromed.* **2014**, *19*, 5–14. [CrossRef] [PubMed]
6. Cavallo, E.; Langle, T.; Bueno, D.; Tsukamoto, S.; Görücü, S.; Murphy, J.D. Rollover Protective Structure (ROPS) retrofitting on agricultural tractors: Goals and approaches in different countries. *J. Agromed.* **2014**, *19*, 208–209. [CrossRef]
7. Weil, R.; Mellors, P.; Todd, F.; Sorensen, J.A. A Qualitative Analysis of Power Take-Off Driveline Shields: Barriers and Motivators to Shield Use for New York State Farmers. *J. Agric. Saf. Health* **2014**, *20*, 51–61. [CrossRef] [PubMed]
8. Aneziris, O.N.; Papazoglou, I.A.; Konstandinidou, M.; Baksteen, H.; Mud, M.; Damen, M.; Bellamy, L.J.; Oh, J. Quantification of occupational risk owing to contact with moving parts of machines. *Saf. Sci.* **2013**, *51*, 382–396. [CrossRef]
9. Gerberich, S.G.; Gibson, R.W.; French, L.R.; Lee, T.Y.; Carr, W.P.; Kochevar, L.; Renier, C.M.; Shutske, J. Machinery-related injuries: Regional rural injury study-I (RRIS-I). *Accid. Anal. Prev.* **1998**, *30*, 793–804. [CrossRef]
10. Al-bassit, L.; Tricot, N. *Improvement of Manure Spreaders Safety. Feasability Study*; Irstea Report; IRSTEA: Antony Cedex, France, 2013.
11. Purschwitz, M.A.; Stueland, D.T.; Lee, B.C. Feasibility Study of Inspection of Farm Machinery Safety Features. *J. Agromed.* **1994**, *1*, 29–38. [CrossRef]
12. Pickett, W.; Hagel, L.; Dosman, J.A. Safety features on agricultural machines and farm structures in Saskatchewan. *J. Agromed.* **2012**, *17*, 421–424. [CrossRef] [PubMed]
13. Narasimhan, G. *Machinery-Related Operational Factors as Determinants of Injury on Canadian Prairie Farms*; Queen's University: Kingston, ON, Canada, 2009.
14. Narasimhan, G.R.; Peng, Y.; Crowe, T.G.; Hagel, L.; Dosman, J.; Pickett, W. Operational safety practices as determinants of machinery-related injury on Saskatchewan farms. *Accid. Anal. Prev.* **2010**, *42*, 1226–1231. [CrossRef] [PubMed]
15. Narasimhan, G.; Crowe, T.G.; Peng, Y.; Hagel, L.; Dosman, J.; Pickett, W. A Task Based Analysis of Machinery Entanglement Injuries among Western Canadian Farmers. *J. Agromed.* **2011**, *16*, 261–270. [CrossRef] [PubMed]
16. Chinniah, Y. Analysis and prevention of serious and fatal accidents related to moving parts of machinery. *Saf. Sci.* **2015**, *75*, 163–173. [CrossRef]

17. Hartling, L.; Pickett, W.; Brison, R.J. Non-tractor, agricultural machinery injuries in Ontario. *Can. J. Public Health* **1997**, *88*, 32–35. [PubMed]
18. DeRoo, L.A.; Rautiainen, R.H. A systematic review of farm safety interventions. *Am. J. Prev. Med.* **2000**, *18*, 51–62. [CrossRef]
19. Tebeaux, E. Improving tractor safety warnings: Readability is missing. *J. Agric. Saf. Health* **2010**, *16*, 181–205. [CrossRef] [PubMed]
20. Tebeaux, E. Safety warnings in tractor operation manuals, 1920–1980: Manuals and warnings don't always work. *J. Tech. Writ. Commun.* **2010**, *40*, 3–28. [CrossRef]
21. Caffaro, F.; Mirisola, A.; Cavallo, E. Safety signs on agricultural machinery: Pictorials do not always successfully convey their messages to target users. *Appl. Ergon.* **2017**, *58*, 156–166. [CrossRef] [PubMed]
22. Caffaro, F.; Cavallo, E. Comprehension of safety pictograms affixed to agricultural machinery: A survey of users. *J. Saf. Res.* **2015**, *55*, 151–158. [CrossRef] [PubMed]
23. Rasmussen, K.; Carstensen, O.; Lauritsen, J.M. Incidence of unintentional injuries in farming based on one year of weekly registration in Danish farms. *Am. J. Ind. Med.* **2000**, *38*, 82–89. [CrossRef]
24. Poisson, P.; Chinniah, Y. Observation and analysis of 57 lockout procedures applied to machinery in 8 sawmills. *Saf. Sci.* **2015**, *72*, 160–171. [CrossRef]
25. Baker, W.; Day, L.; Stephan, K.; Voaklander, D.; Ozanne-smith, J.; Dosman, J.; Hagel, L. *Making Farm Machinery Safer. Lessons from Injured Farmers*; Publication Number 07/190; Rural Industries Research and Development Corp.: Canberra, Australia, 2008; p. 84.
26. Pelliccia, L. *Il Nuovo Testo Unico Di Sicurezza Sul Lavoro (No. 81-2008)*, 4th ed.; Maggioli Editore: Santarcangelo di Romagna, Italy, 2008.
27. Fargnoli, M.; Laurendi, V.; Tronci, M. Design for safety in agricultural machinery. In Proceedings of the DESIGN 2010, Dubrovnik, Croatia, 17–20 May 2010.
28. Cavallo, E.; Ferrari, E.; Bollani, L.; Coccia, M. Attitudes and behaviour of adopters of technological innovations in agricultural tractors: A case study in Italian agricultural system. *Agric. Syst.* **2014**, *130*, 44–54. [CrossRef]
29. ISTAT-Italian National Statisitical Institute. Farm Structure Survey-Year 2013. 2015. Available online: https://www.istat.it/it/archivio/167401 (accessed on 27 September 2017).
30. ISTAT-Italian National Statisitical Institute. Aziende con Superficie Totale Per Classe di Superficie Totale (Superficie in Ettari). Dettaglio Per Regione—Anno 2013. Available online: http://agri.istat.it/sag_is_pdwout/jsp/dawinci.jsp?q=plSPA0000010000012000&an=2013&ig=1&ct=1121&id=68A%7C98A (accessed on 27 September 2017).
31. ISTAT-Italian National Statisitical Institute. Aziende Agricole e Risultati Economici Per Classi di Fatturato, Composizione Percentuale—Anno 2014. Available online: http://agri.istat.it/sag_is_pdwout/jsp/GerarchieTerr.jsp?id=99A%7C46A&ct=314&an=2009 (accessed on 27 September 2017).
32. Bentley, T.A.; Parker, R.J.; Ashby, L.; Moore, D.J.; Tappin, D.C. The role of the New Zealand forest industry injury surveillance system in a strategic Ergonomics, Safety and Health Research Programme. *Appl. Ergon.* **2002**, *33*, 395–403. [CrossRef]
33. Melemez, K. Risk factor analysis of fatal forest harvesting accidents: A case study in Turkey. *Saf. Sci.* **2015**, *79*, 369–378. [CrossRef]
34. Laschi, A.; Marchi, E.; Foderi, C.; Neri, F. Identifying causes, dynamics and consequences of work accidents in forest operations in an alpine context. *Saf. Sci.* **2016**, *89*, 28–35. [CrossRef]
35. Lundqvist, P.; Gustafsson, B. Accidents and accident prevention in agriculture a review of selected studies. *Int. J. Ind. Ergon.* **1992**, *10*, 311–319. [CrossRef]
36. Heist, A.M.; Ziernicki, R.M.; Railsback, B.T. Analysis of the hazards of wood chipper accidents. In Proceedings of the ASME 2011 International Mechanical Engineering Congress and Exposition, Denver, CO, USA, 11–17 November 2011.
37. OSHA. Hazards of Wood Chippers. Safety and Health Information Bulletin. 2008. Available online: https://www.osha.gov/dts/shib/shib041608.html (accessed on 27 September 2017).
38. Lanning, D.N.; Dooley, J.H.; Lanning, C.J. Shear Processing of Wood Chips into Feedstock Particles. In Proceedings of the 2012 ASABE Annual International Meeting, Dallas, TX, USA, 29 July–1 August 2012.
39. Facello, A.; Cavallo, E.; Magagnotti, N.; Paletto, G.; Spinelli, R. The effect of chipper cut length on wood fuel processing performance. *Fuel Process. Technol.* **2013**, *116*, 228–233. [CrossRef]

40. Karha, K. Industrial supply chains and production machinery of forest chips in Finland. *Biomass Bioenergy* **2010**, *35*, 3404–3413. [CrossRef]

41. Spinelli, R.; Cavallo, E.; Eliasson, L.; Facello, A. Comparing the efficency of drum and disc chippers. *Silva Fenn.* **2013**, *47*. [CrossRef]

42. Poje, A.; Spinelli, R.; Magagnotti, N.; Mihelic, M. Exposure to noise in wood chipping operations under the conditions of agro-forestry. *Int. J. Ind. Ergon.* **2015**, *50*, 151–157. [CrossRef]

43. Colantoni, A.; Mazzocchi, F.; Laurendi, V.; Grigolato, S.; Monarca, F.; Monarca, D.; Cecchini, M. Innovative Solution for Reducing the Run-Down Time of the Chipper Disc Using a Brake Clamp Device. *Agriculture* **2017**, *7*, 71. [CrossRef]

44. Al-bassit, L.; Tricot, N. *Amelioration de la Securite de la Dechiqueteuse Forestiere Etude de Reconception*; Irstea Report; IRSTEA: Antony Cedex, France, 2015.

45. OSHA. Accident Search Results. Available online: https://www.osha.gov/pls/imis/AccidentSearch.search?âĉĉ_keyword=%22Chipper%22&keyword_list=on (accessed on 22 September 2017).

46. Cardillo, C.; Cimino, O.; Gabrieli, G.; Giampaolo, A. *La Meccanizzazione Agricola in Italia: Aspetti Tecnici, Economici, Ambientali e Sociali*; Report INEA; INEA: Roma, Italy, 2013.

47. European Union. Standards in Europe. Available online: http://europa.eu/youreurope/business/product/standardisation-in-europe/index_en.htm (accessed on 5 November 2017).

48. Klembalska, A.; Fancello, G. Increasing the quality of agricultural machinery testing—A comparison between Italian and Polish experience. *Manag. Prod. Eng. Rev.* **2015**, *6*, 14–24. [CrossRef]

49. GSE. Rapporto Statistico Energia da Fonti Rinnovabili. 2015. Available online: http://www.gse.it/it/salastampa/GSE_Documenti/Rapporto%20statistico%20GSE%20-%202014.pdf (accessed on 21 September 2017).

50. Berry, R.S.Y. Collecting data by in-depth interviewing. In *Proceeding of the British Educational Association Annual Conference*; University of Sussex: Brighton, UK, 1999; pp. 1–10.

51. CEN/TC 144/WG. N 175 CEN144 8 Rev EN 13525 Note of the French public authority (ref N170).

52. Purschwitz, M.A. Personal Protective Equipment and Safety Engineering of Machinery. In *Agricultural Medicine*; Lessenger, J.E., Ed.; Springer: New York, NY, USA, 2006; pp. 53–69.

53. Cavallo, E.; Ferrari, E.; Coccia, M. Likely technological trajectories in agricultural tractors by analysing innovative attitudes of farmers. *Int. J. Technol. Policy Manag.* **2015**, *15*, 158. [CrossRef]

54. Mrugalska, B.; Kawecka-Endler, A. Machinery design for construction safety in practice. *Lect. Notes Comput. Sci.* **2011**, *6767*, 388–397. [CrossRef]

55. Julius, A.K.; Kane, B.; Bulzacchelli, M.T.; Ryan, H.D. P. Compliance with the ANSI Z133.1—2006 safety standard among arborists in New England. *J. Saf. Res.* **2014**, *51*, 65–72. [CrossRef] [PubMed]

56. Julius, A.K. Investigation of Compliance with the Ansi Z133.1—2006 Safety Standard in the New England Tree Care Industry. Master's Thesis, University of Massachusetts Amherst, Amherst, MA, USA, February 2014.

57. Lorencowicz, E.; Uziak, J. Repair Cost of Tractors and Agricultural Machines in Family Farms. *Agric. Agric. Sci. Procedia* **2015**, *7*, 152–157. [CrossRef]

58. Hagel, L.; Pahwa, P.; Dosman, J.A.; Pickett, W. Economic worry and the presence of safety hazards on farms. *Accid. Anal. Prev.* **2013**, *53*, 156–160. [CrossRef] [PubMed]

59. Caffaro, F.; Lundqvist, P.; Cremasco, M.M.; Nilsson, K.; Pinzke, S.; Cavallo, E. Machinery-related perceived risks and safety attitudes in senior Swedish farmers. *J. Agromed.* **2017**. [CrossRef] [PubMed]

60. Myers, J.R. Factors Associated with the Prevalence of Non-ROPS Tractors on Farms in the U.S. *J. Agric. Saf. Health* **2010**, *16*, 267–280. [CrossRef]

61. Jenkins, P.L.; Sorensen, J.A.; Yoder, A.; Myers, M.; Murphy, D.; Cook, G.; Wright, F.; Bayes, B.; May, J.J. Prominent Barriers and Motivators to Installing ROPS: An Analysis of Survey Responses from Pennsylvania and Vermont. *J. Agric. Saf. Health* **2012**, *18*, 103–112. [CrossRef] [PubMed]

62. Solomon, C. Accidental injuries in agriculture in the UK. *Occup. Med.* **2002**, *52*, 461–466. [CrossRef]

63. Correa, I.M.; Moreira, C.A.; Filipini, S.R.; Mello, R.d.C.; Pontes, P.S. Assessment of agricultural power take-off (pto) drive shafts guards in field conditions. *Appl. Res. Agrotechnol.* **2016**, *9*, 71–77.

agriculture

MDPI

Article

Analysis of the Almond Harvesting and Hulling Mechanization Process: A Case Study

Simone Pascuzzi *,† and **Francesco Santoro** †

Department of Agricultural and Environmental Science (DiSAAT), University of Bari Aldo Moro,
via Amendola 165/A, 70126 Bari, Italy; francesco.santoro@uniba.it
* Correspondence: simone.pascuzzi@uniba.it; Tel./Fax: +39-0805442214
† The Authors equally contributed to the present study.

Received: 1 November 2017; Accepted: 1 December 2017; Published: 4 December 2017

Abstract: The aim of this paper is the analysis of the almond harvesting system with a very high level of mechanization frequently used in Apulia for the almond harvesting and hulling process. Several tests were carried out to assess the technical aspects related to the machinery and to the mechanized harvesting system used itself, highlighting their usefulness, limits, and compatibility within the almond cultivation sector. Almonds were very easily separated from the tree, and this circumstance considerably improved the mechanical harvesting operation efficiency even if the total time was mainly affected by the time required to manoeuvre the machine and by the following manual tree beating. The mechanical pick-up from the ground was not effective, with only 30% of the dropped almond collected, which mainly was caused by both the pick-up reel of the machine being unable to approach the almonds dropped near the base of the trunk and the surface condition of the soil being unsuitably arranged for a mechanized pick-up operation. The work times concerning the hulling and screening processes, carried out at the farm, were heavily affected by several manual operations before, during, and after the executed process; nevertheless, the plant work capability varied from 170 to 200 kg/h with two operators.

Keywords: almond harvest chain; hulling process; manpower employment

---

1. Introduction

Italy's leading regions in the production of almonds are Sicily and Apulia (Italy), with cultivated areas respectively of about 31,090 and 19,578 hectares and corresponding harvest productions of 4.69×10^7 kg and 2.20×10^7 kg. Sicily and Apulia together provide 92% of the total Italian production [1]. During last decades, Italian almond cultivation has registered a notable, progressive reduction both in terms of assigned surface area and production, despite the fact that Italy has the widest variety of almond cultivars. This dramatic crop reduction can be attributed to different reasons, such as the employment of outdated traditional orchards, competition with more profitable crops, uncertain annual yields due to adverse climatic conditions and/or pest attacks and infectious diseases, and the organization of the almond production chain and market [2]. In this regard, many of the intermediate activities involved in the almond processing (sometimes even the harvest) were taken away from the farmer and consequently have an effect on the financial gain. Furthermore there is also a considerable fragmentation because, on average, more than 40% of Italian farms involved use less than 0.20 ha in almond cultivation, and this percentage is even higher (approximately 50%) in Apulia. Nowadays, the harvesting operation, the most labour-intensive of the growth cycle, is often still carried out manually during hull dehisce by knocking the nuts from the tree by means of long poles, collecting the almonds in nets spread on the ground. Harvest alone accounts for an average of 13–17% of the final commercial value of the almond crop [3], without considering the successive processes of hulling and drying, traditionally carried out outside the farm. The more widespread use of trunk shakers used in olive harvesting suggested that these machines could also be used for almonds [4,5]. The employment of the trunk shaker allows a significant increase in the productivity of the individual worker [6,7]. Productivity increases further with the use of self-propelled shakers, which, in addition to the vibrating element, have a reversed-umbrella interceptor. This last solution appears to be the most interesting for the purpose of rational management of almond orchards, because the work chain is limited to two or three working units, reducing the incidence of this cost item to just 20% [8,9].

The almond harvest takes place in Italy in a different way compared to the practices in California, where the almonds farmers produce over 75% of the world's almonds. Inside Californian almond orchards, the harvest is carried out with the following operative phases: early and suitable arrangement of the soil surface (flattening, weeding, tamping), followed by the use of simple shakers to detach the almonds from the plants, side raking of the product on the soil through swathers, and picking up of the swath by means of sweepers. These sweeping practices, however, influence emissions of PM_{10} (particulate matter $\leq$10 μm in nominal aerodynamic diameter) due to the soil material in the windrow, which may add PM emissions during almond pick-up [10,11].

Conversely, the modern Italian almond production, as all modern fruit cultivation, tends toward cultivation intensification, increasing plant density and reducing tree size. The reasons for this general evolution of fruit-growing systems should be sought primarily in reducing manpower costs due to the mechanization of farming operations, with the added value of increased workplace safety [12–17].

Taking in mind the aforementioned observations, the aim of this paper is the analysis of the almond harvesting system, with a very high level of mechanization, frequently employed in Apulia for the almond harvesting and hulling process. Several tests were carried out to assess the technical-economical aspects related to the machinery and to the mechanized harvest system used, highlighting their usefulness, limits, and compatibility within the almond cultivation sector. The analyzed harvest chain was employed by an Apulian farm in line with the standards recommended for an income almond production, both from a dimensional point of view (agricultural land devoted to almond plants of 40 ha) and an agronomic one (plants placed on irrigated flat cultivable land) with freehold machines. This study may then provide farmers with useful guidelines for machine selection in order to reduce management costs, as well as indications to optimize their use.

2. Materials and Methods

In the 36th week of 2015, experimental tests were carried out in an almond orchard ("Filippo Ceo" variety) of 40 hectares located on a farm (40°28′17.73″ N, 17°38′44.64″ E) in the territory of the Municipality of Oria (Brindisi Province, Southern Apulia, Italy) (Figure 1). The trees were planted with a layout of 5.0 m × 5.0 m, giving a density of 400 trees ha^{-1}. The almond orchard was arranged on flat cultivable land with controlled growth weed and irrigation; the size of the headland access path was about 3.5 m and the main trees' structural characteristics are reported in Table 1.

Table 1. Main geometrical characteristics of the almond trees.

Trees Sizes	m
Trunk circumference	0.25–0.55
First branches height above ground level	0.60–1.00
Tree height	3.00–3.60
Canopy width	2.80–4.00
Canopy height	2.50–3.00

The harvesting chain was carried out using a self-propelled trunk shaker with a reversed-umbrella interceptor and a self-moving picker-separator, whilst the hulling process was performed through a high-capacity production huller. The self-propelled harvester by SICMA Ltd. (manufacturing company placed in Acconia di Curinga, Catanzaro Province, Italy), model "Speedy", was equipped with a 4-cylinder 93 kW diesel engine and 3-traction wheels powered by hydraulic engines. The harvester was formed by a trunk shaker (arm linked to a vibrating steerable head) and a reversed-umbrella interceptor (5 m in diameter) (Figure 2). Furthermore, this machine was equipped with a front net to allow the largest operator's visibility and a harvest tank, able to be opened through a hydraulically operated hatch at the bottom in order to empty the contents.

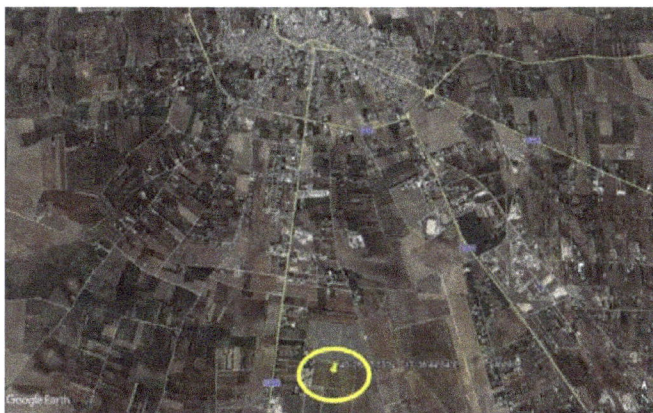

Figure 1. Map of territory of Oria, Italy, with the location of the almond orchard under test.

The self-propelled harvester was driven by a worker whilst another operator knocked the trees with a pole. A third worker was responsible for the cleaning of the product and its transport to the farm (Figure 2).

A hailstorm caused a considerable early drop of almonds just before the harvesting, and this occurrence forced us to also include a mechanized pick-up from ground operation besides the harvest carried out with the trunk shaker. This circumstance also allowed us to evaluate the performance of the mechanized pick-up operation and its feasibility in the harvesting chain.

Figure 2. Self-propelled harvester SICMA Ltd., model "Speedy"; inset shows the manual pole beating for residual product.

This ground pick-up harvesting was carried out by the articulated self-propelled harvester by De Masi Construction Ltd. (manufacturing company placed in Gioia Tauro, Italy), the model "SHA19" picker-upper machine, equipped with a 3-cylinder diesel engine of 12.5 kW. Its 1.5-m working width front gatherer had a pick-up reel with six brushes, and a hopper with a perforated bottom to allow the expulsion of any thin impurities (Figure 3). A worker operated the picker-upper machine, while a further employee attended to the cleaning of the product and its transport to the farm.

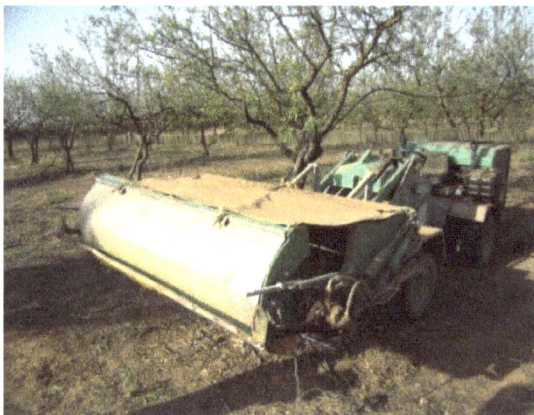

Figure 3. Self-propelled picker-upper machine De Masi Construction Ltd., model "SHA19".

The hulling process was performed through a crafted hulling machine made up by a horizontal cylindrical cage (length 2.87 m, diameter 0.30 m), manufactured by a mean of equally spaced steel rods, containing the hulling device, i.e., a rotating shaft equipped with stiff bodies (molded steel rods) able to separate the hull from the shell. The machine was driven by an electric motor of 1.5 kW. A worker controlled the process and took care of cleaning the product, the hopper filling, the conveyor belt activation, and periodic maintenance of the machine. A further employee took care of the quality control and the dimensional classification of the almonds (Figure 4).

Figure 4. Crafted hulling machine.

The flow chart of Figure 5 summarizes the operations chain performed during the harvesting phase; conversely, the hulling process, carried out outdoors at the farm, was organized as shown in Figure 6.

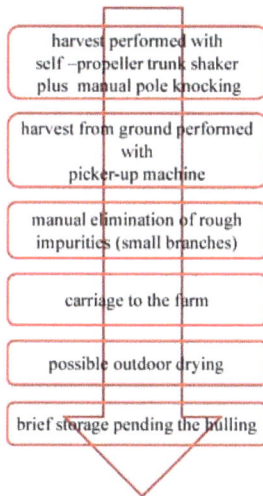

Figure 5. Almond harvesting process performed from the tree and ground.

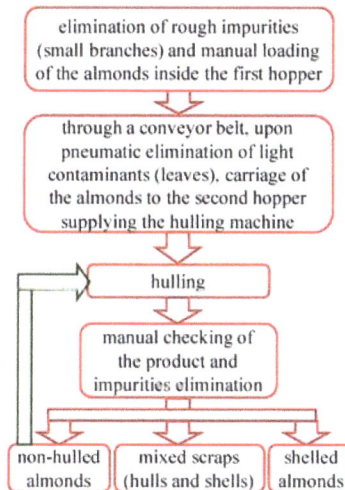

Figure 6. Flow of the almond hulling and screening processes.

3. Results and Discussion

The performance of the self-propelled harvester SIGMA "Speedy", summarized in Table 2, confirms results already found with similar machines used for mechanized harvest in olive orchards [4,5]; altogether, the mechanized harvesting of each tree required less than 2 min with a harvesting capacity within the range 32–36 trees·h^{-1}, corresponding to more than 11 h·ha^{-1}, and the harvesting chain productivity was affected by the amount of the hanging product (9–12 kg·tree^{-1}), equal to 250–400 kg·h^{-1}.

Table 2. Almond mechanized harvesting chain and manual beating average productivity (average hanging production: 10.6 kg·tree^{-1}, corresponding to 4240 kg·ha^{-1}).

Harvesting time	s·tree^{-1}	102
Harvesting capacity	h·ha^{-1}	11.3
Harvesting chain and labor productivity	kg·h^{-1} number of trees·h^{-1} worker hour·ha^{-1}	3.7 35 34

A more detailed analysis of the harvest times highlighted that the tree-shaking operation required only a few seconds (3–6 s), whilst the remaining time was taken up by: (i) operations such as the approach of the machine to the tree, the trunk grippling and release; (ii) the opening and closing of the reverse-umbrella interceptor; (iii) the manual beating in order to harvest almonds that did not fall from the tree; (iv) the first manual sorting operation to eliminate the largest impurities such as twigs before conveying the harvested product to the farm. Mechanized harvesting followed by manual beating allowed a detaching rate greater than 98% of the whole product on the tree. Conversely, the workers' productivity, affected by the amount of the hanging product, was on the average 0.80 worker hours (100 kg)$^{-1}$, i.e., 2.5 to 3.5 times that required for the manual harvest (Table 3) [2,3].

Table 3. Machines and labor productivity for the manual and mechanical almonds harvesting.

Operations	Machine-Hours/100 kg	Worker-Hours/100 kg
Manual beating and product recovery through nets	-	2.0–2.7 [1]
Mechanical harvest through shaker with interceptor and manual beating	0.27	0.80
Mechanical ground pick-up harvesting	0.37	0.75
Total	0.64	1.55
Hulling	0.60	1.2

[1] The average values reported for the production of 10 kg per hectare of almond plants are reduced to less than half in the case of productions of 2.5 to 3 kg/plant.

The articulated self-propelled harvester De Masi "SHA19" allowed for the pick-up of almonds placed on the ground, both those that dropped for natural reasons and due to the hailstorm (approximately the 14% of the total available product) and those not picked up by the umbrella interceptor (almost the 12% of the total hanging product).

The tests pertinent to the mechanized ground pick-up harvesting pointed out a high level of productivity (1.5 h·ha^{-1}) obtained by the aforementioned self-propelled harvester, even with high levels of impurities. On the other hand, the harvester had a low productivity in reference to the picked-up almonds from a single tree (only 1 kg of picked-up product per 3 kg dropped). The main reason for this poor performance is the falling of the almonds in a region very close to the tree trunk base. Those dropped almonds could not be intercepted by the umbrella due to its poor sealing around the trunk, nor by the ground harvesting machine as the ground surface was not well-flattened. Furthermore, the mechanized ground pick-up harvesting required an amount of labor (0.75 worker-hours/100 kg) that was almost the same as that necessary for the mechanized harvesting followed by the manual beating (0.80 worker-hours/100 kg) (Table 3). Therefore, these three operations all together reduced significantly the advantage of the mechanical harvesting compared to the traditional manual harvesting (1.55 vs. 2.0–2.7 worker-hours/100 kg in Table 3).

The harvest testing carried out highlighted the suitability of the mechanized process of almond harvesting from the trees, despite some burdensomeness in the ground picking-up phase if no adequate arrangement of the ground itself had been carried out and in the wrapping collar dimensional adjustment of the intercepting umbrella (Figure 7).

The work times concerning the hulling and screening processes (Figure 6), carried out at the farm, were heavily affected by several manual operations before, during, and after the executed process. Within the hulling process, these operations can be classified in chronological order as preparatory, parallel, and succeeding.

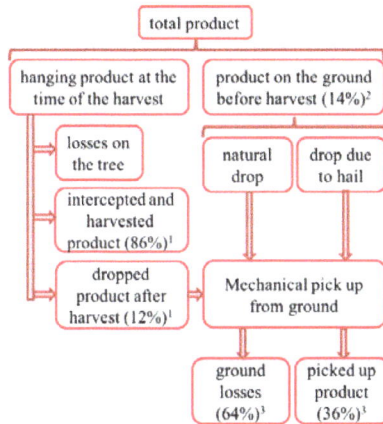

Figure 7. Average percentages referring to crop harvested from the tree, picked-up from the ground, and losses observed. ([1] referred to the whole hanging crop; [2] referred to the total available crop; [3] referred to the fallen crop).

The preparatory operations were related to further impurities separation, manually for the rough ones and pneumatically for lightest ones, as well as the uneven feeding of the hopper and hulling machine; conversely, the hulling process control and the cylindrical cage cleanliness were the main parallel operations; finally, the succeeding operations included sorting the final product from impurities and re-inserting non-hulled almonds back into the hulling machine. The plant work capability varied from 170 to 200 kg/h with two operators, and the product characteristics at the input and output of working chain are reported respectively in Figure 8a,b. The hulled product features are shown in Figure 8c.

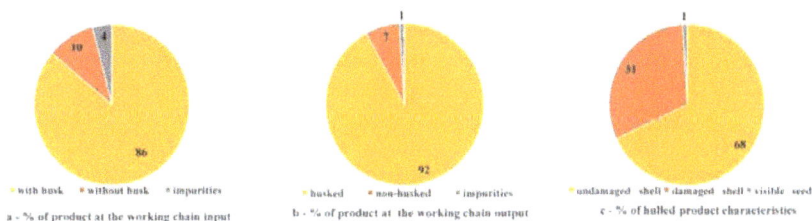

Figure 8. Average characteristics of sampled product before and after the hulling process (% values in weight).

4. Conclusions

Although limited to just one year of tests carried out within an almond orchard at the harvesting time, this research provides some useful evaluation elements related to the efficiency of the used machines and of the harvest chain under test. It has been clearly verified that, even if the almonds can be easily detached from the tree, the total harvesting time is not as low as could be expected because only the tree-shaking time is reduced, not the time necessary for the umbrella positioning and the manual tree beating. Conversely, the ground harvesting machine highlighted a poor productivity in reference to the picked-up almonds from a single tree due to the not well-flattened ground surface and the poor performance of the machine in picking up the almonds very close to the trunks.

The hulling and screening processes were executed at the farm and influenced by a lot of manual operations before, during, and after the performed process.

In agreement with the result obtained, some actions may be proposed:

- to supply guidelines to farmers for the choice of machines, which take into account their optimized employment and cost restraint;
- to study the setup of umbrella interceptors dimensionally consistent with the diameter of the trunks and the plant canopy;
- to encourage farmers to adopt the Californian almond harvesting modalities, founded on the preliminary smoothing of the ground surface and the use of simple shredders to detach the almond from the trees followed by the employment of ground harvesters.

Acknowledgments: The Authors wish to thank C. Gidiuli, V. Marzano and D. Sfregola of the Department of Agricultural and Environmental Science of the University of Bari Aldo Moro, for their helpfulness and commitment in conducting the experimental tests.

Author Contributions: The authors equally contributed to the present study.

Conflicts of Interest: The authors declare no conflict of interest.

References

1. Italian National Institute of Statistics (ISTAT). Area (Hectares) and Production (Quintals) of Hazelnuts, Almonds, Pistachio Nuts, Figs. 2016. Available online: http://agri.istat.it/jsp/dawinci.jsp?q=plC190000010000011000&an=2016&ig=1&ct=270&id=15A\T1\textbar{}21A\T1\textbar{}30A (accessed on 10 June 2017).
2. Briamonte, L. Il Comparto Della Frutta in Guscio. In *I Quaderni Dell'Ortofrutta*; INEA: Roma, Italy, 2007; pp. 1–132. (in Italia)
3. Schiril, A. *Analisi Economiche Della Produzione e del Mercato del Mandorlo e del Nocciolo in Sicilia*; Coreras: Catania, Italy, 2005; pp. 1–141.
4. Manetto, G.; Cerruto, E. Vibration risk evaluation in hand-held harvesters for olives. *J. Agric. Eng.* **2013**, *44*, 705–709. [CrossRef]
5. Vivaldi, G.A.; Strippoli, G.; Pascuzzi, S.; Stellacci, A.M.; Camposeo, S. Olive genotypes cultivated in an adult high-density orchard respond differently to canopy restraining by mechanical and manual pruning. *Sci. Hortic.* **2015**, *192*, 391–399. [CrossRef]
6. Manetto, G.; Cerruto, E.; Pascuzzi, S.; Santoro, F. Improvements in citrus packing lines to reduce the mechanical damage to fruit. *Chem. Eng. Trans.* **2017**, *58*, 391–396.
7. Bianchi, B.; Tamborrino, A.; Santoro, F. Assessment of the energy and separation efficiency of the decanter centrifuge with regulation capability of oil water ring in the industrial process line using a continuous method. *J. Agric. Eng.* **2013**, *44*, 278–282. [CrossRef]
8. Clodoveo, M.L.; Camposeo, S.; de Gennaro, B.; Pascuzzi, S.; Roselli, L. In the ancient world virgin olive oil has been called "liquid gold" by Homer and the "great healer" by Hippocrates. Why is this mythic image forgotten? *Food Res. Int.* **2014**, *62*, 1062–1068.
9. Cecchini, M.; Contini, M.; Massantini, R.; Monarca, D.; Moscetti, R. Effects of controlled atmospheres and low temperature on storability of chestnuts manually and mechanically harvested. *Postharvest Biol. Technol.* **2011**, *61*, 131–136. [CrossRef]
10. Faulkner, W.B.; Downey, D.; Ken Giles, D.; Capareda, S.C. Evaluation of Particulate Matter Abatement Strategies for Almond Harvest. *J. Air Waste Manag. Assoc.* **2011**, *61*, 409–417. [CrossRef] [PubMed]
11. Faulkner, W.B. Harvesting equipment to reduce particulate matter emissions from almond harvest. *J. Air Waste Manag. Assoc.* **2013**, *63*, 70–79. [CrossRef] [PubMed]
12. Pascuzzi, S. A multibody approach applied to the study of driver injures due to a narrow-track wheeled tractor rollover. *J. Agric. Eng.* **2015**, *46*, 105–114. [CrossRef]
13. Pascuzzi, S. The effects of the forward speed and air volume of an air-assisted sprayer on spray deposition in "tendone" trained vineyards. *J. Agric. Eng.* **2013**, *3*, 125–132. [CrossRef]

14. Pascuzzi, S.; Santoro, F. Evaluation of farmers' OSH hazard in operation nearby mobile telephone radio base stations. In Proceedings of the 16th International Scientific Conference "Engineering for Rural Development" Proceedings, Jelgava, Latvia, 24–26 May 2017; Latvia University of Agriculture-Faculty of Engineering: Jelgava, Latvia, 2017; pp. 748–755. [CrossRef]
15. Pascuzzi, S.; Santoro, F. Exposure of farm workers to electromagnetic radiation from cellular network radio base stations situated on rural agricultural land. *Int. J. Occup. Saf. Ergon.* **2015**, *21*, 351–358. [CrossRef] [PubMed]
16. Pascuzzi, S.; Anifantis, A.S.; Blanco, I.; Scarascia Mugnozza, G. Hazards assessment and technical actions due to the production of pressured hydrogen within a pilot photovoltaic-electrolyzer-fuel cell power system for agricultural equipment. *J. Agric. Eng.* **2016**, *47*, 88–93. [CrossRef]
17. Pascuzzi, S.; Santoro, F. Analysis of Possible Noise Reduction Arrangements inside Olive Oil Mills: A Case Study. *Agriculture* **2017**, *7*, 88. [CrossRef]

agriculture

MDPI

Article

Definition of a Methodology for Gradual and Sustainable Safety Improvements on Farms and Its Preliminary Applications

Sirio Rossano Secondo Cividino [1,*], Gianfranco Pergher [1], Rino Gubiani [1], Carlo Moreschi [2], Ugo Da Broi [2], Michela Vello [3] and Fabiano Rinaldi [4]

[1] Department of Agricultural, Food, Environmental and Animal Sciences, Agricultural Engineering Section, University of Udine, Via delle Scienze 208, 33100 Udine, Italy; gianfranco.pergher@uniud.it (G.P.); rino.gubiani@uniud.it (R.G.)

[2] Department of Medical Area, Forensic Medicine Section, University of Udine, Piazzale S.Maria della Misericordia 15, 33100 Udine, Italy; carlo.moreschi@uniud.it (C.M.); ugo.dabroi@uniud.it (U.D.B.)

[3] Sofia & Silaq Corporate Spin-Off, University of Udine, Via Zanon 16, 33100 Udine, Italy; michela.vello@uniud.it

[4] Centro Ricerche Studi dei Laghi, Corso di Porta Vittoria 31, 20122 Milano Italy; info@crslaghi.net

* Correspondence: agricolturasicura@gmail.com; Tel.: +39-0432-558655; Fax: +39-0432-558603

Received: 18 October 2017; Accepted: 22 December 2017; Published: 1 January 2018

Abstract: In many productive sectors, ensuring a safe working environment is still an underestimated problem, and especially so in farming. A lack of attention to safety and poor risk awareness by operators represents a crucial problem, which results in numerous serious injuries and fatal accidents. The Demetra project, involving the collaboration of the Regional Directorate of INAIL (National Institute for Insurance against Accidents at Work), aims to devise operational solutions to evaluate the risk of accidents in agricultural work and analyze the dynamics of occupational accidents by using an observational method to help farmers ensure optimal safety levels. The challenge of the project is to support farmers with tools designed to encourage good safety management in the agricultural workplaces.

Keywords: safety; occupational accidents; agriculture

1. Introduction

To contextualize and define the occurrence of accidents involving farms we need to identify the main risk factors of specific work activities. It is important to remember that sectors such as farming are difficult to standardize and hence various risk types are often underestimated [1–8].

As reported in the literature, there are only two main macro risk categories for accidents involving agricultural work: environmental risks and health risks [1–8].

Among agricultural risk factors, the following three main areas of specific risks have, according to national and regional government data, a significant impact in terms of accidents and occupational pathologies:

1. Mechanical risks (about 60% of serious and fatal injuries);
2. Biomechanical risks due to repetitive movements and postural issues (in recent years there has been a significant increase in claims related to occupational injuries especially in those sectors with low levels of mechanization such as horticulture and floriculture),
3. Interference risks, serious and fatal workplace injuries due to poorly qualified or inexperienced farm workers who may also be employed on several farms.

The Demetra project therefore aims to analyze:

- Those farms where machinery is used;
- Farms where manual processes are still common (pruning and harvesting represent a particularly high biomechanical risk);
- Farms that carry out activities where they interact with other farms [9,10].

Following the preliminary investigative analysis, the initial phase of the project clearly showed that:

- As far as employment is concerned, the farming sector does not follow standard patterns and each individual farm may well demand specific solutions if safety levels are to be improved;
- Improvement pathways and tools need to be devised that work in association with governance models for SMEs (small and medium-sized enterprises) and family-run agricultural enterprises;
- We need to define operational procedures for two main areas: work organization and production, which require dedicated safety solutions for machinery and equipment;
- It is also essential to pay attention to the evolution of both production facilities and reception facilities, and so identify a set of innovative organizational and technical solutions to safety issues.

In the sampling and observational phases of the Demetra project described below we aim to develop a multifactorial procedure which promotes improvements in safety conditions by means of a guideline protocol based on the introduction of gradual, prioritized changes to working practices.

2. Materials and Methods

The preliminary observational and investigative steps of the Demetra model were divided into four evaluation phases:

I.   Definition of the productive and organizational components of the farm;
II.   Creation of a specific pyramidal matrix for each farm;
III. Validation of the model through field tests;
IV.   Building of specific pathways to improve safety levels.

To analyze a farm, the following algorithm was used:

$$\text{Demetra model} = F(Lo) \times F(Lp)$$

where:

- Lo = organizational level. Defines and analyzes safety management from the point of view of farm organization.
- Lp = the operational and productive level and defines different occupational areas.

The Italian agricultural sector is mainly founded on SMEs (small and medium-sized enterprises) characterized by a very small number of employees and production flows which are often poorly standardized, especially in the case of small farms where work is often seasonal and conducted out in the fields.

This representation of the farm defines the intervention levels of the Demetra model. It analyzes the farm as an open system including all its interactions with other external factors (Figure 1).

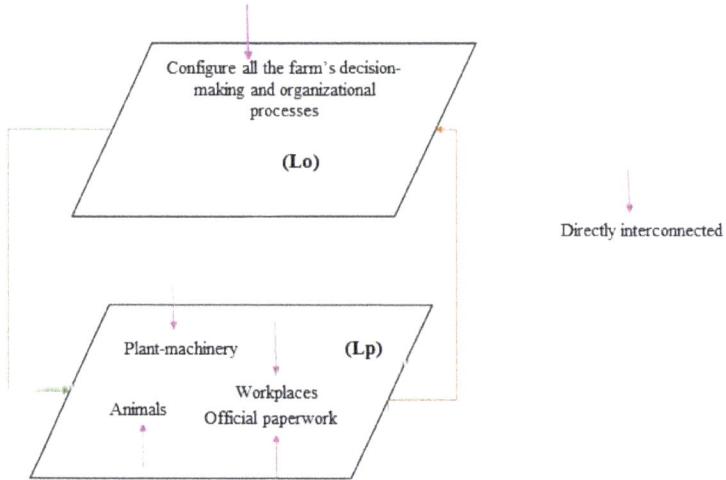

Figure 1. Representation of a farm's risk factors.

As shown in Figure 2, a plan to improve safety on a farm should consider all farming and non-farming factors that affect both production and organization and involve all specific activities.

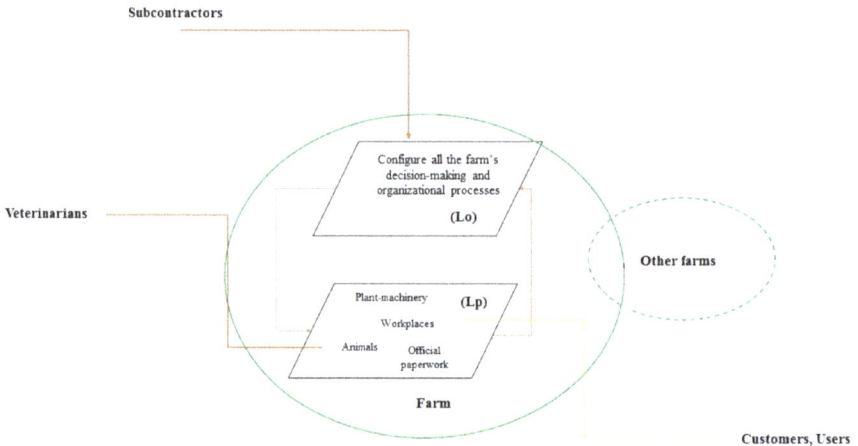

Figure 2. Representation of risk factors of a farm and external non-farming factors.

Demetra considers a series of variables for the management of farm safety planning, such as:

- The constant presence of people not involved in farming activities (veterinarians, technicians, National Health System employees, etc.).
- The presence of visitors, children and school groups (this occurs normally on educational and social farms).
- Productive and organizational activities carried out on one farm by other directly interconnected farms.
- Personnel working on more than one farm.

The logic behind the project and its analytical methodology is shown in Figure 3 through the construction of two distinct elements (pyramids), both characterized by safety levels and color bands which go from red (a serious degree of risk) up to green (an optimal safety situation). This methodology focuses on the position of each farm within the five color bands in the pyramids, so that targets can be set in a program of gradual safety improvements.

As shown in Figure 3 and Table 1, equating a farm's safety performance to the different color-coded safety levels on the pyramids is part of a dynamic, observational process. Farm management and employees therefore have to be engaged in continual dialogue, and evaluate and optimize safety levels, in order to maintain its standards. A color code is assigned to each level: the positioning within a specific color band is represented both in Figure 3 and Table 1 as a result of the analysis carried out on a set of components characterizing the farm.

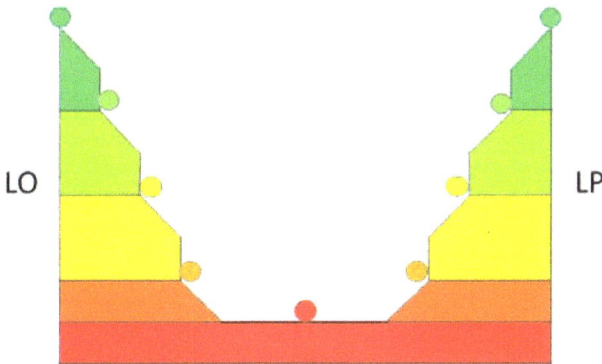

Figure 3. Schematization of the two-pyramid Demetra methodology (where LO = organizational level and defines and analyzes safety management from the point of view of farm organization, whereas LP = represents the operational and productive level and defines different occupational areas. Red Colour = represents maximum risk level; Green Colour = represents optimal safety conditions).

Table 1. Color bands used to build the pyramid.

LO		LP	
Color Code	Meaning	Color Code	Meaning
	Farm organized in an optimal way, which goes beyond the minimum safety levels, with regular internal audits and a safety management system		Production is carried out in a safe and correct manner, above the standards defined by the legislation
	Farm complies with statutory obligations, with proper management of all organizational aspects		Production and operating conditions comply with the regulations
	Farm with deficiencies at an organizational or management level which fails to meet its statutory obligations		Farm has deficiencies that can lead to risk scenarios in work activities
	Farm that has serious deficiencies and criticalities in the organization and management of the farm		Farm has deficiencies and criticalities that can lead to significant risk scenarios in work activities
	Farm without any organization or safety management system covering operational, productive and statutory areas.		Farm which lacks any internal system of risk assessment or safety management at operational, productive and statutory level.

LO = organizational level and defines and analyzes safety management from the point of view of farm organization;
LP = represents the operational and productive level and defines different occupational areas.

To define a standardized methodology which is applicable in all scenarios and which takes into account the specific characteristics of agriculture and SMEs, a procedure has been defined whereby the four points below are analyzed for each farm. In this way, we can assign the correct position for the farm in the pyramid and so identify a pathway of gradual improvement:

(1) Organizational management aspects: how the farm manages production and safety in the workplace.
(2) Documentation: set of technical documents required by the government's health and safety standards.
(3) Operational aspects: how the farm organizes its production according to the specific context in which it operates.
(4) Interface: this defines the receptivity level of the farm in terms of work processes (subcontractors, mobility of the workforce between farms in the network, educational and social farms)

3. Preliminary Results

In our preliminary investigation eleven farms operating mainly in the vine-growing/winemaking (5), zootechnical (livestock/cereal or livestock) (4) sectors were evaluated. The cereal and zootechnical sectors are often closely interconnected in Italy because cereals are mostly used to feed livestock.

The Demetra model was used to analyze these 11 farms. They all showed a high level of specialization, except for one farm, which had different types of non-interconnected production (Table 2).

Our decision to investigate farms employing people on permanent and fixed-term contracts and family farms, was motivated by three factors:

- Gradual changes to Italian law which affect the employment of people on non-standard contracts (agistment, sharecropping, workforce employed on a network of farms);
- Situations in which family farms, in compliance with specific Italian laws, employ family members as subordinate workers;
- The propensity of farms, especially the newly established ones, to offer certain types of contracts which allow them to hire people who are qualified to drive farm tractors (the driver must be a skilled worker who cannot be paid by voucher) (Table 2).

The main results were as follows:

1. The farms we studied mainly specialize in one specific type of production in order to attain greater sustainability.
2. The surface-area of land farmed ranged from a few hectares up to over a hundred.
3. This area was not proportional to the farm's income or the number of people employed. In fact, in order to determine the real productive level of a farm, we need to consider certain key factors: the degree of mechanization, planting distances, production philosophy (organic production is more labour-intensive).
4. The farms we selected were representative, in terms of their characteristics and size, of the average Italian farm (Table 2).

According to data in the sixth census of the Italian Ministry of Agriculture, most personnel were employed in a family context, although in some cases farms employed workers on a permanent contract (on average, family farms employed at least one or two people on a permanent contract) and/or fixed-term contract (Table 2).

The preliminary results from the application of the Demetra model allowed researchers to identify the different risk levels of agricultural accidents and to design appropriate, innovative ways of improving safety, as had been anticipated during the development phase of the Demetra methodology.

The main risk factors identified by the Demetra model are shown in Table 3.

Table 2. Sample characteristics.

Farm Number	Farm 1	Farm 2	Farm 3	Farm 4	Farm 5	Farm 6	Farm 7	Farm 8	Farm 9	Farm 10	Farm 11
Sector	Vine-growing and cereals	Livestock-cereals	Vine-growing and winemaking	Livestock	Vine-growing and winemaking	Fruit and vegetables	Livestock and cereals	Vine-growing and winemaking	Vine-growing	Horticulture	Livestock
Surface area (hectares)	100	50	15	60	50	5	1500 managed, 30 under ownership	5	50	2	20
Employees	yes	no	yes	yes	yes	yes	yes	no (family members only)	no (family members only)	yes	Yes
Outsource or make use of subcontractors	yes	yes	yes	yes	yes	yes	yes	yes	yes	yes	Yes
Payment by voucher	yes	no	yes	no	yes	yes	yes	no	yes	yes	No
Direct sales	yes	no	no	yes	yes	yes	no	yes	no	yes	Yes
Farm restaurant or shop; social or educational farm	yes	no	yes	yes	yes	yes	no	yes	yes	yes	No
Family farmhouse	no	yes	no	yes	yes	no	yes	yes	no	yes	Yes
Accidents	yes	yes	yes	yes	yes	yes	yes	yes	yes	yes	Yes
Governance	Structured family-run enterprises	Family-run enterprises	Co-operative	Family-run enterprises	Structured family-run enterprises	Social farm	Subcontractors	Family-run enterprises	Family-run enterprises	Multiple owners and employers, family-run enterprises	Multiple owners and employers, family-run enterprises

Table 3. Summary of main risk factors.

Risk	Activities/Settings	Sector
Environmental risks		
Explosion	Presence of explosive atmospheres (Biogas plants, autoclaves, storage of granular and dusty material, presence of flours)	livestock, vine-growing and winemaking, cereal sector
Drowning	Streams, irrigation channels, ponds, lakes	all sectors
Fire	Presence of flammable substances, high fire load (e.g., barns), possibility of combustion, high-temperature fermentation	all sectors
Fall from height	Use of simple ladders, silo maintenance activities, fall while using agricultural machinery	all sectors
Fall from ground level	All open field operations, working in the presence of water and residues on the floor	all sectors
Contact with medium-sized and large animals	Care and management of the farm	livestock sector
Mechanical risk	Use of agricultural machinery, open field operations	all sectors
Working at height	Maintenance of and access to silos, wine tanks, use of aerial platforms, construction of rural buildings	all sectors in occasional way
Health risks		
Exposure to low temperatures	Working outdoors during winter, or working in cold storage units	winegrowing, wine, cereals, livestock sector
Exposure to high temperatures	Manual operations outdoors	horticultural, winegrowing, wine, cereal sector
Risks associated with microclimate in general	Protected crops, dairies, wineries, processing in hot, humid environments where there are considerable amounts of organic material	horticultural, wine, floricultural sector
Biological	Animal care and management, management of livestock excrement, direct contact with organic material (picking/harvesting, manual operations outdoors), irrigation	all sectors
Chemical	Treatment, fertilization, sanitization of production environments, prolonged use of chainsaw, brush cutters and grass trimmers (exhaust gas)	all sectors
Asbestos	Presence of asbestos roofs and manufacturing facilities built using asbestos	all sectors
Organizational and cross-cutting risks		
Powders of organic and inorganic origin	Use of agricultural machinery for soil processing, food distribution, livestock—excrement management, plant logging and sawing	all sectors
Noise	Use of agricultural machinery and equipment, open field operations, transformation operations	all sectors
Electrocution	Use of agricultural machinery and electrical equipment	all sectors
Vibrations	Use of agricultural machinery and equipment, management of green and marginal areas	all sectors
Postural issues (stooping, squatting, etc.)	Management of arboreal and herbaceous crops, in particular harvesting and pruning	horticultural, nursery, winegrowing, wine, livestock sector
Repetitive movements	Management of arboreal and herbaceous crops, in particular manual pruning; Processing of products, nursery operations (planting, transplanting, weeding)	horticultural, nursery, winegrowing, wine sector
Night-time operations	Ploughing, harvesting and soil management operations, animal management	cereals, livestock sector
Working in solitude	Driving farm tractors or other vehicles, in isolated places which emergency services will have trouble reaching quickly in the event of an accident	all sectors
Interference risks	Presence of several farms in one area, with shared equipment and personnel	all sectors

4. Discussion

The Demetra project demonstrated, by means of a new analytical methodology, how a process of gradual and continuous improvement can increase the level of organizational and operational safety [9–13]. The analytical tools developed by the Demetra project for small and micro farms demonstrated that it is possible, with the correct analysis of accident risk levels, to ensure safety in agricultural contexts and confirmed that safety targets should be regarded not only as a cost but also as an investment which improves productivity, through the reduction/elimination of work-related injuries or deaths [2,6,9,10,12]. The project succeeded in raising the awareness and increasing the understanding of the farmers involved in this experiment, especially in cases of family-run farms whose owners decided to raise safety levels above the minimum required by Italian law.

Each level of the Demetra pyramids brings together a series of parameters that determine a matrix which allows us to identify the pathway to improvement. Each level includes the analysis of a series of factors that characterize the farm and identify the production organization and safety profiles together with the improvement pathway that should be followed. This method involves the following five steps:

1. Building a matrix that describes the farm's current safety performance;
2. Positioning the collected data in the matrix and the pyramid;
3. Defining aims according to the type of farm in question;
4. Identifying the technical and operational changes that need to be made in order to attain adequate safety levels;
5. Final positioning and assessment of whether aims have been achieved.

A fundamental aspect of the evaluation of a farm by means of the Demetra model is that, even though some parameters may be positive, the farm's real position in the pyramid is always determined by its lowest positioning within the color bands in the matrix, which corresponds to the highest level of risk.

The final results of the Demetra model applied to the farms studied were essentially the following:

- The farm owners/managers used the positioning of coloured matrices correctly in order to carry out the self-assessment of any critical points on their farm;
- The solutions proposed were not costly because they often involved simple changes to the organization or management of the farm;
- From an administrative point of view a series of easily applicable operational solutions and procedures were identified;
- The model promoted innovative solutions involving third parties, showing that a farm can be an "open workplace" which interacts with other networked farms;
- Structural changes are very often unnecessary for farms; in fact, in some cases, the reorganization of productive activities demanded operational solutions rather than structural;
- Changes to machinery and equipment can often be made by means of existing farm resources;
- The protocol and the improvement pathways designed for each individual farm provided objective feedback on the farm's safety status.

The innovative profile of the Demetra model also confirmed the following:

- Each farm has specific requirements where improvements in safety are concerned, and these are influenced by the nature of its governance, structure and production;
- During the risk analysis phase the farms implemented new knowledge and technical skills which were then transmitted to satellite farms or other family farms;
- This new process of safety improvements is easily adaptable to the typical Italian small and medium-sized enterprises and family farms; in fact, all the farms we analyzed were able to comply with Italian safety standards through the application of innovative processes, including those farms which initially had an extremely low safety rating.

The farms we studied displayed a series of critical issues which affect organization, management and administration. Farms also showed a range of activities involving subcontractors that are not managed in compliance with the specific Italian regulations [9–11,14–16].

The Demetra model puts forward an innovative method that can be used for any future experimentation in order to evaluate risk levels related to occupational accidents and diseases by following these steps:

1. Obtain specific farm data;
2. Apply the screened parameters to the coloured pyramid matrices;
3. Analyze the coloured pyramid matrices;
4. Classify farms according to their organizational structure;
5. Set objectives according to their organizational structure;
6. Plan operational and structural decisions in order to improve the safety of the chosen farms;
7. Evaluate the efficiency of new safety plans.

5. Conclusions

To summarize, the Demetra model is a new way to give farms a toolset which can help them define their accident risk levels and in turn increase the safety of all agricultural activities in the near future [10].

Acknowledgments: Our research work was supported by funds received from the National Institute for Insurance against Workplace Accidents and Occupational Disease.

Author Contributions: Sirio Rossano Secondo Cividino, Gianfranco Pergher, Rino Gubiani of the Agricultural Engineering Department evaluated the technical aspects of safety and accidents in farms; Michela Vello and Fabiano Rinaldi of the Sofia & Silaq Corporate Spin-off and Centro Ricerche Studi dei Laghi evaluated the organizational aspects of safety and accidents in farms; Carlo Moreschi and Ugo Da Broi of the Forensic Medicine Department evaluated the medicolegal aspects of accidents in farms.

Conflicts of Interest: The authors declare no conflict of interest.

References

1. Boubaker, K.; Colantoni, A.; Allegrini, E.; Longo, L.; Di Giacinto, S.; Monarca, D.; Cecchini, M. A model for musculoskeletal disorder-related fatigue in upper limb manipulation during industrial vegetables sorting. *Int. J. Ind. Ergon.* **2014**, *44*, 601–605. [CrossRef]
2. Marucci, A.; Monarca, D.; Cecchini, M.; Colantoni, A.; Di Giacinto, S.; Cappuccini, A. The heat stress for workers employed in a dairy farm. *J. Agric. Eng.* **2014**, *44*, 170–174. [CrossRef]
3. Marucci, A.; Monarca, D.; Cecchini, M.; Colantoni, A.; Biondi, P.; Cappuccini, A. The heat stress for workers employed in laying hens houses. *J. Food Agric. Environ.* **2013**, *11*, 20–24.
4. Cecchini, M.; Colantoni, A.; Massantini, R.; Monarca, D. The risk of musculoskeletal disorders for workers due to repetitive movements during tomato harvesting. *J. Agric. Saf. Health* **2010**, *16*, 87–98. [CrossRef] [PubMed]
5. Marucci, A.; Pagniello, B.; Monarca, D.; Cecchini, M.; Colantoni, A.; Biondi, P. Heat stress suffered by workers employed in vegetable grafting in greenhouses. *J. Food Agric. Environ.* **2012**, *10*, 1117–1121.
6. Niskanen, T.; Naumanen, P.; Hirvonen, M.L. An evaluation of EU legislation concerning risk assessment and preventive measures in occupational safety and health. *Appl. Ergon.* **2012**, *43*, 829–842. [CrossRef] [PubMed]
7. Cividino, S.R.S.; Vello, M.; Maroncelli, E.; Gubiani, R.; Pergher, G. Analyzing the manual handling risk in wine growing and wine production sectors. In Proceedings of the Work Safety and Risk Prevention in Agro-Food and Forest Systems, Ragusa, Italy, 16–18 September 2010; Elle Due s.r.l.: Ragusa, Italy, 2010; Volume 1. ISBN/ISSN 97888-903151-6-9.
8. Cecchini, M.; Colantoni, A.; Massantini, R.; Monarca, D. Estimation of the risks of thermal stress due to the microclimate for manual fruit and vegetable harvesters in central Italy. *J. Agric. Saf. Health* **2010**, *16*, 141–159. [CrossRef] [PubMed]

9. Vincenzo, C.; Gubiani, R.; Pergher, G.; Cividino, S.R.S.; Fanzutto, A.; Vello, M.; Grimaz, S. Demetra: A Survey on Work Safety in 103 Agricultural Farms in Friuli Venezia Giulia. *Procedia-Soc. Behav. Sci.* **2016**, *223*, 297–304. [CrossRef]

10. Colantoni, A.; Longo, L.; Biondi, P.; Baciotti, B.; Monarca, D.; Salvati, L.; Boubaker, K.; Cividino, S.R.S. Thermal stress of fruit and vegetables pickers: Temporal analysis of the main indexes by "predict heat strain" model. *Contemp. Eng. Sci.* **2014**, *7*, 1881–1891. [CrossRef]

11. Colantoni, A.; Marucci, A.; Monarca, D.; Pagniello, B.; Cecchini, M.; Bedini, R. The risk of musculoskeletal disorders due to repetitive movements of upper limbs for workers employed in vegetable grafting. *J. Food Agric. Environ.* **2012**, *10*, 14–18.

12. Moreschi, C.; Da Broi, U.; Cividino, S.; Gubiani, R.; Pergher, G. Neck injury patterns resulting from the use of petrol and electric chainsaws in suicides. Report on two cases. *J. Forensic Legal Med.* **2014**, *25*, 14–20. [CrossRef] [PubMed]

13. ProŠrekl, J. Safe behavior and level of knowledge regarding safe work practices on farms. *Res. J. Chem. Sci.* **2011**, *1*, 15–19.

14. Moreschi, C.; Da Broi, U.; Fanzutto, A.; Cividino, S.; Gubiani, R.; Pergher, G. Medicolegal Investigations Into Deaths Due to Crush Asphyxia After Tractor Side Rollovers. *Am. J. Forensic Med. Pathol.* **2017**, *38*, 312–317. [CrossRef] [PubMed]

15. Monarca, D.; Colantoni, A.; Cecchini, M.; Longo, L.; Vecchione, L.; Carlini, M.; Manzo, A. Energy characterization and gasification of biomass derived by hazelnut cultivation: Analysis of produced syngas by gas chromatography. *Math. Probl. Eng.* **2012**, *2012*, 102914. [CrossRef]

16. Monarca, D.; Cecchini, M.; Guerrieri, M.; Colantoni, A. Conventional and alternative use of biomasses derived by hazelnut cultivation and processing. *Acta Hortic.* **2009**, *845*, 627–634. [CrossRef]

agriculture

MDPI

Article

Agricultural Health and Safety Survey in Friuli Venezia Giulia

Sirio Rossano Secondo Cividino *, Gianfranco Pergher, Nicola Zucchiatti and Rino Gubiani

Department of Agricultural, Food, Environmental and Animal Sciences, University of Udine,
via delle Scienze 206, 33100 Udine, Italy; gianfranco.pergher@uniud.it (G.P.);
nicola.zucchiatti@uniud.it (N.Z.); rino.gubiani@uniud.it (R.G.)
* Correspondence: agricolturasicura@gmail.com; Tel.: +39-3281547453

Received: 17 October 2017; Accepted: 18 December 2017; Published: 8 January 2018

Abstract: The work in the agricultural sector has taken on a fundamental role in the last decades, due to the still too high rate of fatal injuries, workplace accidents, and dangerous occurrences reported each year. The average old age of agricultural machinery is one of the main issues at stake in Italy. Numerous safety problems stem from that; therefore, two surveys were conducted in two different periods, on current levels of work safety in agriculture in relation to agricultural machinery's age and efficiency, and to show the levels of actual implementation of the Italian legislation on safety and health at work in the agricultural sector. The surveys were carried out, considering a sample of 161 farms located in the region Friuli Venezia Giulia (North-East of Italy). The research highlights the most significant difficulties the sample of farms considered have in enforcing the law. One hand, sanitary surveillance and workers' information and training represent the main deficiencies and weakest points in family farms. Moreover, family farms do not generally provide the proper documentation concerning health and safety at workplaces, when they award the contract to other companies. On the other hand, lack of maintenance program for machinery and equipment, and of emergency plans and participation of workers' health and safety representative, are the most common issues in farms with employees. Several difficulties are also evident in planning workers' training programs. Furthermore, the company physician's task is often limited to medical controls, so that he is not involved in risk assessment and training. Interviews in heterogeneous samples of farms have shown meaningful outcomes, which have subsequently been used to implement new databases and guidelines for Health and Safety Experts and courses in the field of Work Safety in agriculture. In conclusion, although the legislation making training courses for tractor operators and tractor inspections compulsory dates back to the years 2012 and 2015, deadlines have been prorogued, and the law is not yet fully applied, so that non-upgraded unfit old agricultural machinery is still being used by many workers, putting their health and their own lives at risk.

Keywords: work safety; health and safety; risk prevention; risk assessment document; ROPS; safety belt

---

1. Introduction

In the last decades, the theme of safety at work in the agricultural sector has taken on a fundamental role. Following the 'Tractor Directive' in Italian law on Safety at work (Italian Law 81/2008)—agricultural tractors are currently equated to work machines, the principles of safety at work, ergonomics, and protection of the tractor operator and the other passengers [1,2].

Agricultural tractors in Italy are estimated in 1.7 million units, 35% of which are older than 44 years of age and 50% of which are older than 25 years of age [3]. This is a considerably critical issue in the field of road traffic and safety at work.

Despite many projects and awareness campaigns concerning the issue of safety in agricultural activities, conducted particularly by the National Institute for Insurance against Accidents at Work, vehicles, being non-compliant and potentially fatal in the event of an accident [4–6], are still present in farms and on the market. Many case studies [7–10] show that tractors lacking essential safety requirements-like seat belts and Roll Over Protection Systems-can cause fatal accidents in case of roll-over of the vehicle [1].

Accidents caused by and with tractors are statistically one of the most frequent causes of death in agriculture [1,11,12].

In Italy, the underestimation of this phenomenon has been observed for many years; in fact, only accidents involving farm employees were registered as 'occupational accidents' until 2014, while those involving semi-professional operators were considered as 'domestic accidents' [1].

According to a recent study on serious accidents in agriculture in Friuli Venezia Giulia (North East of Italy), an estimated rate of 30% cases are not surveyed or investigated [1,3]. Considering only the deadly accidents in agriculture and forestry operations, concern arises, as 51% of these accidents happened while workers were operating tractors (75% located on field and 25% while driving on roads) [4,13,14].

As far as accident dynamics are concerned, machine rollover represents 77% of accidents, while accidents involving the cardan shaft account for 0.7% only, but 66% of cases result in the death of the operator [15,16].

According to the reconstruction of 60 fatal accidents with tractors (northeast Italy) [2,11], the origin of these accidents can be categorized into three types:

- Technical causes (set of lacking safety elements)
- Causes of a human or behavioral nature (improper use of the tractor). In this regard it should be stressed that the legislation does not provide the private use of the tractor, it must always be linked to the cultivation or the forest; this is a factor that is often missing in the use of such equipment, in fact, as shown in the analyzed data in five cases the tractor was used in non-agricultural contexts and with playful purposes (race of tractors, carnival parade, loading and unloading of building material, and transport with tractor of building vehicles) [2,11]
- Structural failures (within the analyzed cases, some of them are related to the failure of embankments, bridges or ditches) [2,11].

However, it should be stressed that in the reconstruction of the dynamics, often there is not only one cause but the fatal accident is derived from a human error combined with the use of an unsafe vehicle. [2,11,12].

Within the European Community and according to Italian norms, there is currently a decisive indication by the legislator to make the use of agricultural tractors more professional and more responsible in considering other sectors as the plants to energy conversions and agro industrial [11,12,17–21].

Since 2012, with the 'Technical Law' bill, a specific professional training for the use of this type of machinery has been implemented as mandatory—an obligation that is still to be fully extended within the Italian territory [15,19,22].

In the light of such considerations, this study is meant to investigate a representative sample of the real conditions of the tractors within farms, aiming at bringing to light the main criticalities and proposing effective systems of analysis that can be used by the agricultural entrepreneurs themselves, to improve the present situation.

2. Materials and Methods

A first-level analysis was conducted to assess safety levels on a sample of 103 agricultural farms, with a prevalence of dairy farms and farms with vineyard and/or horticultural crops (Table 1).

Table 1. The sample farms in the first survey.

Type of Farm	No.	%	Average Size (ha)
Dairy farms	36	35.0	67.5
Other livestock	17	16.5	89.9
Vineyard and winery	24	23.3	55.6
Horticulture and nursery	12	11.7	9.4
Other	7	6.8	14.9
Mixed	4	3.9	240.5
Cereal crops	3	2.9	42.3
All farms	103	100.0	63.9

These farms were located in all of the six Health Districts in Friuli Venezia Giulia, each controlled by the respective District Agency. Part of these farms (56.3%) employed hired personnel, while 43.7% were family farms, allowed by the law to use a simplified safety management scheme.

Each farm was visited by one evaluator, and all data were recorded following a specific questionnaire. This questionnaire covered two main areas of interest (Figure 1):

- area A, including general information about the farm;
- area B, which varied depending on farm specialization, and was further divided into three profiles:

  ○ B1: farm machinery;
  ○ B2: personal protective equipment (PPE);
  ○ B3: specific risks.

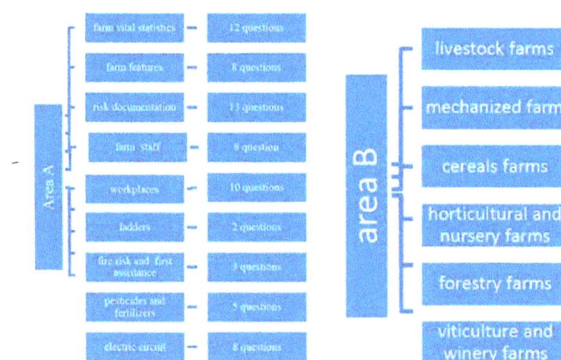

Figure 1. Specific questionnaire.

A second-level analysis was performed on a sample of 58 agricultural tractors (Table 7), employed in 11 selected farms, with the objective of further analyzing the presence or absence of legally required protective items. All main protective equipment and safety systems', as mentioned at point 2.4 part II of Annex V of the Italian Law number 81/2008, were checked and evaluated for compliance with the law. This included roll-over protective structures, safety belts, protections of moving parts, and other items (reported in Table 9).

3. Results

To the purpose of the first study, we analyzed:

- whether official documents and records were actually present at the farm;

- how safety management was organized;
- the working environment in the farm (useful element to correlate machine use and safety);
- the presence of protection devices on tractors;
- the use of prevention and protection equipment.

3.1. First Level Analysis

Table 2 includes only 58 farms with external personnel, which are subjected to full application of Italian Law 81/08, including official documentation. The main document required, i.e., the Risk assessment document, was absent or inadequate in 34.5% of the farms; other required documents were even more often missing, including a scheme for medical surveillance of workers (34.5%), the scheme for emergency procedures (41.4%), and the record of periodic inspection of lifting equipment (44.8%).

Table 2. Official documents at the farm.

Type of Document	Missing or Inadequate (% of Farms)
Risk assessment document	34.5
Risk assessment update	44.8
Medical watch	34.5
Emergency procedures	41.4
Regular inspection record (lifting equipment)	44.8
Compliance certificate of equipment	10.3
Book of use and maintenance	8.6
Pesticide license	24.1
Pesticide safety sheet	25.9
Equipment maintenance plan	36.2

Only those documents provided by third parties were mostly present, such as the Compliance certificate (lacking in 10.3% of farms), the Book of use and maintenance of equipment (8.6%), the Pesticide safety sheet (25.9%), or those required for purchasing pesticides (Pesticide license: 24.1%). Particularly remarkable was the absence of a plan for machinery and equipment maintenance (in 36.2% of farms), because of its great importance for accident prevention.

The Italian law also requires every farm with hired personnel to officially appoint a number of figures in charge of the different protection and prevention services (Table 3). While a safety manager (or head of the prevention and protection service, PPS) was mostly present (82.8% of the farms), other figures were often missing, including a doctor designated for periodic medical surveillance (48.1% of farms), or the supervisors for fire prevention (33.3%), first aid (34.6%) and workers' safety during work (63.0%). Additionally, 38.3% of the farms were not providing the workers with sufficient training and information services, while 44.3% did not have any special training for the various managers and supervisors.

Table 3. Managers and services.

	Not Present (% of Farms)
Safety manager	17.2
Medical doctor	48.1
Fire prevention manager	33.3
First-aid manager	34.6
Workers' supervisor	63.0
Training and information service (workers)	38.3
Special training service (managers)	44.3

Most of the farms had adequate toilet and shower services and dressing rooms for the workers (Table 4). The width of the main entrance to the farm (minimum: 5 m) was mostly in line with the law.

However, protections on gaps or trenches were missing in 28% of the farms. Most remarkable was the absence of any Interference risk analysis, i.e., a plan to avoid risks owing to the presence at the farm of external personnel, especially contractors for cereal or grape harvesting. Only 8.6% of farms had conducted a proper analysis of such risks.

Table 4. Situation of buildings in the farm.

Building Services	Yes (%)	No (%)
Toilets	93.9	6.1
Showers	87.2	12.7
Dressing room	86.4	13.5
Main entrance to farm > 5 m	81.3	18.6
Railing on hole, trench	71.8	28.1
Interference risk analysis	8.6	91.4

In approx. one half of the farms, a specific analysis was made to assess the main features of the tractors (Table 5). The average nominal power was 63 kW, and the average age was 20.9 years. The average annual usage (328 h/year) was related to the small average land area (63.9 ha, Table 1), and was far from the level suggested for profitable management (at least 600 h/year). These data offer some clues as to the current difficult economic situation in most of the farms: the reasons are many, and they cannot be fully discussed here. Nonetheless, this makes it even more difficult for these farms to bear the costs involved by current requirements for risk prevention and protection.

Table 5. Tractors at the farms.

	No. of Tractors	Power (kW)	Age (years)	Usage (h)	Usage (h/year)
Dairy farms	54	76.6	20.7	7339	355
Other livestock	18	65.6	21.6	7078	328
Viticulture	62	57.1	15.3	4444	290
Horticulture and nursery	26	50.3	27.8	3610	130
Other	2	40.4	24.8	1750	71
Mixed	29	61.8	26.1	15329	588
Cereal crops	5	64.7	20.0	6958	348
All farms	196	63.1	20.9	6873	328

In fact, missing protection devices are mostly related to the tractor's old age. In most of the sample farms, tractors were equipped with roll over protection structures (ROPS), protection of moving parts, such as belts and fans, and of hot surfaces (Table 6). However, a safety belt was missing at the driver's seat in 55.1% of the tractors—even though it has been declared mandatory since 2005. PTO (power take off) guards were also missing in 24.7% of the tractors (the study has analyzed in a different and specific way the part of the PTO, as it is often the cause of fatal accidents).

The relationship between the presence of protective items and the tractor's age is shown in Figure 2. All of the new tractors were in line with legal requirements, the only exception being the safe access to the driver's seat; Italian law requires the presence of two handles and stairs for tractors that have a distance exceeding 0.55 m from the ground (Annex V of the Italian Law 81/08) but this is often difficult to attain especially in small tractors for viticulture. This means that the main problem for these farms is the low investment capacity, which makes it difficult for farmers to replace old tractors with new ones.

Figure 3 shows the percentages of farms providing their workers with personal protection equipment (PPE). In general, only basic PPE were present (like cotton overalls and mechanical protection gloves), while specific PPE were seldom found (such as ear muffs, safety foot-ware, chemical resistant clothing and gloves and chemical resistant respirators).

Table 6. Protective devices.

Protective item	Missing (% of Tractors)
ROPS	5.2
Safety belt	55.1
Protection of belts & fans	7.6
Protection of hot surfaces	10.8
Safe access to driver seat	13.6
PTO guards	24.7
CE marking (European Conformity)	37.6
Owner handbook	8.0

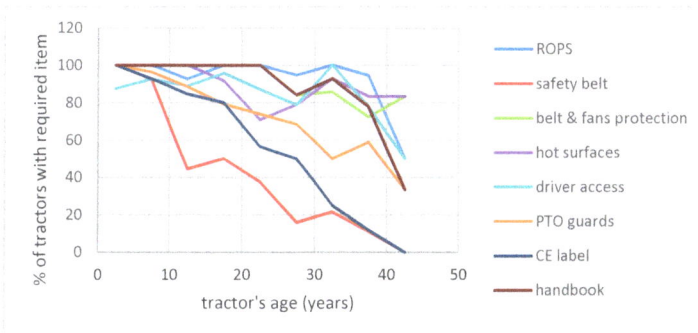

Figure 2. Tractors with required protective items in place.

The main findings from the survey suggested that several agricultural farms were sufficiently aware of the risks associated either with their specific production systems, or with the machinery used, to some extent, particularly in order to avoid the related economic costs. More importantly, information about legal obligations was generally poor, as was the understanding of the possible cost, in terms of fines, damage compensations etc. which failure to comply with the rules might cause.

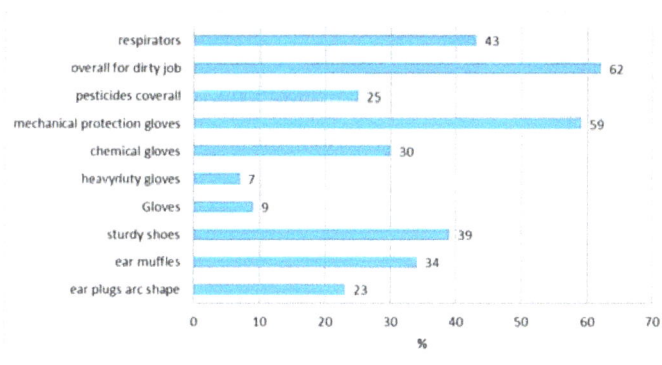

Figure 3. Types of personal protective equipment (PPE) provided at the farms (% of farms where present).

This suggested that most farms would take advantage of some simple informative tool, e.g., in the form of a software, to quickly detect the most critical situations. This software, based on a Microsoft Excel® worksheet (Figure 4a), enables the farmer to check all of the legal requirements for tractors and the main agricultural implements, and suggests how to amend possible defects.

For instance, it is possible to examine the existing ROPS on a tractor (Figure 4b), and understand whether it fulfills legal requirements or it needs changes or replacement; furthermore, indications may be given on how to install a ROPS on an old tractor.

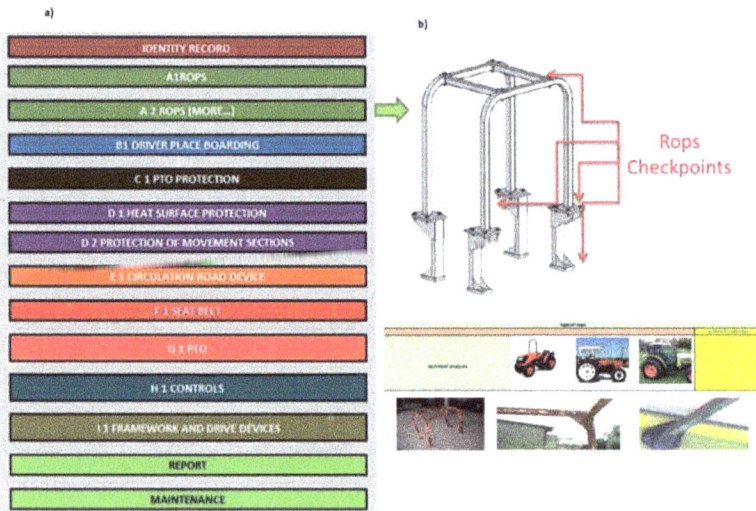

Figure 4. Extract from the software designed for the analysis of the tractors (**a**), (**b**) specific ROPS (rollover protection system) control areas.

3.2. Second-Level Analysis of the Sample Based on 11 Farms

The second survey investigated three groups of farms: vineyard farms, cattle and cereal farms, and a third group of mixed farms (Table 7). This involved an overall number of 58 agricultural tractors.

Table 7. The sample farms in the second survey.

Farm	Type	Own Area, ha	Managed Area, ha	Managed Area, ha
1	Vineyard	180	180	18
3	Vineyard	15	30	4
5	Vineyard	50	50	4
8	Vineyard	5	5	2
	Vineyard farms			28
2	Cattle and Cereals	50	200	5
4	Cattle and Cereals	60	80	5
11	Cattle and Cereals	20	70	3
	Cattle & Cereals Farms			13
7	Cereals and Contractor	300	450	7
6	Orchard	5.8	5.8	2
9	Mixed	50	250	5
10	Market garden	2	3	3
Total		738	1324	58

In the vineyard sector, the mean age of tractors was lower (5728 total h and 14.2 years) compared with both the Cattle % Cereals group (8046 h and 25.2 years) and with the average of the remaining farms (8557 h and 24.3 years) (Table 8). On the other side, the annual use is higher in the vineyard sector (502 h/year, versus 351–370 h/year).

Table 8. Second survey: Tractors power, age and usage.

Type of Farms	Vineyard	Cattle & Cereals	Other	All
Rated power, kW	56	59	65	59
Age, h	5729	8046	8557	7078
Annual usage, h/year	502	351	370	429
Age, years	14.2	25.2	24.3	19.6

The fact that agricultural tractors in the vineyard sector are generally of a younger age implies that they have minor problems in terms of safety and efficiency. In fact, this particular agricultural sector is generally more proactive and prone to invest financial resources, mainly because companies have a higher profitability but also because they are normally larger and therefore more structured and less family-owned. This shows a clearer perception and higher awareness of safety issues and needs (Table 8). Nonetheless, the present study also highlights some extremely important negative features, which can be of paramount importance in implementing corrective measures for the upgrading of current agricultural machinery inventories. In fact, farms dealing with working areas of more than 50 ha extension have been recognized as having the oldest agricultural tractors-with an average age of 25 years.

The main lack is in the power take-off guards (34.5%), followed by driver's seat belts (24.1%), together with lack of hot parts protective shields (32.8%). Lack of moving parts protections (20.7%) and ROPS (19%) has also been highlighted by the study (Table 9).

Table 9. Compliance with safety requirements (% of all tractors). Study-derived technical analysis.

Item	Evaluation	Yes	No
Documents	Compliant	69.0	31.0
PTO guards	Compliant	65.5	34.5
Moving parts, protections	Compliant	79.3	20.7
Hot surfaces, protection	Compliant	63.8	32.8
ROPS	Compliant	81.0	19.0
	Present	98.3	1.7
Driver's seat	Compliant	50.0	50.0
	Type conform	69.0	31.0
	Undamaged	75.9	24.1
Handles	Compliant	72.4	27.6
	Present	77.6	22.4
	Type compliant	74.1	25.9
	Size compliant	72.4	27.6
Stairs	Compliant	84.5	15.5
	Present	94.8	5.2
	Size compliant	84.5	15.5
Safety belt	Compliant	67.2	32.8
	Present	75.9	24.1
	Own installation	22.4	
	Own installation, certified	8.6	
Tires	Compliant	69.0	31.0
	Type compliant	82.8	15.5
	Undamaged	77.6	22.4
Mirrors	Compliant	69.0	31.0
Lights	Compliant	75.9	24.1

Tractor's compliance with the law (% of required items that were indeed present) was analyzed versus the tractor's age (in years) and the type of farm personnel (farms with and without external, hired workers, respectively). Both regressions in Figure 5 were statistically significant ($R^2 = 0.383$ and $R^2 = 0.453$, respectively), showing that: in general, the percentage of compliant items decreased with increased age of the tractors; in particular, tractor compliance was lower in family farms independently of the tractor's age. This can be explained by lower perception of risks in family farms, which certainly represents a failure of awareness campaigns conducted so far, but may also be related to the smaller economic size of these farms, and to the difficulty of bearing the costs involved by extensive equipment updating so as to meet the current requirements for risk prevention and protection.

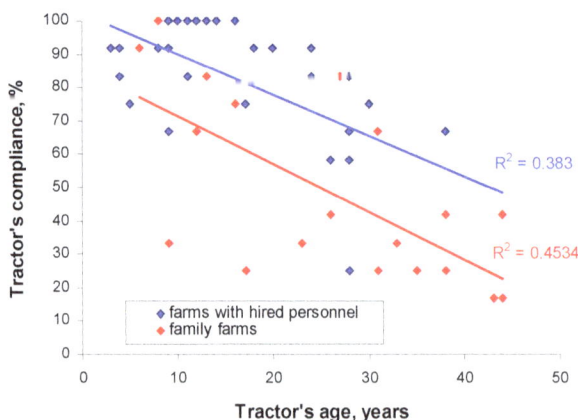

Figure 5. Tractor's compliance (% of compliant items) vs. tractor's age (years) in farms with or without hired personnel.

4. Conclusions

This research work shows that the situation of safety in the agricultural sector is still a real cross-cutting issue, mainly due to three aspects:

- low perception and awareness of the issue of safety at work by the workers in the agricultural sector; in fact, even if the machines are technically obsolete and unsafe, there is the tendency not to sell or adjust the machine, which is considered as a potentially useful vehicle or one that can be used in case of emergency.
- non-economical difficult interventions to upgrade machines having mainly a non-productive, affective value; in fact, very often the agricultural entrepreneur does not want to discard his own machine since the tractor is linked to a memory of a missing family member or parents.
- ineffective control system (e.g., [3,8,10,13,20–22]).

With the introduction of a rapid and efficient control system, this study wants to propose an operational instrument enabling the farmer to analyze the farm machinery and to put into practice simple modifications or installations that in the case of an accident or tipping of the machine could mean saving his/her life. The instrument could also be an excellent guideline not only for the agricultural world but also for the workshops that are only currently approaching the problem of the adjustment of agricultural machinery.

Moreover, the study only concentrated on the harmonized safety standard for the tractors improved safety for operators, such as physical exposure to ergonomic hazards. In fact, operators loading the machines are prone to musculoskeletal disorders since they often undergo awkward postures, repetitive movements and frequent lifting of loads. For this reason, further studies in this matter in particular could prove useful.

In particular, this study highlights the fact that, although the legislation making training courses for tractor operators and tractor inspections compulsory dates back to the years 2012 and 2015, deadlines have been postponed and the law is not yet fully applied.

Acknowledgments: Our research work was supported by funds received from the National Institute for Insurance against Workplace Accidents and Occupational Disease.

Author Contributions: Sirio Rossano Secondo Cividino and Rino Gubiani evaluated the technical aspects of safety and accidents in farms; Gianfranco Pergher conceived and designed the experiments and methodology, Nicola Zucchiatti followed the graphic design. Gainfranco Pergher and Sirio Rossano Secondo Cividino wrote the paper.

Conflicts of Interest: The authors declare no conflict of interest.

References

1. Thelin, A. Fatal accidents in Swedish farming and forestry, 1988–1997. *Saf. Sci.* **2002**, *40*, 501–517. [CrossRef]
2. Cividino, S.; Gubiani, R.; Vello, M.; Pergher, G.; Grimaz, S. Sicurezza sul lavoro: Criticità nell'azienda Agricola. *L'inf. Agrar.* **2016**, *24*, 30–33.
3. Fargnoli, M.; Laurendi, V.; Tronci, M. Design for Safety in Agricultural Machinery. In Proceedings of the International Design Conference—Design, Dubrovnik, Croatia, 17–20 May 2010.
4. INAIL. Istituto Nazionale per L'assicurazione Contro Gli Infortuni Sul Lavoro. Available online: www.inail.it (accessed on 1 May 2015).
5. Etherton, J.R.; Myers, J.R.; Jensen, R.C.; Russel, J.C.; Braddee, R.W. Agricultural Machine-Related Deaths. *Am. J. Public Health* **1991**, *81*, 766–768. [CrossRef] [PubMed]
6. Fulcher, J.; Noller, A.; Kay, D. Framing tractor fatalities in Virginia: An 11-year retrospective review. *Am. J. Forensic Med. Pathol.* **2002**, *33*, 377–381. [CrossRef] [PubMed]
7. Rete Rurale Nazionale. Prevenzione e Sicurezza Sul Lavoro in Agricoltura: Conoscenze e Costi per le Aziende Agricole. Ministero Delle Politiche Agricole Alimentari e Forestali, Dipartimento Delle Politiche Europee ed Internazionali e Dello Sviluppo Rurale. Available online: www.reterurale.it (accessed on 1 May 2016).
8. Seifert, A.L.; Santiago, D.C. Preparation of professionals in the area of agrarian sciences regarding safety in rural work. *Cienc. Agrotecnol.* **2009**, *33*, 1131–1138. [CrossRef]
9. Dogan, K.H.; Dermici, S.; Sunam, G.S.; Deniz, I.; Gunaydin, G. Evaluations of Farm Tractor-Related Fatalities. *Am. J. Forensic Med. Pathol.* **2010**, *31*, 64–68. [CrossRef] [PubMed]
10. Prošrekl, J. Safe behavior and level of knowledge regarding safe work practices on farms. *Res. J. Chem. Sci.* **2011**, *1*, 15–19.
11. Moreschi, C.; Da Broi, U.; Fanzutto, A.; Cividino, S.; Gubiani, R.; Pergher, G. Medicolegal Investigations into Deaths Due to Crush Asphyxia after Tractor Side Rollovers. *Am. J. Forensic Med. Pathol.* **2017**, *38*, 312–317. [CrossRef] [PubMed]
12. Rees, W.D. Agricultural tractor accidents: A description of 14 tractor accidents and a comparison with road traffic accidents. *Br. Med. J.* **1965**, *2*, 63–66. [CrossRef] [PubMed]
13. Cole, H.P.; Myers, M.L.; Westneat, S.C. Frequency and severity of injuries to operators during overturns of farm tractors. *J. Agric. Saf. Health* **2006**, *12*, 127–138. [CrossRef] [PubMed]
14. Monarca, D.; Cecchini, M.; Guerrieri, M.; Colantoni, A. Conventional and alternative use of biomasses derived by hazelnut cultivation and processing. *Acta Hortic.* **2009**, *845*, 627–634. [CrossRef]
15. Kelsey, T.W.; May, J.J.; Jenkins, P.L. Farm tractors, and the use of seat belts and roll-over protective structures. *Am. J. Ind. Med.* **1996**, *30*, 447–451. [PubMed]
16. Hyland-Mcguire, P. Farm accidents involving power take-off devices. *J. Accid. Emerg. Med.* **1994**, *11*, 121–124. [CrossRef] [PubMed]
17. Monarca, D.; Cecchini, M.; Colantoni, A. Plant for the production of chips and pellet: Technical and economic aspects of an case study in the central Italy. In Proceedings of the ICCSA 2011 International Conference on Computational Science and Its Applications, Santander, Spain, 20–23 June 2011; Volume 6785, pp. 296–306.

18. Moscetti, R.; Saeyes, W.; Keresztes, J.C.; Goodarzi, M.; Cecchini, M.; Monarca, D.; Massantini, R. Hazelnut Quality Sorting Using High Dynamic Range Short-Wave Infrared Hyperspectral Imaging. *Food Bioprocess Technol.* **2015**, *8*, 1593–1604. [CrossRef]

19. Monarca, D.; Colantoni, A.; Cecchini, M.; Longo, L.; Vecchione, L.; Carlini, M.; Manzo, A. Energy characterization and gasification of biomass derived by hazelnut cultivation: Analysis of produced syngas by gas chromatography. *Math. Probl. Eng.* **2012**, *2012*, 102914. [CrossRef]

20. Colantoni, A.; Ferrara, C.; Perini, L.; Salvati, L. Assessing trends in climate aridity and vulnerability to soil degradation in Italy. *Ecol. Indic.* **2015**, *48*, 599–604. [CrossRef]

21. Marucci, A.; Monarca, D.; Cecchini, M.; Colantoni, A.; Cappuccuni, A. The heat stress for workers employed in laying hens houses. *J. Food Agric. Environ.* **2013**, *11*, 20–24.

22. Cutini, M.; Forte, G.; Maietta, M.; Mazzenga, M.; Mastrangelo, S.; Bisaglia, C. Safety-Critical Manuals for Agricultural Tractor Drivers: A Method to Improve Their Usability. *Agriculture* **2017**, *7*, 67. [CrossRef]